NAMES10: "New Achievements in Materials and Environmental Sciences"

Journal of Physics: Conference Series Volume 416

Nancy, France
26-29 October 2010

ISBN: 978-1-62748-294-3
ISSN: 1742-6588

Printed from e-media with permission by:

Curran Associates, Inc.
57 Morehouse Lane
Red Hook, NY 12571

Some format issues inherent in the e-media version may also appear in this print version.

Printed by Curran Associates, Inc. (2013)

For permission requests, please contact the Institute of Physics
at the address below.

Institute of Physics
Dirac House, Temple Back
Bristol BS1 6BE UK

Phone: 44 1 17 929 7481
Fax: 44 1 17 920 0979

techtracking@iop.org

Additional copies of this publication are available from:

Curran Associates, Inc.
57 Morehouse Lane
Red Hook, NY 12571 USA
Phone: 845-758-0400
Fax: 845-758-2634
Email: curran@proceedings.com
Web: www.proceedings.com

TABLE OF CONTENTS

NANOMATERIALS AND NANOTECHNOLOGIES

FUNCTIONAL MATERIALS AND COATINGS

ENVIRONMENT AND CLEAN TECHNOLOGY

ITER RELATED TOPICS

NATURAL FIBROUS MATERIALS

CONTINUOUS MEDIA MECHANICS

MATERIALS FOR OPTICS AND OPTOELECTRONICS

Journal of Physics: Conference Series **416** (2013) 011001 doi:10.1088/1742-6596/416/1/011001

History of NAMES Conferences

Franco–Russian NAMES Seminars are held for the purpose of reviewing and discussing actual developments in the field of materials science by researchers from Russia and from the Lorraine Region of France. In more precise terms, as set down by the organizers of the seminar (the Moscow Institute of Steel and Alloys and the Institut National Polytechnique de Lorraine), the mission of the seminars is as follows:

- the development of scientific and academic contacts, giving a new impulse to joint fundamental research and technology transfer
- the development and consolidation of scientific, technical and business collaboration between the regions of Russia and Lorraine through direct contact between the universities, institutes and companies involved

The first Seminar took place on 27–29 October 2004, at the Institut National Polytechnique de Lorraine (on the premises of the Ecole Européenne d'Ingénieurs en Génie des Matériaux, Nancy, France). The number, variety and quality of the oral presentations given and posters exhibited at the first Seminar were of high international standard. 30 oral presentations were given and 72 posters were presented by

- 19 participants from five universities and three institutes of the Russian Academy of Sciences
- participants from 11 laboratories of three universities from the Lorraine region
- three industrial companies, including the European Aeronautic Defence and Space Company—EADS, and ANVAR (Agence Nationale de Valorisation de la Recherche)

From 2005 onwards, it was decided to organize the Seminar every other year. The second Seminar convened on the occasion of the 75th Anniversary of the Moscow Institute of Steel and Alloys on 10–12 November 2005 in Moscow, Russia. The seminar demonstrated the efficiency of the scientific partnership founded between the research groups of Russia and France during the first Seminar. High productivity of the Franco–Russian scientific cooperation on the basis of the Research-Educational Franco–Russian International Centre was demonstrated. By the high standards of the reports presented, as well as by its overall organization, the second Seminar met the standards of an international conference.

Reviews of state-of-the-art developments in materials science were given by leading scientists from Moscow and from the Lorraine region. The three days of the seminar were structured into four main themes:

- Functional Materials
- Coatings, Films and Surface Engineering
- Nanomaterials and Nanotechnologies
- The Environment

and three Round Table discussions:

Journal of Physics: Conference Series **416** (2013) 011001 doi:10.1088/1742-6596/416/1/011001

- Defining practical means of carrying out Franco–Russian collaborations in technology transfer and innovation
- Materials science
- ARCUS: Lorraine–Russian collaboration in materials science and the environment

32 oral and 25 poster presentations within four sections were given by a total of 110 participants. NAMES 2007, the 3rd Franco–Russian Seminar on New Achievements in Materials and Environmental Sciences, took place in Metz, France on 7–9 November 2007. The conference highlights fundamentals and development of the five main themes connected to the Lorraine–Russia ARCUS project with possible extension to other topics.

The five main subjects included in the ARCUS project are:

- Bulk–surface–interface material sciences
- Nanomaterials and nanotechnologies
- Environment and natural resources
- Plasma physics—ITER project
- Vibrational dynamics

The first, second and third NAMES conferences were financially supported by the following organizations:

- Ambassade de France à Moscou
- Communauté Urbaine du Grand Nancy
- Région Lorraine
- Conseil Général de Meurthe et Moselle
- Institut National Polytechnique de Lorraine
- Université de Metz
- Université Henry Poincaré
- CNRS
- ANVAR
- Federal Agency on Science and Innovations of the Ministry of Education and Science of the Russian Federation
- Moscow Committee on Science and Technologies
- Moscow Institute of Steel and Alloys (Technological University)

The 4th conference is supported by the Ministry of Foreign Affairs of France and the Lorraine Region Council. The conferences have indicated directions for future research and stimulated the possibilities of cooperation between scientists from Lorraine and Russian universities and academic institutions. The participants of the conferences reviewed the remarkable worldwide progress with numerous breakthroughs in areas of fundamental research and industrial applications, specifically in the fields of nanomaterials and nanotechnologies, surface engineering, biomaterials and multifunctional coatings, functionally graded materials, new materials for microelectronics and optics, nanostructured thin films and nanodispersion strengthening coatings, combustion synthesis, new micro- and nanosystems and devices, natural resources, environmental sciences, clean technology, and recently, natural fibrous materials, etc. The participants consider that new fundamental knowledge, new materials, and industrial production methods generated as a result of international cooperation between both countries will be of interest to the industrial sector in Lorraine and Moscow, France and Russia.

Professor Lev O Filippov

Coordinator of NAMES conferences

Journal of Physics: Conference Series **416** (2013) 011001 doi:10.1088/1742-6596/416/1/011001

IV France-Russia Conference

NAMES'10

New Achievements in Materials and Environmental Sciences

Nancy, FRANCE - October 26-29, 2010

Organized by

The Institut National Polytechnique de Lorraine
The National University of Science and Technology "MISIS"

Journal of Physics: Conference Series **416** (2013) 011001 doi:10.1088/1742-6596/416/1/011001

INTERNATIONAL ORGANISING and SCIENTIFIC COMMITTEE

COORDINATOR OF NAMES CONFERENCES CHAIR of NAMES'10

Lev O. FILIPPOV
LEM UMR 7569 CNRS Nancy Université and NUST "MISIS"

INTERNATIONAL SCIENTIFIC COMMITTEE

P. ARCHAMBAUT, *Institut Jean Lamour, Nancy Université, Nancy, France*
J. Ph. BAUER *Institut Jean Lamour, Nancy Université, Nancy, France*
P. BOURSON *University Paul Verlaine and Supelec, Metz, France*
V. CHANTOURIA *Institute of Comprehensive Exploitation of Mineral Resources RAS, Moscow, Russia*
P. CHARPENTIER *ENSTIB Ecole Nationale Supérieure des Technologies et Industries du Bois, Nancy, France*
A. CELZARD *Institut Jean Lamour, ENSTIB,Epinal, France*
T. GROSDIDIER *LETAM, UMR CNRS 7078, University Paul Verlaine, Metz, France*
K. FEDOTOV *Irkutsk state technical university, Irkutsk, Russia*
I. FILIPPOVA *LEM UMR 7569 CNRS Nancy Université, Vandoeuvre-lès-Nancy, France*
S. HEURAUX *Institut Jean Lamour, Nancy Université, France CNRS UMR 7198, Vandoeuvre, France*
E. LEVASHOV *National University of Science and Technology "MISIS", Moscow, Russia*
T. LYUBIMOVA *Institute of Continuous Media Mechanics UB RAS, Perm, Russia*
E. McRAE *Institut Jean Lamour, Nancy Université UMR CNRS 7198, ,Vandoeuvre-lès-Nancy, France*
A. MAILFERT *LEM UMR 7569 CNRS Nancy Université, Vandoeuvre-lès-Nancy, France*
S. NIKULIN *National University of Science and Technology "MISIS", Moscow, Russia*
E. PAYAN *CNRS Delegation Centre Est, Vandoeuvre-lès-Nancy, France*
V. PETROV *IGEM Russian Academy of Sciences, Moscow, Russia*
D. ROIZARD *LRGP CNRS3349 - Nancy-Université, ENSIC, Nancy, France*
V. SHUR *Ferroelectric Laboratory, Ural State University, Ekaterinburg, Russia*
F. THOMAS *LEM UMR 7569 CNRS Nancy Université, Vandoeuvre-lès-Nancy, France*
J. YVON *LEM, UMR 7569 CNRS, Nancy-University, ENSG, Vandoeuvre-lès-Nancy, France*
E. ZHARIKOV *Mendeleev University of Chemical Technology of Russia, Moscow, Russia*

LOCAL ORGANISING COMMITTEE

I. FILIPPOVA, F. DIOT, R. JOUSSEMET, E. GAUTIER, D. MARTIN
M. GRANDJEAN, V. SEVEROV, A. DUVERGER, A. MATININ, P. PIRIOU

The Conference is organized at the ENS Géologie location by Laboratoire Environnement et Minéralurgie, LEM UMR 7569 CNRS-INPL.

DAY 1 – Tuesday 26 October 2010 – AFTERNOON SESSION		
12.00 – 14.00	**WELCOME**	
14.00 – 14.30	**Lorraine-Russia Collaborations from 1900**	**L. FILIPPOV**, Coordinator of NAMES Conferences **A. MAILFERT**, Coordinator ARCUS program **M. BERNS-COQUILLAT,** Directeur de l'Enseignement Supérieur de la Région Lorraine
Session 1 – Natural Fibrous Materials **Session Chair : P. Charpentier**		
14.30 – 15.00	Wood : a multifunctional material	**M.C. Triboulot** *ENSTIB, France*
15.00 –15.20	Integrated transformation of plant biomass to valuable chemicals, biodegradable polymers and nanoporous carbons	**B.N. Kuznetsov**[1,2], N.V.Chesnokov[1,2,] ,V.E. Taraban'ko [1,2], S.A. Kuznetsova[1,2], A.V. Petrov[1] *[1]Institute of Chemistry and Chemical Technology SB RAS,Russia* *[2]Siberian Federal University, Krasnoyarsk Russia*
15.20 – 15.40	Multifunctional porous solids derived from tannins	**A. Celzard**[1],V.Fierro[1], A. Pizzi[2], W. Zhao[1] *[1]IJL – UMR UHP - CNRS 7198, ENSTIB, France* *[2]LERMAB, EA 4370, ENSTIB, France*
15.40 – 16.00	New families of carbon gels based on natural resources	**A. Szczurek**[1], G. Amaral[1], V. Fierro[1], A. Pizzi[2], A. Celzard *[1]IJL – UMR UHP - CNRS 7198, ENSTIB, France* *[2]LERMAB, EA 4370, ENSTIB, France*
16.00 – 16.20	**COFFEE BREAK**	
Session 2 – Environment and clean technology **Session Chair : D.Roizard**		
16.20 – 16.40	Synthesis of alkali ferrates by using a waste as raw materials"	**N. Kanari**[1], L. Filippov,[1] F. Diot,[1] J. Mochón,[2] I. Ruiz-Bustinza,[2] E. Allain[1] *[1] LEM, UMR 7569 CNRS, Nancy-University, ENSG, France* *[2]CENIM – CSIC, Spain*
16.40 – 17.00	Preparation and Optimization of TiO2/b-SiC Foam For Industrial Photocatalytic Water Treatment	**N.A. Kouamé**[1], D. Roberta[1], V. Kellerb[2], N. Keller[2] *[1]Antenne de Saint-Avold, UPV-Metz, France* *[2] Ecole Européenne de Chimie, Polymères et Matériaux, France*
17.00 – 17.20	Carbonation of residual brines produced by ammonia-soda process	**P. Piriou**, I.V. Filippova, L.O Filippov, M. Grandjean *Laboratoire Environnement et Minéralurgie, UMR 7569 CNRS, ENSG-INPL, France*
17.20 –17.40	Tailoring of analytical performances of urea biosensors using nanomaterials	**W. Nouira**[1], H. Barhoumi[1], A.Maaref [1], N. Jaffrézic Renault[2], M. Siadat[3] *[1] Laboratoire de physique et chimie des interfaces, Tunisia* *[2]Laboratoire des Sciences Analytiques : Université Claude Bernard, Lyon 1, France* *[3] LASC – ISEA - Université de Metz, France*
EVENING 19.00	**GUIDED TOUR OF CITY OF NANCY**	

DAY 2 – Wednesday 27 October 2010 – MORNING SESSION

9.00 – 9.50	Official opening ceremony of NAMES	**François LAURENT**, Président INPL – Nancy Université **Dmitry V. LIVANOV**, Rector – NUST "MISIS" **Christophe CHOSEROT**, Conseiller régional de Lorraine - Région Lorraine
Session 3 – Functional coatings Session Chair : J.-Ph. Bauer		
9.50 – 10.10	Nanostructured Materials for Electrospark Deposition and Disperse-Strengthening by Nanoparticles Diamond Tools	**E.A. Levashov** [1], A.E. Kudryashov [1], E.I. Zamulaeva [1], V.V. Kurbatkina [1], V.A. Andreev [2], A.A. Zaitsev [1], D.A. Sidorenko [1] *[1] National University of Science and Technology "MISIS",Russia* *[2] Kermet Company, Russia*
10.10 – 10.30	Characterization of Cr-B-N films deposited by pulsed DC magnetron sputtering from targets with various B/Cr ratios	**D. Pilloud**[1], Ph. V. Kiryukhantsev-Korneev[2], J.F. Pierson[1], J.P. Bauer[1] *[1]Institut Jean Lamour (UMR CNRS 7198), France* *[2]National University of Science and Technology "MISIS", Russia*
10.30 – 10.50	Structure and properties of amorphous Cr-Al-Si-B-(N) coatings	**Ph.V. Kiryukhantsev-Korneev**[1], J.F. Pierson[2], I.A.Bashkova[1],J.P Bauer[2], E.A. Levashov[1], D.V. Shtansky[1] *[1]National University of Science and Technology "MISIS", Russia* *[2] Institut Jean Lamour (UMR CNRS 7198), France*
10.50 – 11.20	**COFFEE BREAK**	
Session 4 – Environment and Clean Technologies Session Chair : V. Petrov		
11.20 – 11.50	Impact of membranes and membranes processes in environmental issues	**V.V. Teplyakov** *A.V. Institute of Petrochemical Synthesis, Russian Academy of Sciences (TIPS RAS), Russia*
11.50 – 12.10	Use of residual materials for manufacturing lightweight expanded granulates	**J. Yvon**[1], N. Kanari[1], J. Banse[1], R. Joussemet[1], F. Diot[1], E. Allain[1], F.X. Faure[2] J.L. Seveque[2] J.P. Chambard[3] *[1] LEM, UMR 7569 CNRS, Nancy-University, ENSG, France* *[2] VBC 3000,France* *[3] GEM, France*
12.10 – 12.30	Development of vinylic and acetylenic functionalized structures bases on high permeable glassy polymers as membrane materials for gas mixture separation	**D.Roizard**[2], Y.Kiryukhina[1], A. Masalev[1], V. Khotimskiy[1], V.Teplyakov[1], D. Barth[2] *[1] A.V.Topchiev Institute of Petrochemical Synthesis Russian Academy of Sciences, Russia* *[2]Laboratoire Réactions et Génie des Procédés (CNRS3349 - Nancy-Université), ENSIC, France*
12.30 – 12.50	Size effects in combustion of hydrocarbons over Pt/Al2O3 catalysts for abatement of car exhausts.	**V.I. Bukhtiyarov**, I.E. Beck, A.Y. Stakheev *Boreskov Institute of Catalysis, Russia*
12.50 – 14.00	**LUNCH BREAK**	

DAY 2 – Wednesday 27 October 2010 – AFTERNOON SESSION

14.00 – 15.00	**POSTER SESSION**	
Session 5 – Nanomaterials and Nanotechnologies Session Chair : E. Levashov		
15.00 – 15.30	Domain nanotechnology in ferroelectrics	**V.Ya.Shur** *Ural State University, Russia*
15.30 – 15.50	The synthesis of carbon nanostructures by pyrolysis of ethanol using a new Ni/($NiO+Y_2O_3$) catalyst	**E. V. Zharikov**[1], E. McRae[2], X. Devaux[3], S.Yu.Tsareva[1], A. N. Kovalenko[1], P.P. Faikov[1], B. Vigolo[2], K.S. Zaramenskhih[1]. *[1]D. Mendeleev University of Chemical Technology of Russia, Russia* *[2,3] Institut Jean Lamour UMR CNRS 7198, France*
15.50 – 16.10	On the effect of sulphur on the formation of carbon nanotubes and multibranched carbon nanostructures	**E. McRae**[1], X. Devaux[2], S. Y. Tsareva[3], A. N. Kovalenko[3], K. S. Zaramenskih[3], E. V. Zharikov[3] *[1]Institut Jean Lamour, UMR 7198 CNRS - Nancy-Université - UPV-Metz, France* *[2] Institut Jean Lamour UMR CNRS 7198, Nancy Université, Ecole des Mines, France* *[3] D. Mendeleev University of Chemical Technology of Russia, Russia*
16.10 – 16.30	About quantum-mechanical properties of the Block Elements in Nanomaterials	**V.A. Babeshko**, O.V. Evdokimova, O.M. Babeshko *Centre of joins using of Heavy Vibroseismic Sources Kuban State University, South Scientific Center of RAS, Russia*
16.30 – 16.50	**COFFEE BREAK**	
Session 6 – New Materials synthesis Session Chairs : E. McRae		
16.50 – 17.20	Pertinence of new communicating material paradigm: A first step towards wood mass marking.	**J. Jover**[1], A. Thomas[1], J.M.Leban[2] [1] Centre de Recherche en Automatisme de Nancy (*CRAN, UMR 7039),France* *[2]INRA Nancy,France*
17.20 – 17.40	Study of micro and nano structures of lithium niobate by Raman scattering	**P.Bourson**[1], P.Zelenovskiy[2], M.D.Fontana[1], S. Margeuron[1], V.Shur[2] *[1]LMOPS, université Paul Verlaine, Metz France* *[2] Ural State University, Russia*
17.40 – 18.00	Combustion synthesis in mechanical and structural activated systems	**A.S.Rogachev** *Institute of Structural Macrokinetics and Materials Science, Russia*
18.00 – 18.20	Atomic and electronic structure of the (001) surface of the Al_9Co_2 complex metallic alloy	**S. Alarcón Villaseca**[1], E. Gaudry[1], J. Ledieu[1], V. Fournée[1], P. Gille[2], M.-C. De Weerd[1], J.-M. Dubois[1] *[1]Institut Jean Lamour, UMR 7198 (CNRS – Nancy Université – UPV Metz), École des Mines de Nancy, France* *[2]Department of Earth and Environmental Sciences, Crystallography Section, LMU,Germany*
EVENING 19.00	**WELCOME BY MAIRIE DE NANCY**	

DAY 3 – Thursday 28 October 2010 – MORNING SESSION		
Session 7 – Nanomaterials and Nanotechnologies **Session Chairs : V. Shur**		
9.00 – 9.45	Highly spatially-resolved microscopy	**M. Dossot** *LCPME - UMR 7564 CNRS-Nancy Université, France*
9.45 – 10.15	Towards control of the diameter of single wall carbone nanotube in a CVD process	**X. Devaux**[1], S.Y. Tsareva[2] [1]*Institut Jean Lamour, Nancy Université UMR CNRS 7198, France* [2] *Mendeleev University of Chemical Technology of Russia*
10.05 – 10.30	Development of new concepts of wear resistant coatings by laser cladding	**N.V. Mityushina**[1,2] I. Movchan,[1] E.A. Levashov[2] [1]*ENISE-DIPI, Maison des Pôles, 74 rue des Aciéries, Saint-Etienne* [2] *National University of Science and Technology "MISiS", Russia*
10.30 – 11.00	**COFFEE BREAK**	
Session 8 – ITER related topics **Session Chairs : E. Zharikov**		
11.00 – 11.30	Materials for nuclear reactors : evolutions and revolutions	**P. Dubuisson**, P. YVON *Nuclear Materials Department, Nuclear Energy Division CEA Saclay, France*
11.30 – 11.50	Numerical modeling of tokamak plasma micro turbulence wave number spectrum reconstruction using radial correlation reflectometry data	**N.Kosolapova**[1,2], E.Gusakov,[1] S.Heuraux[2] [1]*Ioffe Institute, Russia* [2] *Institut Jean Lamour, UMR 7198 CNRS France*
11.50 – 12.10	Automatic video tracking and statistical analysis of dust particles trajectories from all tungsten ASDEX Upgrade	**S. Bardin,**[1] F. Brochard,[1] J.L. Briançon,[1] J. Bougdira,[1] N. Endstrasser,[2] T. Lunt,[2] V. Rohde,[2] R. Neu,[2] [1]*Institut Jean Lamour, UMR 7198 CNRS, Département CP2S, France* [2]*Max-Planck-Institut für Plasma Physik, Germany*
12.10 – 12.30	Reflectometry simulation as a tool to evaluate new concepts of diagnostics to extract fusion plasma turbulence characteristics	**S. Heuraux**[1], E. Gusakov[2], A. Popov[2], F da Silva[3], N. Kosolapova[1,2], K. Syisoeva[2] *1 IJL CNRS UMR 7198-UHP, France* *2 Ioffe Institute ,Russia* *3 Associação EURATOM/IST Centro de Fusão Nuclear, Instituto Superior Técnico, Portugal*
12.30 – 14.00	**LUNCH BREAK**	

DAY 3 – Thursday 28 October 2010 – AFTERNOON SESSION		
14.00 – 15.00	**POSTER SESSION**	
Session 9 – Environnement and Clean Technologies **Session Chairs : V. Teplyakov,**		
15.00 – 15.20	High-performance carbonaceous adsorbents for hydrogen storage	**W. Zhao**[1], V. Fierro[1], C. Zlotea[2], E. Aylon[3], M. T. Izquierdo[3], M. Latroche[2], A. Celzard[1] *[1]IJL – UMR UHP - CNRS 7198, ENSTIB, France* *[2]ICMPE, UMR 7182, CNRS, France* *[3]Instituto de Carboquímica (CSIC), Zaragoza, Spain*
15.30 – 15.50	Dry and steam biofermentation products reforming into hydrogen containing gas using innovative inorganic porous membrane catalytic systems	**A. Fedotov**, M.V.Tsodikov *Institute of Petrochemical Synthesis, Russian Academy of Sciences (TIPS RAS), Russia*
15.50 – 16.10	Stress-time context of fault permeability at the Krasnokamensk Area, SE Transbaikalia	**V. Petrov** *Institute of Geology of Ore Deposits, Petrography, Mineralogy and Geochemistry (IGEM), Russian Academy of Sciences, Russia*
16.10 – 16.30	Morphology of calcite particles precipitated from saline waste solution	**M. Grandjean**[1], I.V. Filippova[1,2], L.O.Filippov[1,] *[1] LEM, UMR 7569 CNRS, Nancy-University, ENSG, France* *[2] National University of Science and Technology "MISiS", Russia*
16.30– 16.50	**COFFEE BREAK**	
Session 10 – Processing and mechanical aspects of materials **Session Chairs : A. Celzard**		
16.50 – 17.10	Endurance analysis of protective coatings for constructions	**Y.I. Romanov**, M.G. Ryumin *Moscow State University of Railway Engineering (MIIT),Russia*
17.10 – 17.30	Structure formation in submicrocrystalline low carbon steek during equal channel angular pressing and heating	**S. Dobatkin**[1], G. Raab[2], M. Pankova [3] and S. Shagalina[1] *[1]A.A.Baikov Institute of Metallurgy and Materials Science, RAS, Russia* *[2] Ufa State Aviation Technical University, Russia* *[3]I.P. Bardin Central Research Institute for Ferrous Metallurgy,Russia*
17.30 – 17.50	Particle interaction in a vibrational flow	**T. Lyubimova**[1], A. Cherepanova[1], M. Shardin[1], L. Filippov[2] *[1] Institute of continuous media mechanics UBRAS, Russia* *[2] LEM, UMR 7569 CNRS, Nancy-University, ENSG, France*
17.50 – 18.10	Economical interest of the use of an X-ray scanner in a planing mill	**B. Almecija**[1], P. Charpentier[2], R. Baquitaine, J-M Leban *[1]CRAN , Epinal, France* *[2]LERMAB, EA 4370, ENSTIB, France*
EVENING	**GALA DINER**	

DAY 4 – Friday 29 October 2010 – MORNING SESSION		
9.00 – 9.30	**ROUND TABLE OPENING CEREMONY** **Francois Laurent President INPL** **Representative of the Russian Ministry of Education and Science** **Representative of the European Commission (DG Research & DG Informational Society and Media)**	
9.30 – 12.30	**SESSION 1 : Strategy of Development of Large-Scale Research Infrastructures of the Russian Federation & Cooperation with the European Union** **Chairmens : Academician Vladimir Babeshko, Prof. Evgeny Levashov**	
9.30 – 10.00	International Research Networks – GDRI a tool for scientific cooperation	**Martine Bonin** *Head of Direction Internationals Relations,CNRS, Paris* ***Centre national de la recherche scientifique, France***
10.00 – 10.30	"Research Infrastructure = Russia and Cooperation with EU. EuroRIs-Net FP7 project"	**Evgeny Levashov** *Dr. Sc., Prof., Head of Russian NCP "Research Infrastructure",* *Head of SHS-Center and Department "Powder Metallurgy and Functional Coatings"* ***National Research Technological University "MISIS, Russia***
10.30 – 11.00	Organisation de la recherche coopérative dans le domaine des matériaux en France : le réseau des pôles de compétitivité	**Olivier Bonnet** Head of direction Materalia pole **Materalia Pole, Metz, France**
11.00 – 11.30	Joint Use Center for Vibroseismic Sources	**Vladimir Babeshko** *Dr. Sc., Prof., Academician of RAS,* *Head of the Joint Use Center for Vibroseismic Sources* ***Kuban' State University, Russia***
11.30 – 12.00	BIC Research Complex for Physical-Chemical Characterization of Functional Nanomaterials.	**Valery Bukhtiyarov** *Dr. Sc., Prof., Member Correspondent of RAS,* *Deputy Director and Laboratory Head of the Boreskov' Institute of Catalysis, Siberian Branch of Russian Academy of Science.* ***Boreskov' Institute of Catalysis, Siberian Branch of Russian Academy of Science, Russia***
12.00 – 12.30	Siberian Center of Synchrotron and Terahertz Radiation	**Valeriy Pindyurin** *PhD, Senior Research Worker* ***Siberian Joint Use Center of Synchrotron and Terahertz Radiation,*** ***Budker Institute of Nuclear Physics,*** ***Siberian Branch of Russian Academy of Science***
12.30 – 14.00	**LUNCH BREAK**	

DAY 4 – Friday 29 October 2010 – AFTERNOON SESSION

14.00 – 17.30	**SESSION 2 :** **Strategy of Development of Large-Scale Research Infrastructures of the Russian Federation & Cooperation with the European Union** **Chairmen :** **Member-Correspondent, Valery Bukhtiyarov and Professor Jean Steinmetz**	
14.00 – 14.30	Pole de compétitivité « Fibre naturelle »	
14.30 – 15.00	"Research Infrastructure = Russia and Cooperation with EU. EuroRIs-Net FP7 project"	**Pavel Todua** *Dr. Sc., Prof., Director of Research Center for Surface and Vacuum (Russia), Dean of the State Technical University "Moscow Physical-Technical Institute" (Russia)*
15.00 – 15.30	"Scientific Research and Development Center for Structural and Functional Nanomaterials"	**Pavel Kuznetsov** *Dr. Sc., Deputy General Director for Research in Nanotechnology of the Federal State Unitary Enterprise "Central Research Institute of Structural Materials "PROMETEY", Russia*
15.30 – 16.00	"Scientific-Educational Center "Nanomaterials and Nanotechnologies" of Moscow Institute of Steel and Alloys. Facilities and Activities"	**Mikhail Filonov** *Dr. Sc., Prof., Vice-Rector for Science and Innovations, Head of Scientific-Educational Center "Nanomaterials and Nanotechnologies", National Research Technological University "MISIS", Russia*
16.00 – 16.30	"Radioastronomical study of turbulent solar wind"	**Igor Chashey** *Dr. Sc., Prof., Deputy Director of the Pushchino Radio Astronomy Observatory, Russian Academy of Science, Russia*
16.30 – 17.00	"Present Research Activities and Future Trends of Ural Joint Use Center "Modern Nanotechnologies"	**Vladimir Shur** Dr. Sc., Prof., Director of Ural Joint Use Center "Modern Nanotechnologies", Ural State University, Russia
17.00 – 17.30	Russian-French Associated Laboratory : Experience in membranes and smart separation processes	**Denis Roizard and Vladimir Teplyakov**
17.30 – 18.00	**DISCUSSION**	
18.00	**CLOSING CEREMONY**	
	**** *Wine and Cheese* ****	

Poster Board Number	Name	Institution	Title of presentation
		Nanomaterials and Nanotechnologies	
1	**M. Aillerie**[1], S.M. Kostritskii[2], B. Chick-Bled[1,3] and E.P. Kokanyan[4]	[1]LMOPS, University Paul Verlaine of Metz and Supelec, France [2]RPC Optolink, Proton, Zelenograd, , Russia [3]URMER, University Abou Bakr Belkaïd, Tlemcen, Algeria [4]Institute for Physical Research, National Academy of Sciences of Armenia,	Light-induced effects in reduced and oxidized iron-doped lithium niobate crystals
2	**N. Allain-Bonasso**[1], V.P. Rotshtein[2], E. Bouzy[1], Yu. F. Ivanov[2], T. Grosdidier[1],	[1]Letam ISGMP Université Paul Verlaine-Metz, Ile du Saulcy, Metz, France [2]Institute of High-Current Electronics, Tomsk, Russia	Characterizations of Ti-Al surface alloys obtained by pulsed electron beam melting of film-substrate-system
3	**X. Devaux**[1], S.Yu. Tsareva[2], A.N. Kovalenko[2], E. McRae[3], E. Zharikov[2]	[1]Institut Jean Lamour, Nancy Université UMR CNRS 7198, Ecole des Mines, Nancy, France [2] Mendeleev University of Chemical Technology of Russia, Moscow, Russia [3] Institut Jean Lamour, Nancy Université UMR CNRS 7198, Faculté des Sciences et Techniques, Vandoeuvre-lès-Nancy, France	Structure and growth scenario of multibranched thick carbon nanotubes
4	S.Yu. Tsareva[1], **X. Devaux**[2], A.N. Kovalenko[1], K. S. Zaramenskih[1], E. McRae[3], E. Zharikov[1]	*[1] Mendeleev University of Chemical Technology of Russia, 9, Miusskaya sq., Moscow 125047, Russia* *[2] Institut Jean Lamour, Nancy Université UMR CNRS 7198, Ecole des Mines, 54042 Nancy cedex, France* *[3] Institut Jean Lamour, Nancy Université UMR CNRS 7198, Faculté des Sciences et Techniques, B.P. 70239, 54506 Vandoeuvre-lès-Nancy Cedex, France*	Influence of CS_2 on the growth of carbon nanotubes
5	S.M. Kostritskii[1], S Margueron[2], M Aillerie[2], **P. Bourson**[2]	[1]RPC Optolink, Proton, Russia [2]LMOPS, University Paul Verlaine of Metz and Supelec, Metz, France	Two-photo luminescence of small polarons in reduced $LiNbO_3$ crystals
6	**E. McRae**[1], X. Devaux[2], S. Yu. Tsareva[3], A. N. Kovalenko[3], K. S. Zaramenskih[3], E. V. Zharikov[3]	[1] Institut Jean Lamour UMR CNRS 7198, Nancy Université, Faculté des Sciences et Techniques, Vandoeuvre-lès-Nancy, France [2] Institut Jean Lamour UMR CNRS 7198, Nancy Université, Ecole des Mines, Nancy, France [3] D. Mendeleev University of Chemical Technology of Russia., Moscow, Russia	On the effect of sulphur on the formation of carbon nanotubes and multibranched carbon nanostructures

Poster Board Number	Name	Institution	Title of presentation
Environment and Clean Technologies			
7	**I.V. Filippova**[1,2], L.O. Filippov[1,2], V.A. Chantyria[3], M.V. Ryazantseva[1,3]	[1] Laboratoire Environnement et Minéralurgie, ENSG-INPL, Vandoeuvre-lès-Nancy, France [2] National university of technology and sciences NITU "MISIS", Moscow, Russia [3] Research Institute of Comprehensive Exploitation of Mineral Resources, Russian Academy of Sciences (IPKON RAS), Moscow, RUSSIA	The effect of high power nanosecond electromagnetic pulses on the selectivity of pyrite flotation from arsenopyrite
8	**O.N Krivolapova**., V.P.Tarasov	National University of Science and Technology "MISIS", RUSSIA	Research of Li_5AlO_4 synthesis
9	L.O. Filippov[1,2], **A.S. Matinin**[1,2], V.D. Samiguin[2], I.V.Filippova[1,2]	[1] Laboratoire Environnement et Minéralurgie, UMR 7569 CNRS Nancy Université, Vandoeuvre-lès-Nancy, France [2] National University of Science and Technology «MISIS» (MISIS), Moscou, Russia	Effect of ultrasound on flotation selectivity in the reactor-separator
10	**V.V. Severov**[1,2] L.O. Filippov[1,2], I.V. Filippova[1,2],	[1] Laboratoire Environnement et Minéralurgie, ENSG-INPL, Nancy-Université, France [2] National University of Science and Technology "MISIS", Russia	Flotability of Fe-bearing silicates in the presence of starch - Adsorption and spectroscopic studies.
11	**M. Siadat**, M. Lumbreras	LASC/Université de Metz, France	Discrimination between different nature of painting and substrates with an electronic nose
12	**L. Toniazzo**, V. Fierro, F. Braghiroli, G. Amaral, A. Celzard	IJL – UMR UHP - CNRS 7198, ENSTIB, Epinal, France	Biosorption of model pollutants in liquid phase on raw and modified rice husks
13	**E. Zelinskaya**	Irkutsk State Technical University, Russia	Freezing as the non-traditional method of hydromineral raw enrichment

Poster Board Number	Name	Institution	Title of presentation
		Processing And Mechanical Aspects Of Materials	
14	**T. Lyubimova**[1], V. Konovalov[1], Ya. Parshakova[1], N. Shumilova[1], A. Lepikhin[2]	[1]Institute of Continuous Media Mechanics UB RAS, Perm, Russia [2]Mining Institute UB RAS, Perm, Russia	Three-dimensional numerical modeling of admixture transport in turbulent flow at river confluence
15	**D. Lyubimov**, S. Mingalev	[1]InsPerm State University, Perm, Russia	Pressure-driven peristaltic flow of biofluid in a channel with elastic walls
16	**T. Lyubimova**[1], A. Mailfert[2]	[1]Institute of Continuous Media Mechanics UB RAS, RUSSIA [2]INPL-ENSG – LEM, France	Thermal convection in a closed cavity in zero-g space conditions with stationary magnetic forces
17	**A.A. Komissarov,** Yu.B. Sazonov	The Federal State Educational Institution of the Higher Professional Education "National University of Science and Technology "MISIS", RUSSIA	Ultra-fine grained structure formation in steels by heat treatment in the range of subcritical temperatures
18	**A.A. Komissarov**	The Federal State Educational Institution of the Higher Professional Education "National University of Science and Technology "MISIS", RUSSIA	Improving reliability of structural steels by heating at subcritical temperatures
19	**P-J. Méausoone**, S. Auchet	LERMAB-ENSTIB, EPINAL, France	Quality of wood machining with bandsaws. Blade - bandmill interactions and vibrations
20	**T.Tarasova**	Bauman Moscow State Technical University, RUSSIA	Influence of laser radiation on structure and properties of steels and alloys
21	**V.Y. Turilina**[1], S.A. Nikulin[1], S.V. Dobatkin[2]	The National University of Science and Technology "MISIS, RUSSIA "	High - Strength Steel with the Reverted Austenite
		Functional Coatings Coatings	
22	**Ph.V. Kiryukhantsev-Korneev**, E.A. Levashov, D.V. Shtansky	Scientific-educational center of SHS, National University of Science and Technology "MISIS", RUSSIA	Chemical analysis of nanostructural and nanolayer coatings using radiofrequency glow discharge optical emission spectroscopy

Journal of Physics: Conference Series **416** (2013) 011003 doi:10.1088/1742-6596/416/1/011003

Peer review statement

All papers published in this volume of *Journal of Physics: Conference Series* have been peer reviewed through processes administered by the proceedings Editors. Reviews were conducted by expert referees to the professional and scientific standards expected of a proceedings journal published by IOP Publishing.

Journal of Physics: Conference Series **416** (2013) 012001 doi:10.1088/1742-6596/416/1/012001

Photorefractive Damage in congruent $LiNbO_3$. Part I. Zinc doped Lithium Niobate Crystals

M Aillerie, P Bourson, M Mostefa, F Abdi & M D Fontana
LMOPS, University Paul Verlaine of Metz and Supelec, 2 rue E. Belin, 57070 Metz, France

email: aillerie@metz.supelec.fr

Abstract. The experimental measurements of the photorefractive damage in a series of zinc doped congruent lithium niobate are reported and discussed. The first experimental technique used is based on the measurement with time of the photoinduced distortion of an HeNe beam spot at 632.8 nm. The second method is based on the direct measurement of the photoinduced birefringence variation with time, using a powerful laser beam at 514.5 nm as gating radiation, and a laser weak beam at 632.8 nm as probe. The dependences of the photorefractivity and of the photosensitivity on the power and dopant concentration have been investigated. The photorefractive damage is highlight by the electro-optic behaviour with the zinc concentration and the defect structure of the doped crystals. We conclude that doping above a second threshold concentration with divalent Zn ions leads to a significant decrease of the photorefraction with respect to pure congruent crystals. In this concentration range, $LiNbO_3$:Zn are strongly photorefractive damage resistant and thus present an interesting alternative for modulating devices to the stoichiometric lithium niobate. Link to the high electro-optic coefficients, these crystals are very suitable for Q-switching applications.

1. Introduction

Lithium niobate (LN) crystals have good optical quality and uniformity, an excellent electro-optic, piezoelectric, acoustic and nonlinear optical properties. Generally, due to their physical properties but also because of their growth facilities, LN crystals with congruent composition, corresponding to a ratio [Li]/[Li+Nb]=48.6%, are generally used in a lot of optical and acoustic devices. Nevertheless, the main drawback of congruent lithium niobate for successful applications in devices is its relatively low optical damage resistance compared to other competing oxide crystals such as $LiTaO_3$ (LT) or BaB_2O_4 (BBO) [1,2]. The optical damage corresponds to local modifications of the refractive indices of the medium induced by light beam [1]. The main process in LN crystals is the photorefractive damage due to the presence of space charge field induced by the optical beam, which generate mobile charge carriers in the bright region and migrate toward the dark zones. Consequently, induced via the electro-optic effect, the space charge field thus creates modifications of refractive indices. For general considerations, another origin of the optical damage could be due to thermal effects, such as the thermo-optic effect. In this effect, the power density of the laser beam induces a local heating in the crystal that changes its refractive indices. Nevertheless, in LN crystal, for relatively high power densities of the beam, these effects are negligible compare to the photorefractive effect [3].

In order to reduce the photorefractive damage in LN or to improve some non-linear optical properties, different approaches have been developed. One of them consists of decreasing the number of intrinsic defects by growing LN crystals with the ratio R=[Li]/[Nb] closer to the one corresponding to the stoichiometric composition (R=1 and R=0.946 for stoichiometric and congruent compositions, respectively). But stoichiometric crystals cannot be grown by conventional Czochralski technique even in melt with ratio R far above one. The growing process is delicate and stoichiometric lithium niobate cannot be yet produce in large quantities and constant qualities as necessary for applications. The second approach consists in doping congruent lithium niobate crystals by appropriate doping ions like divalent ions such as Mg^{2+} [4,5] and Zn^{2+} [6] or trivalent ions such as Sc^{3+} and In^{3+}, which are known, for specific concentrations, to improve the optical damage resistance of LN crystals. Moreover, it has been shown that doping with hafnium also reduces the optical damage in congruent LN crystals [7]. Significant reduction of optical damage was observed at doping concentrations about 5.5 mol.% MgO [4,5], 7 mol.% ZnO [6], 1.5 mol.% Sc_2O_3 [8], 1.5 mol.% In_2O_3 [9] and 4 mol. % HfO_2 [7].

The purpose of this paper is to investigate the optical damage in four congruent crystals doped with increasing ZnO concentrations. The techniques used for these characterizations were already described in previous papers [10,11]. We present the behavior of the photoinduced birefringence change at saturation and the photoconductivity as a function of the power densities of an Argon pump beam for the two sets of crystals. In order to clarify our analysis, the results will be confronted with those obtained in electro-optical characterizations made on the same crystal samples.

It is to be of note that the optical damage investigations in congruent lithium niobate are presented in two publications. The present part is dedicated to the photorefractive behavior of zinc doped congruent LN, which is analyzed after the presentation of the two methods used for the characterizations. Then, the second part follows with the presentation of the photorefractive damage of a series of magnesium doped congruent LN done within the same experimental processes and which finishes with a general discussion on the behavior and its comparison of these two doped LN families. Both publications were written to be self-consistent and to be read independently.

2. Materials and Methods

2.1. Crystal presentation

We have performed photorefractive measurements in four congruent zinc doped crystals. In order to discard the influence of other involuntary kinds of impurity, crystals have been prepared in the same conditions. This set of crystals is made up of LN crystals doped with different concentrations of zinc and provided by T. Volk [6]. These crystals were grown using the Czochralski method from congruent melt to which Zn was added in its oxide form. The concentrations of doping ions in the crystals are listed in Table 1; they are based on the concentration in the melt using the empirical equation proposed by Schlarb et al. [13].

The optical quality of all crystals was checked by an optical transverse mapping, which shown in the visible range a good optical uniformity without polishing or growth striations, respectively at the surface or inside crystals. The nonuniformity of the birefringence along the axis does not exceed 10-5 cm-1. For a suitable comparison, Y-cut plates with the same thickness equal to 0.68±0.02 mm were especially prepared for the laser-induced birefringence measurements presented in this paper.

To determine the role of Zn and LN structural defects in the discussion, we have introduced experimental results obtained with a pure congruent LN crystal; unfortunately, this crystal was not grown in same conditions than the Zn series and additional effects due to impurities (as iron at various valence states) have to be considered.

Table 1. Labels and concentrations of Zinc doped LN crystals

Crystal	Me-oxide in the crystal (mol.%)
Zn-1	2.90
Zn-2	3.78
Zn-3	6.40
Zn-4	7.75

2.2. Experimental methods

Photorefractive damage measurements were carried out by means of the two methods presented in details in Ref. [10]. We just summarize here, the fundamentals of both methods necessary for a good understanding of the relevant parameters measured in the experimental work to characterize the photorefraction of the samples.

The first method consists of the direct measurements of the self-defocussing of a laser beam photo-induced in the crystal. This method is based on the implementation of the pseudo Z-scan method where the sample is fixed at the waist position of the laser beam and measurements of transmitted intensity are done through the sample and a pinhole.

The curves of variation of transmission as a function of time can be fitted by a first-order exponential function:

$$T(t)(\%) = I/I_0 = T_{min} + \Delta T \exp(-t/\tau) \quad (1)$$

where T_{min} is the minimal transmission at saturation, $\Delta T=100-T_{min}$, I_0 and I, the transmitted beam intensities at t=0 and at a time corresponding to the saturation of the scattering respectively and τ is the dielectric relaxation time which is linked to the dielectric constant and photoconductivity by:

$$\tau = \varepsilon_0\varepsilon_{33} / \sigma_{ph} \quad (2)$$

In this method the relevant characteristics are the variation ΔT of the transmitted laser beam and σ_{ph}.

The second method was implemented for the direct determination of the photoinduced birefringence change $\delta\Delta n$, using a very sensitive set-up, based on a modified FDEOM method detailed in Ref. [14] and presented in its adaptation form for photorefractive measurements in Ref. [10]. Within this method, the optical damage is measured by monitoring the change in the magnitude of the induced birefringence as a function of irradiation time. Two aligned and concentric laser beams with different wavelengths are used for the pump and for the probe. The focused pump cw laser beam induces in the crystal an internal space charge electric field due to the displacement of carriers between the bright and the dark zone. This field involves a modification of the refractive indices via to the electro-optic effect and as a consequence, the output phases of both optical beams are modified. The induced change of the probe beam phase is directly monitored on an oscilloscope screen, which yields to the value of the light-induced birefringence changes $\delta\Delta n$, via the well-known equation

$$\delta\Delta n = \delta\Gamma\lambda/2\pi L \quad (3)$$

where $\delta\Gamma$ is the induced phase shift, λ is the wavelength of the probe beam and L is the width of the sample. The changes were monitored as a function of irradiation until saturation. The curves obtained within this method can be fitted with a first order exponential time response function given by

$$\delta\Delta n(t) = \delta\Delta n_s\,[1 - \exp(-t/\tau)] \qquad (4)$$

where τ is the characteristic time of the PR effect (as in Eqs. 1 and 2) and $\delta\Delta n_s$ the saturated value of the birefringence change. In this method the relevant characteristics are $\delta\Delta n_s$ and σ_{ph}.

3. Results and discussion

3.1. Self-defocusing measurements

We have measured the self-defocussing of an He-Ne spot beam at a wavelength of 633nm passing through the crystal and a power density at the focal plane equal to 475 W/cm^2. For each crystal, we have recorded as a function of the illuminated time, the two beam intensities I and I_0, to determine the characteristic T(t) (Eqs. 1). From the fits of the T(t), we deduce the maximum light-induced scattering ΔT corresponding to the saturation obtained after few minutes in the change of T(t), $\Delta T(t)$ and finally, via Eq. 2, the value of the photoconductivity σ_{ph}.

We have plotted in Fig 1, experimental results obtained as a function of Zn concentration in crystals. We can see, in Fig. 1 that the light-induced scattering ΔT decreases monotonically with increasing Zn contents. No concentration threshold seems to appear in this concentration range. The behaviour of the photoconductivity is opposite among that of ΔT corresponding to a monotonous increase.

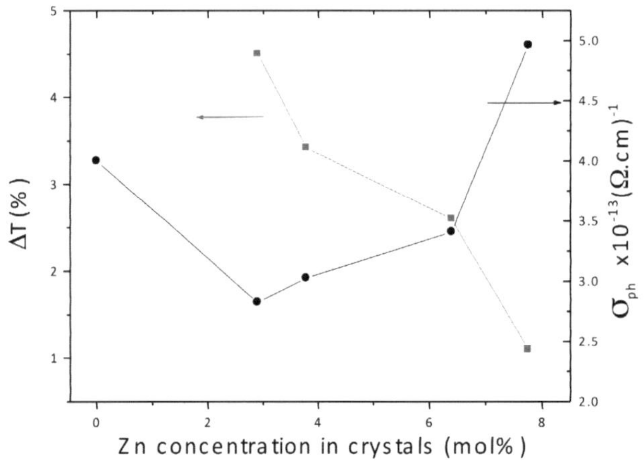

Figure 1. Variations of the transmittance ΔT and photoconductivity σ_{ph} as a function of Zn concentration in LN:Zn crystals for a laser beam intensity equal to 475 W/cm^2 at λ = 633nm.

An interpretation of these behaviors will be suggested below in the discussion with considerations of electro-optic experimental results obtained on these crystals

3.2. Photoinduced birefringence change

We have measured the photoinduced birefringence changes as a function of time, $\delta\Delta n(t)$ with a focused low-intensity He-Ne laser at a wavelength of 633nm for the probe beam and a focused cw argon laser beam laser oscillating at 514.5nm with a power density of 1.2 kW/cm^2 for the pump beam. For each crystal we have determined the light-induced phase shift as a function of the illuminated time, $\delta\Gamma(t)$ and we have calculated, using Eq. 3, the value of the maximum birefringence changes at saturation $\delta\Delta n_s$ and by Eq. 4, via the dielectric relaxation change, the photoconductivity σ_{ph}. At this power density and wavelength for the pump beam, we have report in Fig. 2 the photoinduced birefringence change at saturation and the photoconductivity versus dopant concentration. In our measurements, $\delta\Delta n_s$ is the sum of birefringence changes caused by the photorefractive effect and that caused by the thermo-optic effect, both due to the Ar-laser irradiation. Nevertheless when the Ar-laser irradiation is stopped after illumination, no significant birefringence changes occurred even some hours later. That means that the birefringence change due to thermal effects is negligible and that the measured $\delta\Delta n$ in this samples is predominantly induced by the photorefractive effect, which confirm results obtain by Furukawa et al. in pure LN crystals [3]. We can also consider that the relative stability after the end of illumination indicates a weak dark conductivity σ_d in all these crystals compared to the photoconductivity σ_{ph}, which is also confirmed by other authors in pure and doped crystals [2]. Thus, the observed effect is mainly due to the photovoltaic space charge field generated along c-axis. We emphasize that under these experimental conditions, a low intensity probe and high intensity pump focus beams, photorefraction is only due to the bulk photovoltaic effect generated by the pump beam.

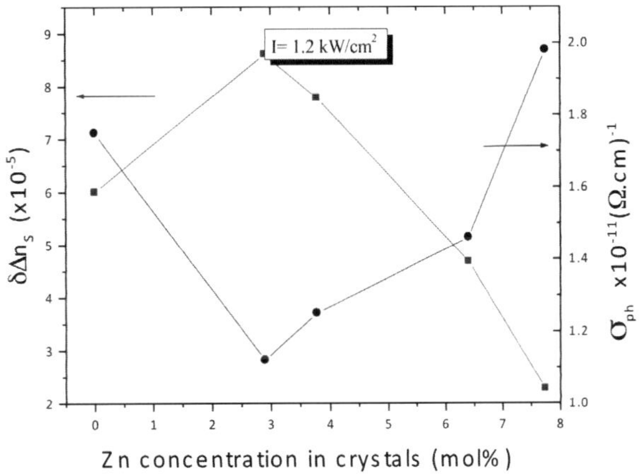

Figure 2. Photoinduced birefringence change at saturation and photoconductivity as a function of Zn concentration in LN:Zn crystals for a laser beam intensity equal to 1.2kW/cm^2 at λ = 514.5 nm.

We can see in Fig. 2 that $\delta\Delta ns$ have the same behavior versus Zn concentration than the intensity variations, ΔT, measured with the first method and reported in Fig. 1. Moreover, these measurements done with a different method as those presented above confirm the values and the behaviour of the photoconductivity.

The $\delta\Delta ns$ dependence on the specific value of the optical damage resistant doping concentration in crystals was earlier connected with the changes of photoconductivity σ_{ph} [14]. Doping with an optical damage resistant impurities has the same effect on the optical damage resistance as increasing with the composition ratio R of pure crystal. This dependence between $\Delta\delta n_s$ and σ_{ph} with the doping concentration could be also correlated to the amount of intrinsic defects.

4. Discussion

The optical damage resistance of LN crystals doped with specific optical damage-resistant ions has been widely discussed (see by example Ref. [15]). According to the simple photovoltaic model writing by expression [16]

$$\delta\Delta n \approx r_{eff} K\alpha\, I/(\sigma_{ph}+\sigma_d) \qquad (5)$$

where r_{eff} is the effective electro-optic coefficient, k the Glass constant and α the optical absorption coefficient. We can explain a decrease in photoconductivity results in an increase in the birefringence change.

In order to understand the role in the variation of photorefraction with the amount of optical damage resistant doping ions in lithium niobate, we have considered the photorefractive sensitivity, δS like defined by Jermann et al. in Ref. [15]. In ferroelectric crystals, as the main mechanism that controls the electron transport in the absence of an external electrical field is the photovoltaic effect, δS is directly connected to the bulk photovoltaic current. The photorefractive sensitivity is considered as a figure of merit for the comparison of the photorefractive behaviour of crystals and is defined as:

$$\delta S=\Delta n_s/I\tau= r_{eff}(j_{ph}/\varepsilon_0\varepsilon_r) \quad , \qquad (6)$$

where I is the light beam intensity and j_{ph} is the photovoltaic current density. Fig. 3 shows the variations of δS versus Zn ions concentration.

The photosensitivity can be suitably used to determine and discuss the PR properties and their origin regardless to the model. Indeed a large resistance to PR optical damage requires large photoconductivity and low photoinduced birefringence change $\delta\Delta n$ related to the PV current. This current and the photoconductivity have generally a complex dependence with the laser intensity, the composition and the concentration of optical damage resistant ions like shown in Fig. 3.

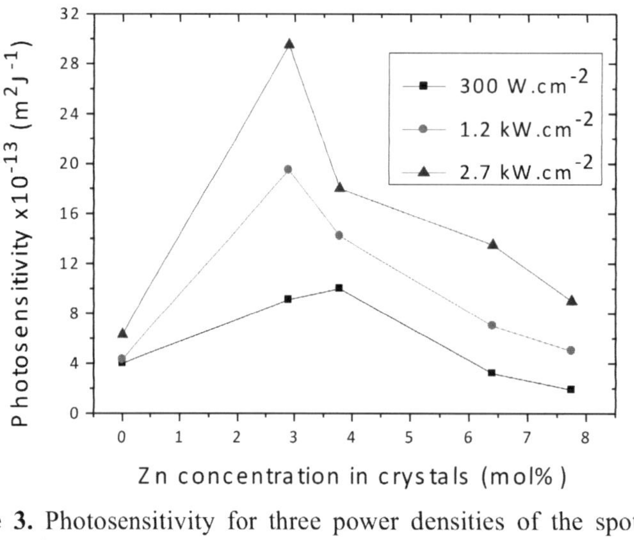

Figure 3. Photosensitivity for three power densities of the spot beam as a function of Zn concentration in LN:Zn crystals.

We can hence deduce from these curves of δS versus Zn dopant concentration, that the highly doped crystal is the most damage resistant.

The characteristic photorefractive time is link to the dielectric constant ε_0 and the relative dielectric constant ε_r, so we can express, from Eq. 6, the photovoltaic current as

$$j_{ph} =\Delta n_s \, \sigma_{ph}/r_{eff} \qquad (7)$$

To calculate j_{ph} and to define parameters that occur in the photovoltaic properties of LN crystals, we have also performed measurements of the effective electro-optic (E/O) coefficient involved in the optogeometric configuration corresponding to the y-cut samples i.e. $r_{eff}=r_c=r_{33}-(n_0/n_e)r_{13}$. These samples are obtained by the same growing process and in the same conditions that samples used for PR measurements. On the other hand, we have disposed for this part of the study more samples than for the PR measurements. The experimental method implemented for the E/O measurements, based on a Sénarmont optical arrangement, is the FDEOM method largely detailed in [20]. This method, from which is originate the MFDEOM method [14] used for the photorefractive characterization in this study, offer a great accuracy in the determination of the dc electro-optic coefficients, better than 5%, in the determination of the dc electro-optic coefficients. The dc EO phase retardation in the crystals under test allows the calculation of the coefficients r_c [21] and its dependence versus the dopant concentrations, which is reported in Fig. 4. The electro-optic coefficient, r_c versus Zn doping concentration, plotted on figure 4, presents a large amplitude of a non monotonous behaviour.

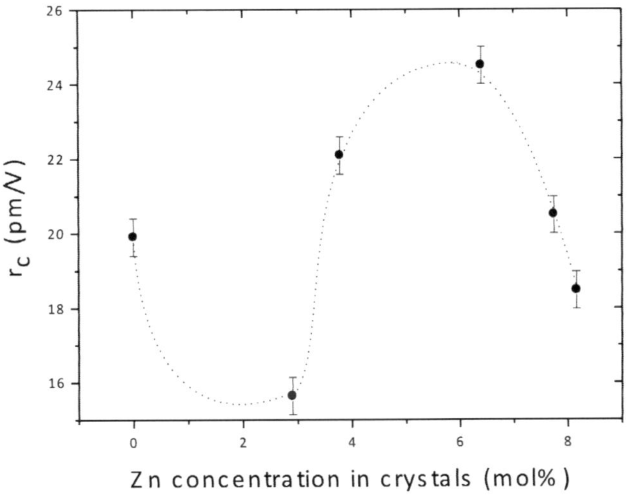

Figure 4. EO coefficient rc versus Zn concentration in the crystal (The line is just a guide for the eyes).

In this Fig. 4, three concentration ranges may be distinguished. In the low concentration range up to 2-3 mol% Zn r_c strongly decreases down to a minimum; in an intermediary range up to around 7.5 mol% Zn, r_c abruptly grows reaching a maximum; finally, in high concentration range, r_c sharply decreases.

Finally, using these experimental values in the Eq. 7, the bulk photovoltaic current can be calculated and its behaviour is plotted in Fig. 5 in function of Zn concentration.

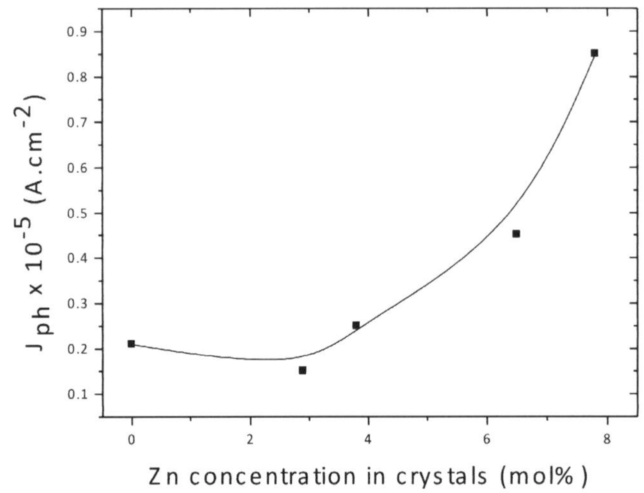

Figure 5. Photovoltaic current versus Zn concentration in LN:Zn crystals for a laser beam intensity equal to 1.2kW/cm2 at l = 514.5 nm (The line is just a guide for the eyes).

The bulk photovoltaic current of Zn doped LN samples presents quite the same behaviour that the photoconductivity in spite of the large non-monotonous variation in the electro-optic coefficients with the doping concentration. Thus, these data clearly indicate that the main contribution to the variation of photorefraction comes from variations in $1/\sigma_{ph}$ with the concentration.

Now, we try to correlate the dependencies of the photorefractive and electro-optic properties on Zn concentrations to variations in the defect structure. In literature, the low concentration range, below the first extremum at 2-3 mol% Zn, is often regarded as a specific range in the concentration dependence of some properties such as photoconductivity and photorefraction [22] and the concentration of 7 mol% Zn is usually referred to as a concentration threshold and manifests itself as a second extremum in the concentration dependence of other optical properties of LN:Zn such as refractive indices, IR absorption, phase matching temperature of SHG, etc... [16,17]. An analysis of the influence of Zn doping on electrooptical properties and structure parameters of LN crystals was proposed in Ref [21].

The two proposed models of defect structure in LN doped by divalent ions, based on Li [15-17] or Nb [21,23] vacancies for the starting pure congruent LN agree that in low doping concentration the divalent ions introduced in the crystal replace niobium on lithium site, Nb_{Li} so-called niobium antisites. In a recent modelling of the various phases of the substitution process in Mg doped LN crystals [23], based on the Nb vacancies model suggested by Abdi et al. for pure undoped LN from congruent to stoichiometric compositions [24], we have shown that in this low concentration range, this substitution process is accompanied by a decrease of Nb vacancies and a change in the Li vacancies to ensure the condition of electrical neutrality of the cell. The first extremum is reached when the total amount of vacancies is minimum. It is to be of note that at this extremum position and above, all cited models describe the substitution and compensation process by the same way. Above this extremum by analogy with phenomena that occurred in Mg doped LN crystals [19], we suggest that Zn ions replace Li ions in their own natural sites up to the threshold concentration corresponding to the lowest limit Li/Nb of the ZnO-$LiNbO_3$ phase diagram. This replacement is accompanied by the creation of lithium vacancies for charge compensation. Finally, when we consider the incorporation of Zn in the high-concentration range above the threshold, the structural data have shown that the most probable compensation mechanism in this case is the self-compensation of $[Zn_{Nb}]^{3-}$ by $[Zn_{Li}]^{+}$ which required a concentration ratio Zn_{Nb}/Zn_{Li} =33% to conserved the charge neutrality [15]. In this range, the incorporation of Zn ions induces a huge diminution of the Li vacancies as confirmed by various authors [15,22,23]. Therefore, the threshold concentration corresponds to a maximum concentration of Li vacancies and it is known that the photocurrent and the photorefractive properties depend on the concentration on lithium vacancies.

After this description, we can correlate the dependencies of the photorefractive properties on the defect structure of LN:Zn. Nevertheless, we have first to consider the concentration of Zn ions in crystals. The crystal Zn-1 has a dopant concentration, which corresponds to the extremum concentration herein mentioned, whereas the highly doped crystal, Zn-4 is at the threshold concentration. For low light intensities ($I < 1MW/m^2$), as Zn ions only occur as Zn^{2+}, it is allowed that they do not directly contribute to charge transport processes responsible of the photovoltaic effect and of the bulk photoconductivity, which is mainly dominated by Fe^{2+} ions. Nevertheless, at high light intensities ($I \geq 1MW/m^2$), as used for this study, the previous one center model [15] is not yet available and additional contributions to the photoconductivity and to the bulk photovoltaic effect due to Zn ions and to charge compensating vacancies occur. Thus, the photoconductivity is due to competition effects originate from niobium antisites and lithium vacancies, which means a major influence of substitution phenomena that occurs in the lithium site. At low Zn concentrations, a decrease of the photoconductivity is linked to the decrease of niobium in antisite and we can observe in this concentration range an increase of the photorefraction and of the photosensitivity. For medium concentration range (2-3 mol% $\geq$ [Zn] $\geq$ 7 mol%), due to the disappearance of niobium antisite and the increase of lithium vacancies, it appears a large increase of the photoconductivity accompanied by a decrease of the photosensitivity. It was shown [25] that at low impurity concentrations, i.e. in the two described ranges above, the photoconductivity is of n-type whereas above the threshold, it becomes dominantly p-type and increases with the increase of the contribution from the hole component with Zn at the expense of a decrease of the concentration of hole traps i.e. V_{Li} [17].

Like observed in undoped $LiNbO_3$ crystals, the E/O properties is a competition between the deformation and the polarizability of NbO_6 octahedra and the conductivity due to Li vacancies. Therefore, we can analyze the role of Zn, as of Mg in Mg-doped $LiNbO_3$ on the EO properties, by the consideration of Nb, Li and vacancies amounts in doped crystals, which is similar to a variation of the composition from the congruent up to the stoichiometric ones for pure LN crystals. At low concentration, the E/O coefficient is linked to diminution of the total amount of vacancies and to the decrease of Nb_{Li}. Above this extremum, the creation of new Li vacancies induces an increase of the electro-optic coefficients up to the threshold. In the high-concentration range, the sharply decrease of the lithium vacancies is accompanied by a new decrease of the electro-optic coefficient.

5. Summary

The photorefractive damage of zinc-doped congruent lithium niobate crystals agrees with the existence of a first extremum concentration for Zn doped crystals around 2-3 mol.% and a threshold about 7 mol%. We have shown that doping above the first extremum concentration with divalent Zn ions leads to a significant decrease of the photorefraction with the increase of Zn concentration.

In Zn doped LN crystals with concentrations above 7 mol%, we observed a large photorefractive damage resistant and thus, due to lower growing difficulties, these crystals present a very interesting alternative in some optoelectronic applications compared to the stoichiometric lithium niobate one. Moreover, the electro-optic coefficients in this concentration range are high and thus, these crystals are very suitable for Q-switching applications.

Acknowledgments

We are grateful to Tatyana Volk from Russian Academy of Sciences (Russia), which provide the zinc doped crystals studied in this work.

Journal of Physics: Conference Series **416** (2013) 012001 doi:10.1088/1742-6596/416/1/012001

References

[1] Krätzig E and Schirmer O F 1988 *Photorefractive Materials and Their Applications I*, ed P. Günter and J.P. Huignard (Springer-Verlag, Berlin) p 131-166.
[2] Volk T and Wöhlecke M 1998 *Ferroelect. Rev.* **1** 195.
[3] Polgar K, Kovacs L, Foldvari I and Cravero I 1986 *Sol. Stat. Comm.* **59 (6)** 375.
[4] Furukawa Y, Sato M, Kitamura K, Yajima Y and Minakata M 1992 *J. Appl. Phys.* **72, 8**, 3250.
[5] Bryan D A, Gerson R and Tomaschke H E 1984 *Appl. Phys. Lett.* **44** 847.
[6] Volk T R, Pryalkin V I and Rubinina N M 1990 *Optics Letters* **15** 996.
[7] Razzari L, Minzioni P, Cristiani I, Degiorgio V and Kokanyan E 2005 *Appl. Phys. Lett.* **86**.
[8] Volk T R and Rubinina N M 1992 *Ferroelectrics Letters* **14** 37.
[9] Yamamoto J K, Kitamura K, Iyi N, Kimura S, Furukawa Y and Sato M 1992 *Appl. Phys. Lett.* **61** 2156.
[10] Fontana M, Chah K, Aillerie M, Mouras R and Bourson P 2001 *Opt. Mater.* **16** 111.
[11] Schlarb U, Wöhlecke M, Gather B, Reichert A, Betzler K, Volk T and Rubinina N 1995 *Opt. Mat.* **4** 791.
[12] Aillerie M, Abdi F, Fontana M D, Theofanous N and Abarkan E 2000 *Rev. Sci. Instrum.* **71-4** 1627.
[13] Volk T R, Maximov B, Sulyanov S, Rubinina N and Wöhlecke M 2003 *Opt. Mat.*, **23-1**, 229.
[14] Volk T, Rubinina N and Wöhlecke M 1994 *J. Opt. Soc. Am. B* **11 9** 1681.
[15] Jermann F, Simon M and Krätzig E 1995 *J. Opt. Soc. Am. B* **12** 2066.
[16] Theofanous N, Aillerie M, Fontana M D and Alexakis G 1997 *Rev. Sci. Instrum.* **68-5** 2138.
[17] Donnerberg H, Tomlinson S M, Catlow S R A and Schirmer O F 1989 *Phys. Rev.* B **40** 11909.
[18] Schlarb U and Betzler K 1994 *Phys. Rev. B* **50-2** 751.
[19] Grabmaier B C, Wersing W and Koestler W 1992 *J. Cryst. Growth* **110** 340.
[20] Donnerberg H, Tomlinson S M, Catlow S R A and Schirmer O F 1991 *Phys. Rev.* B **44** 4877.
[21] Abdi F, Aillerie M, Fontana M D, Bourson P, Volk T R, Maximov B, Sulyanov S, Rubinina N and Wöhlecke M, 1999 *Appl. Phys.* B **68** 795.
[22] Zhang Y 2004 Thesis of Metz University.
[23] Abdi F, Aillerie M, Bourson P, Fontana M D 2009 *J. Appl. Phys.* **106** 033519.
[24] Abdi F, Fontana M D, Aillerie M, Bourson P 2006 *Appl. Phys.* A **83-3** 427.
[25] Deng J C, Wen J K, Wu Z K, Wang H F 1994 *Appl. Phys. Lett* **64** 2622.

Photorefractive Damage in congruent $LiNbO_3$. Part II. Magnesium doped Lithium Niobate Crystals

M Aillerie, P Bourson, M Mostefa, F Abdi and M D Fontana
LMOPS, University Paul Verlaine of Metz and Supelec, 2 rue E. Belin, 57070 Metz, France

E-mail: aillerie@metz.supelec.fr

Abstract. The photorefractive effect in lithium niobate (LN) crystals is one the main drawback for its integration in optoelectronic devices using high light intensity. Doping congruent LN crystals by appropriate dopants like divalent ions such as Mg^{2+} are known, for specific concentrations, to improve their optical damage resistance. We present experimental measurements of the photorefractive damage in a series of magnesium doped congruent lithium niobate, performed with two experimental technique based on the measurement with time of the photoinduced distortion for the first one and based on the direct measurement of the photoinduced birefringence variation with time for the second. The dependences of the photorefractivity and of the photosensitivity on the power and dopant concentration have been investigated and discussed with the consideration of the electro-optic properties dependence with the defect structure of the doped crystals. We conclude that that doping above a second threshold concentration with divalent Mg ions leads to a significant decrease of the photorefraction with respect to pure congruent crystals. We conclude that doping LN congruent crystals with Mg strongly increase the photorefractive damage resistance and, in association to the interesting electro-optic coefficients, LN:Mg present an interesting alternative for modulating devices to the stoichiometric LN, which is difficult to growth in high quality and in large quantity.

1. Introduction

$LiNbO_3$ (LN) crystals are of great interest for applications in optics thanks to their large electro-optic (EO) and non linear optical coefficients [1,2]. Nevertheless, the integration of non stoichiometric LN crystal in electro-optic modulators, Q-switches or frequency converters is hampered by a relatively small laser-damage threshold (equal to 0.3 GW/cm^2 for the congruent composition that we can compared to BaB_2O_4 crystals (BBO) equal to 10 GW/cm^2) [3]. In LN crystal, the optical or photorefractive damage is one of the numerous properties that depends on the charge transport properties, which is directly linked to the nature and the concentration of intrinsic defects like antisite Nb_{Li}, Li- and Nb-vacancies [4], present in the LN structure. The first origin of the optical damage could be due to thermal effects, such as the thermo-optic effect or pyroelectric effects due to a local heating in the crystal along the light path changing the refractive indices seen by the beam but for relatively high power densities, these effects are negligible compare to the photorefractive effect [5]. So, the main process at the origin of this damage is due to the presence of space charge field induced by the optical beam, which generate mobile charge carriers in the bright region and migrate toward the dark zones. This space charge field induces via the electro-optic effect, modifications of the refractive indices seen by the beam. The growth of crystals with stoichiometric composition is one of the ways explored by growers to reduce the optical damage [6,7]. Indeed, the growing of LN crystals with a ratio R=[Li]/[Nb] closer or equal to one, corresponding to the stoichiometric composition, decrease the number of intrinsic defects. But, even with ratio R far above one in melt, the growth by conventional Czochralski technique cannot produce large quantities and constant qualities of stoichiometric crystals as necessary for applications. These difficulties to obtain large stoichiometric crystals has opened an alternative way to increase the laser threshold consisting in doping congruent LN crystals with divalent or trivalent ions such as Mg, Zn, In or Sc ions. [8,9]. Simultaneously some linear and non-linear optical properties are influenced by these dopant ions [8-10]. Among them, Mg is one of the most frequently used with Zn. Zhong et al [11] showed that the damage is strongly reduced by the introduction of 4.6 mol% Mg in a congruent melt while significant reduction of optical damage was observed at dopant concentrations exceeding 5.5 mol.% MgO by Furukawa et al [7] and Bryan et al [12].

The purpose of this paper is to investigate the optical damage in four congruent crystals doped with increasing MgO concentrations. The techniques used for these characterizations were already described in previous papers [13]. We present the behavior of the photoinduced birefringence change at saturation and the photoconductivity as a function of the power densities of an Argon pump beam for this set of crystals. In order to clarify our analysis, the results will be confronted with those obtained in electro-optical characterizations made on the same crystal samples.

It is to be of note that the optical damage investigations in doped congruent LN are presented in two publications. The present one is dedicated to the photorefractive behavior of magnesium doped congruent LN. It follows a first publication, which discussed the photorefractive properties of Zn doped congruent LN, which were measured within the same experimental processes. Even if both publications were written to be self-consistent and to be read independently, the present publication largely summarizes the experimental techniques and proposes a general discussion and comparison of the photorefractive behavior as a function of the nature and concentration of dopant in these two set of doped LN families.

2. Crystal presentation

We have performed photorefractive measurements in four congruent magnesium doped crystals with different concentrations of magnesium. These crystals were grown by K. Polgar [4] using the Czochralski method with a congruent melt added by Mg in its oxidic form. For the determination of the dopant concentration in crystals as reported in Table 1, we have considered the segregation coefficient k_{eff} for Mg doped crystals as proposed by Tan et al. which have reported that k_{eff} for Mg, during growth changed from 1.3 to 0.95 up to 8 mol% Mg content in the melt [14].

Table 1. Labels and concentrations of Mg doped Crystals

Crystal	Mg-oxide in the crystal (mol.%)
Mg-1	0.12
Mg-2	1.04
Mg-3	2.77
Mg-4	4.61

The optical quality of all crystals was checked by an optical transverse mapping, which shown in the visible range a good optical uniformity without polishing or growth striations, respectively at the surface or inside crystals. The non-uniformity of the birefringence along the axis does not exceed 10^{-5}

cm^{-1}. For a suitable comparison, Y-cut plates with the same thickness equal to 0.68±0.02 mm were especially prepared for the laser-induced birefringence measurements presented in this paper.

3. Experimental methods and results

Photorefractive damage measurements were carried out by means of the two methods presented in details in Ref. [13] and mainly summarize in the previous publication devoted to Zn doped congruent LN, as a part of the present work [15].

The first method, based on the implementation of the pseudo Z-scan method consists of measurements in the far field region and after a pinhole of the dynamic variation of the transmitted intensity, adjusted in our measurements to 475 W/cm^2 at the input face of the crystal placed at the focal plane, of an He-Ne spot beam at a wavelength equal to 633 nm. For each crystal, we have recorded as a function of the illuminated time, the two beam intensities I and I_0, to determine the characteristic T(t)

$$T(t)(\%) = I/I_0 = T_{min} + \Delta T \exp(-t/\tau) \quad (1)$$

where T_{min} is the minimal transmission at saturation, $\Delta T=100-T_{min}$, I_0 and I, the transmitted beam intensities at t=0 and at a time corresponding to the saturation of the scattering respectively and τ is the dielectric relaxation time, which allows the determination of the photoconductivity by the equation:

$$\sigma_{ph} = \varepsilon_0\varepsilon_{33}/\tau \quad (2)$$

The experimental results obtained as a function of Mg concentration in crystals are plotted in Fig. 1.

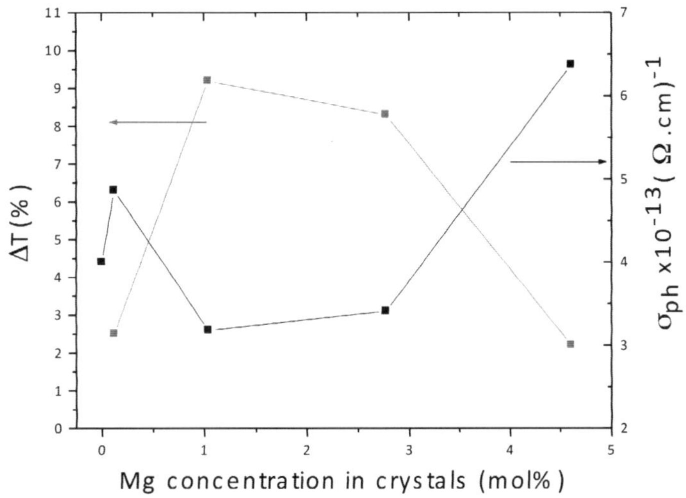

Figure 1. Variations of the transmittance ΔT (see Eq.1) and photoconductivity σ_{ph} as a function of Mg concentration in LN:Mg crystals for a laser beam intensity equal to 475 W/cm^2 at λ = 633nm.

We see that with the concentration of Mg induces an abrupt increase of the defocusing up to a saturation concentration value around 2 mol.% following by a progressive decrease down to the value obtained for the highly doped. On the other hand, the behaviour of the photoconductivity is opposite among that of ΔT with a decrease until Mg concentration around 2 mol% and an increase after this threshold.

The second method allows the direct determination of the photoinduced birefringence change $\delta\Delta n$, using the modified FDEOM [16] adapted for photorefractive measurements. [13] Within this method, the optical damage is measured by monitoring the change in the magnitude of the induced birefringence as a function of irradiation time. The characteristic curve representing $\delta\Delta n$ obtained within this method can be fitted with a first order exponential time response function given by

$$\delta\Delta n(t)= \delta\Delta n_s \left[1 - \exp(-t/\tau)\right] \qquad (3)$$

where τ is the characteristic time of the PR effect and $\delta\Delta n_s$ the saturated value of the birefringence change. We have measured the photoinduced birefringence changes as a function of time, $\delta\Delta n(t)$ with a focused low-intensity He-Ne laser probe beam at a wavelength of 633nm and a focused cw argon laser pump beam oscillating at 514.5nm with a power density of 1.2 kW/cm^2. Fig 2 shows the photoinduced birefringence change at saturation and the photoconductivity versus dopant concentration. $\delta\Delta n$ in these samples is predominantly induced by the photorefractive effect $\delta\Delta n_s$ as no significant birefringence changes occurred after illumination when the Ar-laser irradiation is stopped, which allows us to exclude the contributions related to temperature, such as the thermo-optic effect [5]. The relative stability after the end of illumination indicates a weak dark conductivity σ_d as observed in pure and doped LN crystals [5]. With a low intensity probe and high intensity pump focus beams, we emphasize that photorefraction is only due to the photovoltaic space charge field generated by the pump beam along the c-axis.

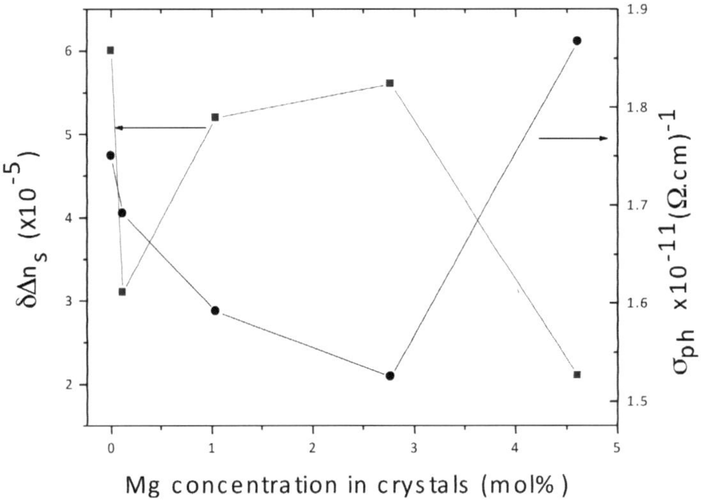

Figure 2. Photoinduced birefringence change at saturation and photoconductivity as a function of Mg concentration for a laser beam intensity equal to 1.2 kW/cm^2 at λ = 514.5 nm

We see in Fig. 2 that $\delta\Delta$ns have the same behaviour versus Mg concentration than the intensity variations, ΔT, reported in Fig. 1. An abrupt increase with the concentration up to a maximum obtained for crystals with intermediary concentration is followed by a progressive decrease down to the value obtained for the highly doped crystal. By else, these measurements done with a different method as those presented above confirm the values and the behaviour of the photoconductivity. The optical damage resistance of crystals labelled Mg-1 and Mg-4 is much higher than that of Mg-2 and Mg-3. We can observe that the concentration threshold is the same for the both properties. These non-monotonous behaviours indicate that the optical damage resistance is lower for both crystals with intermediary Mg concentration and that, even if pure congruent crystal have a high concentration of intrinsic defects, the defocusing is at a level as low as the highly Mg doped.

4. Discussion

The photorefractive sensitivity, δS as defined by Jermann et al. [4] is considered as a figure of merit for the qualification of the photoconduction and the link of the defects in LN crystals. In ferroelectric crystals, δS is directly links to the bulk photovoltaic current and it is defined as

$$\delta S = \Delta n s / I\tau = r_{eff}\,(j_{ph}/\varepsilon_0\varepsilon_r) \ , \qquad (4)$$

where r_{eff} is the effective electro-optic coefficient, I is the light beam intensity and j_{ph} is the photovoltaic current density. Eq. 4 shows that a large resistance to the photorefractive damage requires low photoinduced birefringence change δΔn relative to the PV current.

The sensitivity, δS versus Mg, is plotted in Fig.3, which points out a complicated dependence on the laser intensity and on the concentration of Mg.

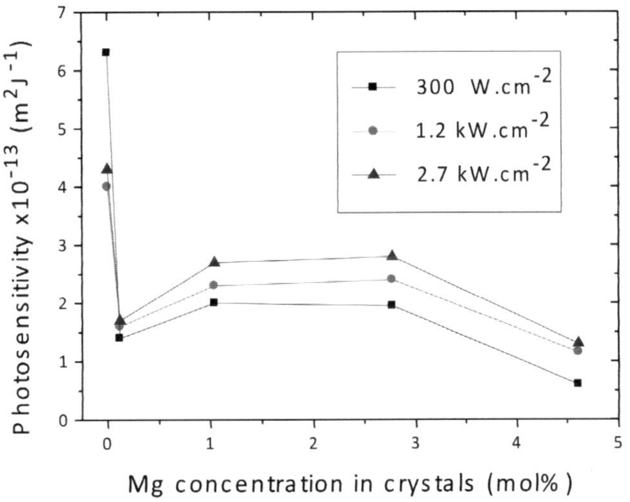

Figure 3. Photosensitivity for three power densities of the spot beam as a function of Mg concentration in LN:Mg crystals.

We can hence deduce from the curves of δS versus Mg dopant concentration, Fig. 3 that the highly doped crystal is the most damage resistant. The behaviour of the sensitivity of Mg doped LN crystals presents the opposite behaviour that the photoconductivity. These data clearly indicate that the main contribution to the variation of photorefraction comes from variations in $1/\sigma_{ph}$ with the concentration.

Various authors [17,18] present the existence of three phases in the substitution and compensation processes involved in LN:Mg doped crystals as a function of the concentration of dopant. We have confirmed [19] by Raman spectroscopy measurements done on the same samples and additional ones with higher Mg concentration above 5 mol%, but unfortunately not convenient for the PR measurements achieved in the present work, the existence of the three phases. In the first phase, when Mg is introduced in the LN lattice up to 3 mol%, it occupy the A sites linked to the natural lithium site and push out the Nb antisites on their "normal" B sites. In the second phase, for Mg doping up to 5 mol%, the site A is strongly affected by Mg ions that incorporate successively instead of Nb antisites, V_{Li} and Li, without any additional disorder induces in the LN structure. On the contrary, the substitution of Li ions by Mg ions in the third phase induces a large disorder on sites A, while the sites

B are not directly affected by Mg incorporation, and which shows only a small change of Nb content along the three phases. The concentration of 5 mol% Mg is often referred to as a concentration threshold and manifests itself as a second extremum in the concentration dependence of other optical properties of LN:Mg such as refractive indices, IR absorption, phase matching temperature of SHG, etc... [20]. Therefore, the threshold concentration corresponds to a maximum concentration of Li vacancies. Thus, in the low range, Mg ions replace the niobium on antisites, Nb_{Li} and as a consequence, we observe a decrease of the photoconductivity with the increase of Mg concentration up to the first extremum. Above this first extremum, the doping by Mg ions is accompanied by an increase of the lithium vacancies. So that, the photoconductivity increase for higher concentration of Mg in crystal.

The two sets of Zn and Mg lithium niobate crystals, presented in part 1 and 2, respectively in this series of contributions came from growers using various powders and methods for preparation, growing and post growing treatments, which can induce additional effects due to the influence of impurities such as iron present in various valence states. Thus the comparison of absolute experimental values presented in this series of contributions should be considered taking into account this fact.

Nevertheless, it is to be of note that all measurements in a series were done on samples providing from the same powders and growing oven and method performed by the same grower. We can consider that the level of involuntary impurities is the same in all crystals. It was shown that the photoconduction process is mainly dominate by the contribution of Fe^{2+} ions for low intensities and that Mg ions they do not directly contribute to charge transport processes. However, this study was done using high light intensities ($I \geq 1MW/m^2$) and additional contributions appear in the photoconduction response attributed to Mg ions and to charge compensating vacancies. The non-linear behaviour of the photoconductivity versus Mg concentration, Fig. 2 and the non proportionality of the sensitivity with the power density of the beam, Fig. 3, confirm this proposal. Thus, the bulk photoconductivity is directly and indirectly influence by the doping concentration [12]. The various processes cited herein that contribute to photorefraction, for Mg doped LN crystals are also involved for crystals doped with Zn ions [15]. Nevertheless, due to the higher electronegativity of Zn ions, the threshold in doping concentration observed in some properties takes place at different levels for Zn than for Mg doped crystals [18].

5. Summary

The photorefractive damage of magnesium-doped congruent lithium niobate crystals agrees with the existence of a first extremum for a doping concentration around 2 mol% and a threshold above 5 mol%.

We have shown that the crystal doped with a concentration close to this threshold is the most damage resistant. The photorefractive behaviour of LN crystal was analyzed considering the substitution process that occurs during the growth of Mg doped LN crystals with the doping concentration. We have analyse the obtain results with those obtain for LN:Zn crystals, which present the same behaviour than LN:Mg crystals when considering the difference of the electronegativity of the two dopant ions. LN:Mg and LN:Zn crystals are strongly photorefractive damage resistant when the doping concentration is closer to the respective threshold concentration associated to each dopant. The Mg and Zn doped congruent LN crystals present an interesting alternative for modulating devices to the stoichiometric LN, which is difficult to growth in high quality and in large quantity.

Acknowledgments

We are grateful to Kataline Polgar from Budapest Research Centre (Hungary), which provide the magnesium doped crystals.

References

[1] Räuber A 1978 *Current Topics in Material Sciences*, ed. E. Kaldis, North Holland, Amsterdam, p 481.
[2] Krätzig E and Schirmer O F 1988 *Photorefractive Materials and Their Applications I*, ed P. Günter and J.P. Huignard (Springer-Verlag, Berlin) p 131-166.
[3] Eimerl D, Velsko S, Davis L and Wang F 1990 *Prog. Cryst. Growth* **20**, 59.
[4] Jermann F, Simon M and Krätzig E 1995 *J. Opt. Soc. Am. B* **12** 2066.
[5] Polgar K, Kovacs L, Foldvari I and Cravero I 1986 *Sol. Stat. Comm.* **59 (6)** 375.
[6] Malovichko G I, Grachev V G, Kokanyan E P, Schirmer O F, Betzler K, Gather B, Jermann F, Klauer S, Schlarb U and Wöhlecke M 1993 *Appl. Phys. A* **56** 103.
[7] Furukawa Y, Sato M, Kitamura K, Yajima Y and Minakata M 1992 *J. Appl. Phys.* **72, 8** 3250.
[8] Volk T, Rubinina N and Wöhlecke M 1994 *J. Opt. Soc. Am. B* **11** 1681.
[9] Volk T, Wöhlecke M, Rubinina N, Reichert A and Razumovski N 1996 *Ferroelectrics* **183**, 291.
[10] Schlarb U and Betzler K 1994 *Phys. Rev. B* **50-2** 751.
[11] Zhong G G, Jian J and Wu Z K 1980 *Proceedings of the Eleventh International Quantum Electronics Conference, IEEE* (New York) 631
[12] Bryan D A, Gerson R and Tomaschke H E 1984 *Appl. Phys. Lett.* **44** 847.
[13] Fontana M, Chah K, Aillerie M, Mouras R and Bourson P 2001 *Opt. Mater.* **16** 111.
[14] Tan H R, Ma Y X, Zhu Q B, Hua W X, Wu Y A and Zhang Y X 1994 *J. Cryst. Growth.* **142, 1-2,** 111.
[15] M. Aillerie, P. Bourson, M. Mostefa, F. Abdi and M. D. Fontana 2010 sub. to *J. Phys: Conf. Series*.
[16] Aillerie M, Abdi F, Fontana M D, Theofanous N and Abarkan E 2000 *Rev. Sci. Instrum.* **71-4** 1627.
[17] Donnerberg H, Tomlinson S M, Catlow S R A and Schirmer O F 1991 *Phys. Rev. B* **44** 4877.
[18] Abdi F, Aillerie M, Fontana M D, Bourson P, Volk T R, Maximov B, Sulyanov S, Rubinina N and Wöhlecke M, 1999 *Appl. Phys. B* **68**, 795.
[19] Abdi F, Aillerie M, Bourson P and Fontana M D 2009 *J. Appl. Phys.* **106** 033519.

On the effect of sulphur on the formation of carbon nanotubes and multibranched carbon nanostructures

E. McRae[1], X. Devaux[2], S. Yu. Tsareva[3], A. N. Kovalenko[3], K. S. Zaramenskih[3], E. V. Zharikov[3]
[1] Institut Jean Lamour UMR CNRS 7198, Nancy Université, Faculté des Sciences et Techniques, B.P. 70239, 54506 Vandoeuvre-lès-Nancy Cedex, France
[2] Institut Jean Lamour UMR CNRS 7198, Nancy Université, Ecole des Mines, CS14234, 54042 Nancy, France
[3] D. Mendeleev University of Chemical Technology of Russia, 9, Miusskaya sq., Moscow, 125047, Russia

Email: Edward.mcrae@lcsm.uhp-nancy.fr

Abstract
While sulphur is a ubiquitous element in the field of carbon nanotube (CNT) synthesis, a full understanding of its role is still open to question. A perusal of the literature reveals that it is sometimes referred to as being a "triggering element" for CNT branching [1], a "growth promoter" [2], a means of fragmenting larger-sized catalyst particles to enhance catalytic activity [3], an element which allows controlling both the number of layers in multi-walled CNTs as well as the overall CNT yield [4]. Sulphur has been put forward as an essential means of enhancing nucleation of CNT growth [5] and as having strong effects on the overall morphology of CVD-synthesised carbon nanostructures [6, 7]. In spite of these obviously beneficial effects, sulphur can also be a catalyst poison, depending on its concentration. All these different potential roles are further complicated by the fact that they depend on the synthesis route and the catalyst used. Using highly sensitive investigation tools is vital since even very minute quantities of sulphur may be of critical importance.
After a brief overview of the above-mentioned background literature, we will present experimental results obtained for catalytic decomposition of benzene on iron nanoparticles in the presence of CS_2. Depending on the exact synthesis conditions, carbon nanoballs, single-walled CNTs, multi-walled CNTs, thick fibres, and/or multibranched nanostructures have been observed. Based on extensive electron microscopy investigations (SEM, TEM, and HRTEM) and on a comprehensive analysis of our CVD process, we propose a phenomenological growth model.

References:

1 J. M. Romo-Herrera et al., *Angew. Chem.. Int. Ed..* **2008**, *47,* 2948-53
2 Q. Liu et al., Carbon **2008**, 46, 1417–23
3 Q/-J. Gong et al., *Compos. Sci. Technol.* **2007**, 67, 2986-89
4 J.Wei et al., *Carbon* **2007**, 45, 2152-58
5 W. Ren et al., J. Phys. Chem. B, **2006**, 110, 16941-46
6 L. Ci et al., *Carbon,* **2003**, 39, 329-35
7 X. Devaux et al., Physica E, **2010,** in press.

Journal of Physics: Conference Series **416** (2013) 012004 doi:10.1088/1742-6596/416/1/012004

Nanostructured Materials for Electrospark Deposition and Disperse-Strengthening by Nanoparticles Diamond Tools

E A Levashov [1], A E Kudryashov [1], E I Zamulaeva [1], V V Kurbatkina [1], V A Andreev [2], A A Zaitsev [1], D A Sidorenko [1]

[1] National University of Science and Technology "MISIS" , Leninsky Prospect, 4, Moscow 119049, Russia, [1]

[2] Kermet Company, Ltd, Moscow

Email: levashov@shs.misis.ru, [2] vvkurb@mail.ru

Three groups of nanostructured electrode materials which are used as anodes in pulse electrospark deposition (PED) and chemical reaction assisted electrospark deposition (CRAPED) were presented: *dispersive-hardening ceramic materials* with effect of simultaneous strengthening of carbide grains and metallic binder result in precipitations; *nanoparticles disperse-strengthened composite materials* with modified structure produced using focused alloying by refractory compounds nanoparticles which are modificators affected to the process of structure formation thought the liquid phase and lock the recrystallization; *nanostructured cemented carbides*.

Recent studies concerning ceramic materials with the effect of dispersion hardening by nanoparticles in order to controlling the distribution of alloying elements within carbide grains are reviewed. Combination of force SHS-pressing followed by annealing makes it possible to synthesize composite materials with desirable structure and properties. Some service properties are increased due to nanosized $Me^{V}C$ or Me^{V}-Ti precipitations to be formed result in concentration separation of supersaturated solid solutions. Composite ceramic materials based on Ti-xC, Ti-Me–xC (Me= Nb, Zr, Ta) systems were produced by SHS. Contents of third elements are varied from 5 to 25 at.%. Composition and structure of the synthesis products before and after annealing at 850°C, 1 hour were studied using optical and SEM, XRD, AES and defined more exactly using synchrotron radiation. Composition and structure of carbide grains and intergranular phase just after combustion process is not equilibrium: supersaturated solid solutions are formed because of high temperature gradient and combustion velocity. Precipitates are appeared result in concentration separation of supersaturated solid solution. Carbide grains content precipitations sized 20-200 nm based on solid solution (Ti- Me). Precipitates are formed as a results of concentration separation of supersaturated solid solution via two possible schemes:

$(Ti, Me)_{x+y}C = (Ti, Me)_xC + y(Me\text{-}Ti)$, where Me- Nb, Ta, Zr (less 6%)

or $(Ti, Zr)_{x+y}C = (Ti, Zr)_xC + y(Me\text{-}Ti) + z(Zr, Ti)_zC$.

Hardness, wear- and heat resistance of the dispersion hardening materials are high as to compare with analogs based on titanium carbide.

Coatings deposited by PED and CRAPED on Ti-, Ni-, Fe- alloy substrates has a raised hardness, elastic recovery, adhesion strength to the substrate, heat resistance, and reduced friction coefficient. Coatings thickness (more than 50 μm) at density till 100%, lower roughness (less than 0.1 μm) were achieved due to high energy expended to erosion of nanostructured anode at high frequency and relatively lower pulse discharge energy. Phase- and structure formation mechanism of coatings are discussed.

Journal of Physics: Conference Series **416** (2013) 012004 doi:10.1088/1742-6596/416/1/012004

Basic research and regularities of the effect of nanosized WC, ZrO_2, Al_2O_3, W additives on sintering process were investigated in order to produce disperse-strengthening metal matrix composites based on cobalt, iron, and copper. Involvement of nanoparticles to metallic binder are used in order to improve working ability and service life of diamond tool. Main purposes of nanosized additives are follow: increasing the chemical potential of interphase between diamond and metallic binder result in alloying by chemically active nanoparticles; disperse strengthening the binder by involvement of nanoparticles inside of grains (hardness and toughness growing); involvement of nanoparticles on the grain boundaries (decrease the friction coefficient in cutting zone). Initial mixtures were prepared using planetary mills followed by ultrasonic treatment of nanoparticles containing mixtures. Obtained mixtures with various content of nanosized additives were sintered by HIP in inert atmosphere. It was established that nanoparticles located on grain boundaries has a constitutive effect on mechanism and kinetic of sintering. By the example of two systems, first - nanoparticles of chemically inert against the matrix in «ZrO_{2nano}-Co» and second - nanoparticles of chemically active against the matrix in «$WC_{нано}$-Co», shown that sintering kinetic strongly depends on thermodynamics singularity of interaction between metal and nanoparticles nature. Significant increasing of hardness, bending strength, wear resistance and decreasing the friction coefficient were observed result in disperse-strengthening by nanoparticles effect.

Nanocrystalline zirconium alloys obtained by severe plastic deformation

S Nikulin[1], **S Dobatkin**[1,2], **S Rogachev**[1]
[1]The National University of Science and Technology "MISIS", Leninsky prospekt 4, 119049 Moscow, Russia
[2]A.A.Baikov Institute of Metallurgy and Materials Science, Russian Academy of Sciences, Leninsky prospekt 49, 119991 Moscow, Russia

nikulin@misis.ru

Abstract. The structure and properties of commercial zirconium alloys Zr-2.5%Nb and Zr-1%Nb-0.3%Fe-1.2%Sn after severe plastic deformation (SPD) by high pressure torsion (HPT) at room-temperature under pressure of 4 GPa to N=5 (revolutions), which corresponds to a true deformation of $\varepsilon \approx 6$, and their thermal stability upon subsequent annealing have been studied. The alloys for HPT were taken in the two states: alloys were quenched from the single-phase β-region and from the two-phase α-β region. The structure formed in the Zr–2.5%Nb and Zr-1%Nb-0.3%Fe-1.2%Sn alloys after HPT is characterized by approximately similar nanograin sizes, 35–55 nm and 30–40 nm, respectively. The HPT of the alloys quenched from the two-phase region results in smaller grain sizes compared to those obtained from the alloys quenched from the single-phase region. It is shown that the HPT increases the microhardness of both alloys by a factor of 2.0–2.5. Annealing at 350°C leads to an insignificant grain growth in the nanocrystalline matrix. The increase in hardness of the alloys after HPT remains after annealing at a temperature of 350–400°C.

1.Introduction

Zirconium alloys are promising materials for the manufacture of medical implants due to their good biocompatibility [1, 2]. However, the use of zirconium alloys for this purpose is limited because of their low strength in comparison, particularly with titanium alloys. Increased strength of zirconium alloys can afford to significantly expand their use in medicine.

One of the promising directions of improving the mechanical properties of zirconium alloys is the formation of nano- and submicrocrystalline structure by severe plastic deformation (SPD) [3, 4].

The purpose of this paper is to study the structure and properties of commercial zirconium alloys after severe plastic deformation by high pressure torsion (HPT). The Zr-2.5%Nb and Zr-1%Nb-0.3%Fe-1.2%Sn alloys, which are widely used to manufacture nuclear reactor core components, were selected for the study [5, 6]. HPT can use high pressure, which leads to the ultimate refinement of the structure and allows to study the possible phase transformation.

2. Experimental Procedure

High pressure torsion (HPT) was performed in Bridgman anvils at room-temperature under pressure of 4 GPa to N=5 (revolutions), which corresponds to a true strain of ~6. For HPT the specimens were cut into disks with a diameter of 10 mm and initial thickness of 0.6 mm. The alloys for HPT were taken in the following two states: the Zr-2.5%Nb alloy was quenched from the single-phase β-region (920°C, holding for 30 min, water-quenching, grain size of ~130 μm) and from the two-phase α+β-region (860°C, holding for 30 min, water-quenching, grain size of ~6 μm); the Zr-1%Nb-0.3%Fe-1.2%Sn alloy was quenched from the single-phase β-region (950°C, holding for 30 min, water-quenching, grain size of ~110 μm) and two-phase α+β-region (860°C, holding for 30 min, water-quenching, grain size of ~5 μm). Analysis of the structure was carried out using a JEM-100CX transmission electron microscope. Microhardness measurements were carried out by Vickers method using test machine MICROMET 5101 under a load of 100 g.

3. Results and Discussion

The structure and properties of commercial zirconium alloys Zr-2.5%Nb and Zr-1%Nb-0.3%Fe-1.2%Sn taken in two initial quenched states and their thermal stability after high pressure torsion (HPT) at room temperature and subsequent annealing at temperature of 550° C (1 hour, air cooling) have been studied.

The scheme of HPT deformation is inhomogeneous: it is minimal in the center of the specimen and maximum at the edges. To estimate the heterogeneity of the strain after HPT for Zr-2.5% Nb and Zr-1%Nb-0.3%Fe-1.2%Sn alloys the distribution of the microhardness values along the diameter of the samples was constructed (Fig. 1). For alloys quenched from the α+β-region, the microhardness along the diameter of the specimen is distributed more evenly. Structural studies and phase analysis were performed on the middle of the radius of the specimen.

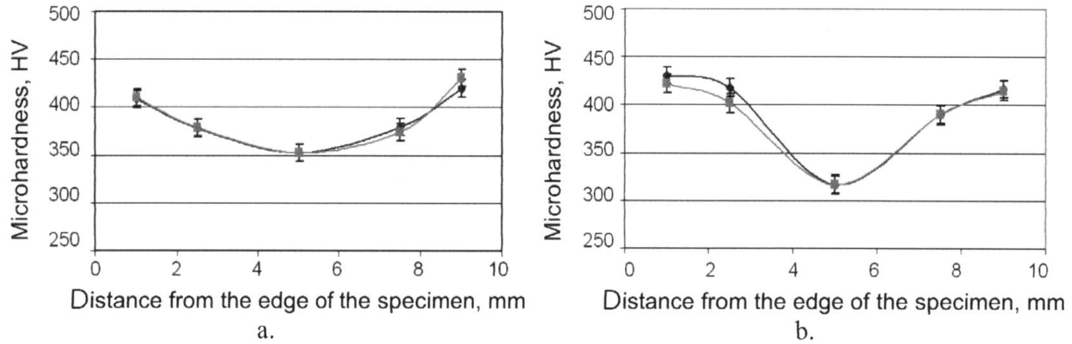

Figure 1. Distributions of the values of Vickers microhardness along the diameter of the specimens of the Zr-2.5% Nb alloy: a - quenched from the two-phase region, 860°C; b - quenched from the single-phase region, 920°C

HPT provides a significant increase the microhardness of the both alloys by a factor of 2.0-2.5 compared to the initial state of the alloys (160-170 HV), but decrease the hardness of the center. Quenching of both alloys from the single phase β-region provides the maximum increase in hardness after HPT due to a higher supersaturation of the solid solution.

Electron microscopic analysis showed that the HPT of the Zr-2.5%Nb and Zr-1%Nb-0.3%Fe-1.2%Sn alloys leads to a nanocrystalline structure which is characterized by approximately similar nanograin sizes, 30–55 nm (Fig. 2, 3). The HPT of the alloys quenched from the two-phase region results in smaller grain sizes compared to those obtained from the alloys quenched from the single-phase region: 30–35 nm and 40–55 nm, respectively. It should be noted that, along with nanosized grains with high-angle boundaries there are small areas of subgrains of the same size with a low-angle boundaries.

Annealing at 350°C (holding for 1 h) leads to an insignificant grain growth in the nanocrystalline matrix. After annealing, the alloys quenched from the two-phase region exhibit a smaller grain size (of 35–45 nm) than that observed in the alloys quenched from the single-phase region, which after HPT exhibit a grain size of 70–75 nm.

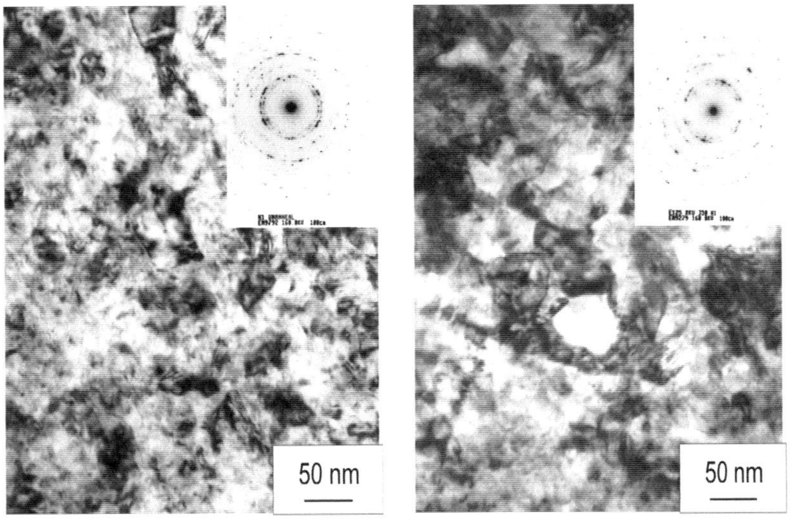

a. b.

Figure 2. TEM micrographs of the Zr-2.5%Nb alloy, (quenched from 860°C) after HPT: a - without annealing; b - after annealing at 350 °C

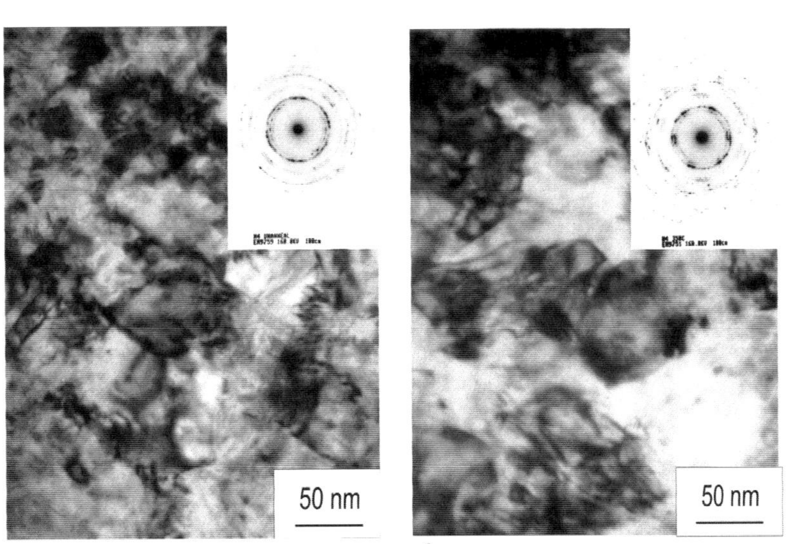

a. b.

Figure 3. TEM micrographs of the Zr-2.5%Nb alloy, (quenched from 920°C) after HPT: a - without annealing; b - after annealing at 350°C

To study the thermal stability of the increase in hardness after HPT the dependence of the microhardness on the temperature of the annealing of the deformed alloys in the temperature range 50-550 °C was measured (Fig. 4). The increase in hardness of the alloys after HPT remains unchanged after annealing at temperature of 350–400°C, despite some increase in grain size (by a factor of 1.2-1.4). With a further increase in temperature above 400°C, the microhardness is significantly reduced due to the growth of grains in the matrix.

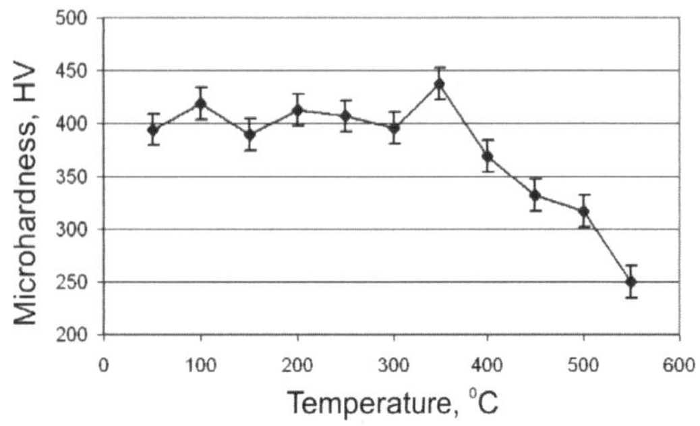

a.

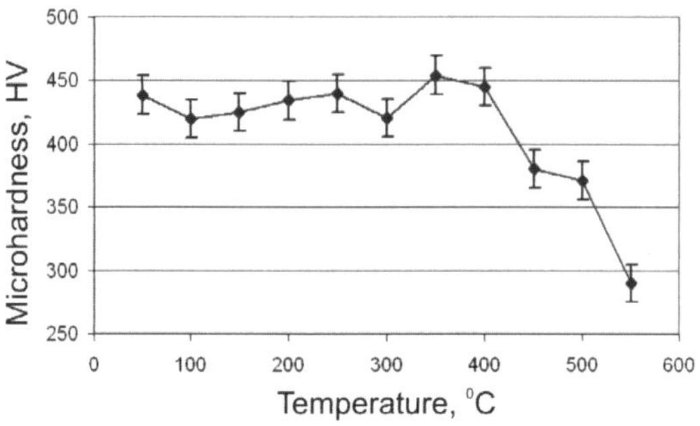

b.

Figure 4. Change in the Vickers microhardness vs. annealing temperature for the Zr-2.5% Nb alloy: a - quenched from the two-phase region, 860 °C; b - quenched from single-phase region, 920°C

HPT leads to the formation of the ω-Zr high pressure phase in α-Zr matrix (α-Zr-phase is partially preserved). During annealing the phase of ω-Zr is transformed into α-Zr. In the HPT-ed Zr-2.5% Nb and Zr-1%Nb-0.3%Fe-1.2%Sn alloys after annealing at 350 °C, some quantity of ω-phase is preserved.

Thus, HPT zirconium alloys leads to the formation of nanocrystalline structure with the grain size of 30-55 nm and to increase of microhardness by a factor of 2.0-2.5 in comparison with the initial undeformed state. The increase in hardness after HPT remains unchanged during annealing at temperatures up to 350-400°C.

4. Conclusions

1. The structure formed in the Zr–2.5%Nb and Zr-1%Nb-0.3%Fe-1.2%Sn alloys after HPT is characterized by approximately similar grain sizes, 35–55 nm. The HPT of the alloys quenched from the two-phase region results in smaller grain sizes compared to those obtained from the alloys quenched from the single-phase region: 30–35 nm and 40–55 nm, respectively.
2. It is shown, that the HPT increases the microhardness of both alloys by a factor of 2.0-2.5. This effect is more pronounced in the alloys quenched from the single-phase region due to a higher supersaturation of the solid solution in spite of somewhat larger grain size. The increase in hardness of the alloys after HPT is remains unchanged after annealing at temperature of 350-400°C.
3. Annealing at 350°C (holding for 1 h) after HPT leads to an insignificant grain growth in the nanocrystalline matrix. After annealing, the alloys quenched from the two-phase region exhibit a smaller grain size (of 35–45 nm) than that observed in the alloys quenched from the single-phase region, which exhibit a grain size of 70–75 nm.
4. The Zr-2.5%Nb alloy upon HPT undergoes the $\alpha'_{Zr} \rightarrow \omega_{Zr}$ transformation. A partial reverse $\omega_{Zr} \rightarrow \alpha_{zr}$ transformation occurs upon annealing.

References

[1] Steinemann S G 1980 *Evaluation of Biomaterials* 1
[2] Fujita M 1993 *J. Jpn. Stomatol. Soc.* **60** 54 (in Japanese)
[3] Perez-Prado M T, Gimazov A A, Ruano O A, Kassner M E, Zhilyaev A P 2008 *Scripta Mater.* **58** 219
[4] Edalati K, Horita K S Yagi S 2009 *Mater. Sci. Eng.* A **523** 277
[5] Zaymovskiy A S, Nikulina A V, Reshetnikov F G 1994 *Zirconium alloys in nuclear industry* (Moscow: Energoatomizdat) p 253 (in Russian)
[6] Nikulin S A 2007 *Zirconium alloys for nuclear power reactors* (Moscow: MISIS) p 92 (in Russian)

Journal of Physics: Conference Series **416** (2013) 012006 doi:10.1088/1742-6596/416/1/012006

Influence of laser radiation on structure and properties of steels and alloys

T Tarasova[1] and E Popova[1]

[1] Department of Material Science, Bauman Moscow State Technical University, Moscow, Russia

E-mail: ttv@bmstu.ru

Abstract. In present study, and laser alloying of different steels and laser cladding of Ti and SiC powders mixtures was carried out, and microstructure, as well as microhardness profile and wear properties were examined. Research of the influence of lasers alloying modes on the elastic and plastic characteristics of the surface was conducted. As a result of chemical reactions in the cladded layer, a new phase (TiC) was synthesized during cladding process. The results showed that, in the clad layer, TiC was solidified to form dendrites in the clad zone. Produced coatings have high microhardness values in the upper and middle clad areas, about two time higher than clad matrix microhardness.

1. Introduction

Different ratios between power density and laser irradiation time are used in different laser applications. By varying power density and laser irradiation time and controlling the ratio between them, different applications are possible, for instance, drilling, welding , thermal hardening and cladding (figure 1). Laser surfacing techniques have been employed to form a ceramic-metal composite on conventional alloys for enhancing wear resistance [5,6]. Laser cladding by powder injection is the surface engineering method to produce high quality, metallurgically bonded and thick coatings on deficient substrates with a minimal heat input into the work piece. Usually, the main aim of laser cladding and alloying is to improve wear, impact and corrosion resistance properties of surfaces, by generating a protective layer. A complete description of the laser cladding and alloying process is rather complex, because of numerous of interactions (laser beam/powder stream, laser beam/substrate surface, powder stream/melt pool, powder stream/solid substrate, etc.) and physical phenomena (mass and heat transfer, fluid flow, phase transformations during rapid solidification, etc.) are present [1]. The present study aims at investigating the in situ synthesis of a TiC/Ti/Fe composite layer on lox-carbon steel by laser cladding using SiC and Ti powders as the starting materials, and analyzing the microstructures formed for different powder mixtures . Ti-based alloys possess high specific strength, excellent corrosion resistance and good high-temperature strength up to 400°C. These attractive characteristics have made Ti alloys important structural materials in the aircraft industry. However it is well known that Ti-based alloys have poor wear resistance with respect to abrasion, which imposes a limitation on these alloys for a wider range of applications [3, 2]. TiC has been widely used as the reinforced phase of composite materials because of its high hardness, high modulus and rather high flexural strength. In the past decade, the in-situ composite method has been extensively studied to produce composite materials . In this method, reinforcements are formed in the matrix by reaction between added pure elements or compounds, so the reinforcements are much more compatible with the matrix and the interface is much cleaner than that of composites made by conventional technique [4].

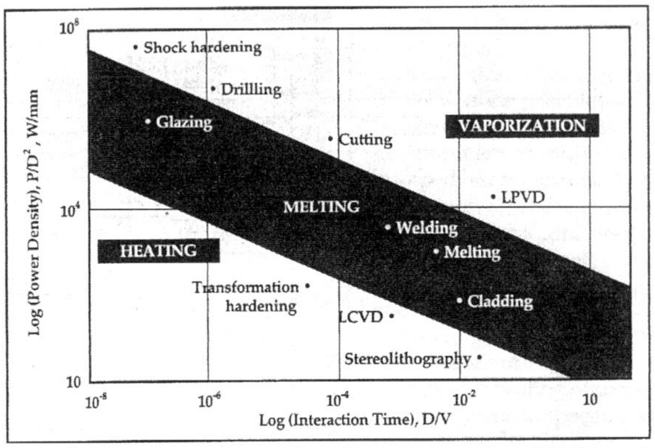

Figure 1. Different ratios between power density, laser irradiation time and different laser applications.

2. Experimental procedures

Laser treatment was conducted on Trumpf DMD 505 industrial-scale machine. A nozzle transports powder coaxially to the CO2 5kW laser beam onto the molten pool created on the workpiece surface. 4 modular powder feeders can be used at the same time for powder mixing *in situ* the process and thus allowing manufacturing multimaterial multifunctional objects. The process is characterized by a minimal heat input into the workpiece. To prevent oxidation of the cladded layers, a protection skirt made of heat-resistant material is used. By comparing samples manufactured with and without protection skirt, we can observe a notable quality improvement. The substrate was low-carbon steel, with nominal chemical composition in wt.% is: 0,17-0,22 C, 0.40–0.60 Mn, Si less 0,05, Ni less 0,03, Cr less 0,03, P less 0,04, Cu less 0,03, balance Fe. Mixtures of SiC powder (135 μm) and Ti powder (60 μm) were thoroughly mixed by ball milling according to the ratios Ti/SiC = 6:4, 6:5 (wt%). The laser output power was varied between 4 and 5 kW, the laser beam scanning velocity was from 500 to 1000 mm/min, powder feeding rate was 75 g/min. During cladding process, monolayer coatings were produced (30x30 mm). Metallographic samples of the composite coatings were prepared using standard mechanical polishing procedures and were etched in $HF:HNO_3:H_2O$ water solution in volume ratio of 1:1:8. Microstructure of the coatings was studied by optical microscopy (OM), scanning electron microscopy (SEM). The phases present in the cladding layer were identified by X-ray diffraction (XRD). The microhardness of the coating was measured with the load of 100 g by using a BUEHLER hardness tester machine.

3. Results and discussion

3.1 Laser alloying

Let us consider peculiarities of phase and structure transformation in steels under laser heating. Laser action zones have layer structure because of the fact that different layers were heated up to different temperatures. Generally, three layers can be distinguished. The first layer is the melted zone and corresponds to the hardening in the liquid state. In most cases, it has dendrite structure. The primary axes of the crystals are perpendicular to the bottom border of the melted zone. Martensite is the main structural element. As usual, carbides dissolve. The second layer corresponds to the hardening in the solid phase (without melting). The bottom border is determined by heating up to the critical temperature AC1. In this case, both complete and partly hardening is realized. The third layer is a transitional zone. Here the metal is heated below the critical temperature AC1. Microstructure and hardness can vary depending on the initial structure of the steel. Figure 2 represents microstructure of steel 20X13 under pulsed laser radiation (E = 18 J, ΔF = 0).

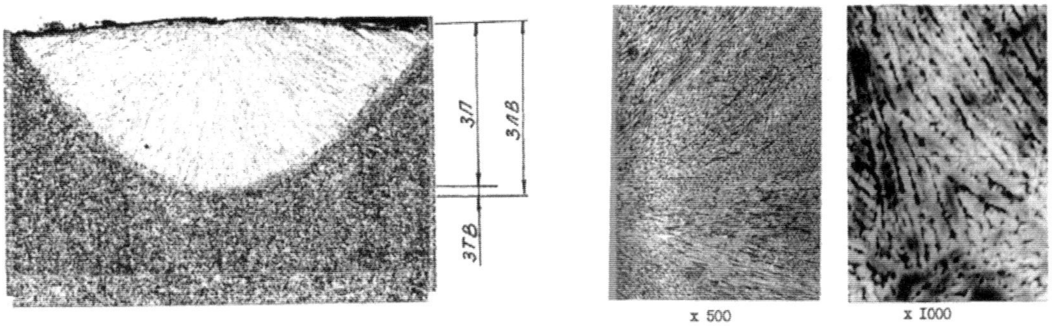

Figure 2. Microstructure of steel 20X13 under pulsed beam irradiation (E = 18 J, ΔF = 0).

Laser treatment with melting is characterized by high cooling rate of the melted metal. Thus, the hardening if the liquid state occurs. The resulting structure has several peculiarities: a) the grains formed at primary crystallization have small sizes because of a large number crystallization centers. The distance between the dendrites axes of the second order substantially reduces.(At the cooling rate of 106 °C/s and more they may disappear. Dendrite liquation reduces due to suppression of the separating diffusion at the crystallization front), b) formation of metastable structures is possible. They are oversaturated solid solutions and metastable intermediary phases, c) as a result of changes in the crystallization mechanism, anomaly structures may form such as stratification of eutectics and quasi-eutectics, amorphous and metal crystal structures.

3.2 The effect of laser alloying on steel hardness

Now our concern will be methods of laser alloying. The table 1 represents the effect of laser alloying on steel hardness. In laser alloying a zone of the surface is heated above the melting temperature, then alloying component are introduced into the melted zone. Figure 3 shows microhardness distribution into the depth of the alloyed zone after laser cementation of steel 12X13 (E = 9 J, δ = 0,4 mm, ΔF = 0).

Table 1. Effect of Laser Alloying on the Microhardness of Carbon Steel (0.4%C, 1%Cr)

Class (Type) Alloying Components	Alloying Components	Microhardness, MPa
	Boron	10000-20000
Nonmetal Elements	Graphite	10000
	Nitrogen	9500
Metals	Chromium	8000
	Titanium	4000-4500
Chemical compounds:		
Carbides	Boron carbide Titanium carbide	9500-11000
Nitrides	Vanadium nitride Aluminum nitride Zirconium nitride	10000-12000

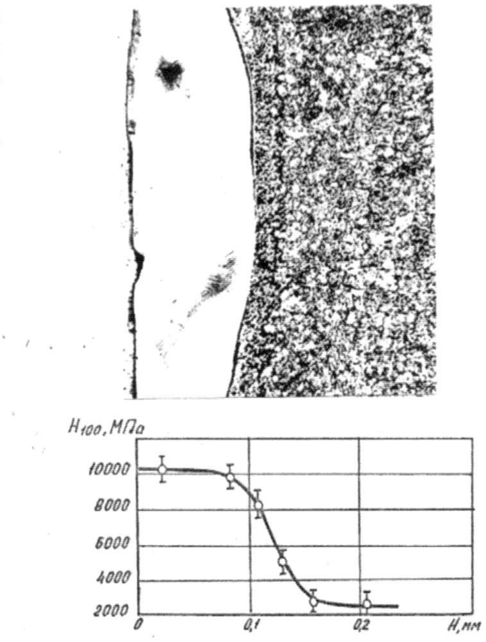

Figure 3. Microhardness distribution into the depth of the alloyed zone after laser cementation of steel 12X13 (E = 9 J, δ = 0,4 mm, ΔF = 0).

Based on experimental research, phase diagrams were constructed. They show existence of zones of structures of various types depending on laser energy and powder layer thickness. The influence of laser cementation conditions on structure and microhardness of melted zones of steel 12X13 (ΔF = 0) is shown on figure 4.

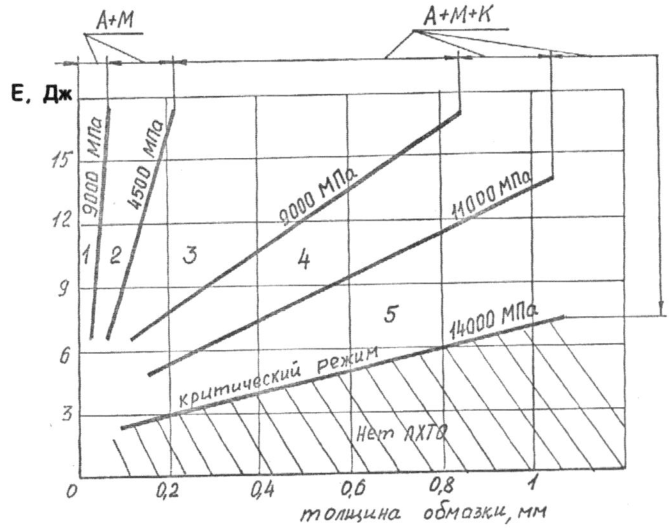

Figure 4. Influence of laser cementation conditions on structure and microhardness of melted zones of steel 12X13 (ΔF = 0).

3.3 Evolution of wear rate of the hardened samples

Wear rate of samples after laser treatment is by 2-2.5 time lower than that of samples from the similar steels after conventional thermal treatment, experiment had been held on SMC-2 machine. Laser alloying leads to much more noticeable decrease of wear rate. This is shown on figure 5.

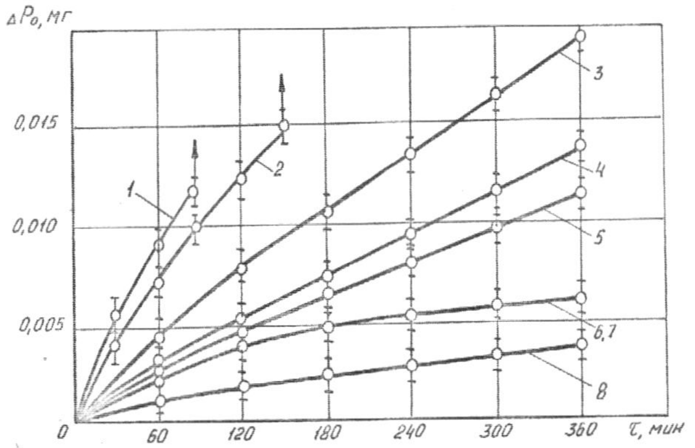

Figure 5. Evolution of wear rate of the hardened samples on SMC-2 machine: 1-volume hardening and tempering, 20X13 steel, 2- volume hardening and tempering, 40X13 steel, 3-LCT with melting, 20X13 steel, E=30J, 4- LCT with melting, 40X13 steel, E=30J, 5- gas cementation, 12X13 steel, 6-LA (C), E=20J, d=0,05, steel 20X13, 7-LA (Si), E=11J, d=0,2, steel 20X13, 8-LA (C), E=8J, d=0,3, steel 20X13.

Wear resistance and abrasive wear versus hardness of the alloyed surface had been measured.

The tests showed than the laser alloying increases microhardness of the samples. Friction coefficient and wear rate decreased. As to abrasive wear, the optimal conditions are those that provide the maximum hardness without cracks and pores. The results are presented on figure 6.

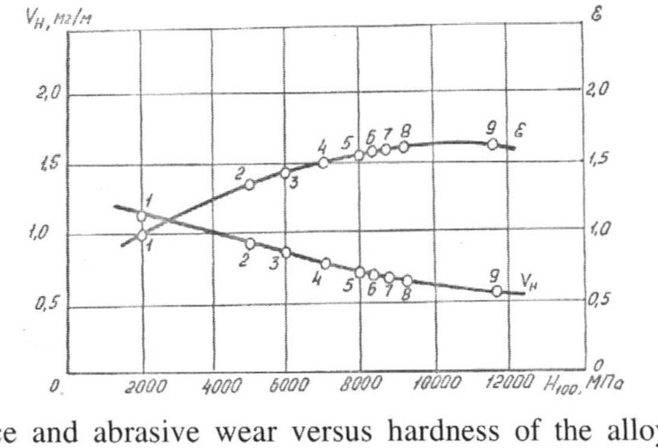

Figure 6. Wear resistance and abrasive wear versus hardness of the alloyed surface: 1-annealing, 20X13 steel, 2- conventional thermal treatment, steel 20X13, 3-traditional heat-treatment, steel 40X13, 4-LCT with melting, 20X13 steel, E=30J, 5- LA with melting, 40X13 steel, E=30J
6- laser cementation, 12X13 steel, 7- gas cementation, 12X13 steel, 8- LA (C), E=20J, d=0,05, steel 20X13, LA (Si), E=11J, d=0,2, steel 20X13, 9-LA (C), E=8J, d=0,3, steel 20X13.

3.4 Laser cladding

The purpose of current study is production of fine and uniformly distributed TiC in the cladd layers. TiC has been widely used as the reinforced phase of composite materials because of its high hardness. The majority of reinforced phases are directly added into coating materials. Figure 7 presents powder characteristics used in our research. The average Ti and SiC powder particles size is about 50 μm. The substrate was low-carbon steel.

Journal of Physics: Conference Series **416** (2013) 012006 doi:10.1088/1742-6596/416/1/012006

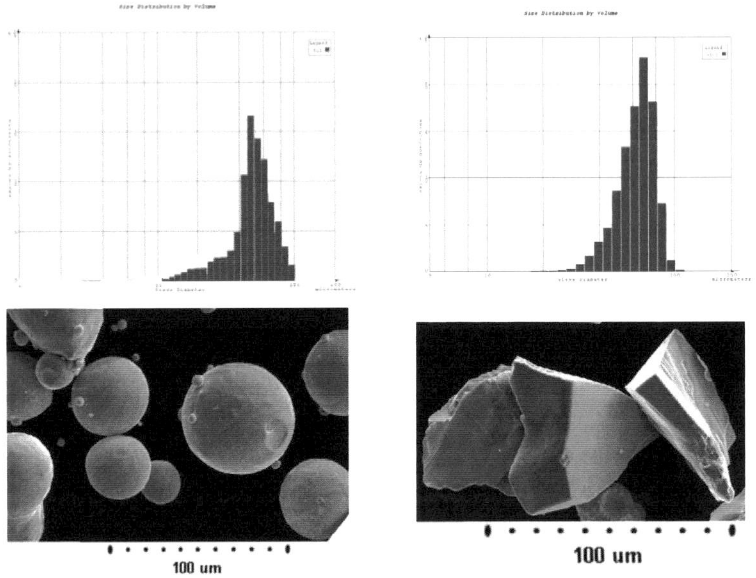

Figure 7. Powder characteristics: a-TiC powder, b-SiC powder.

Phase analysis of the coating indicates that the dark precipitations in the cladded layer are Titanium Carbide. It is shown on figure 8.

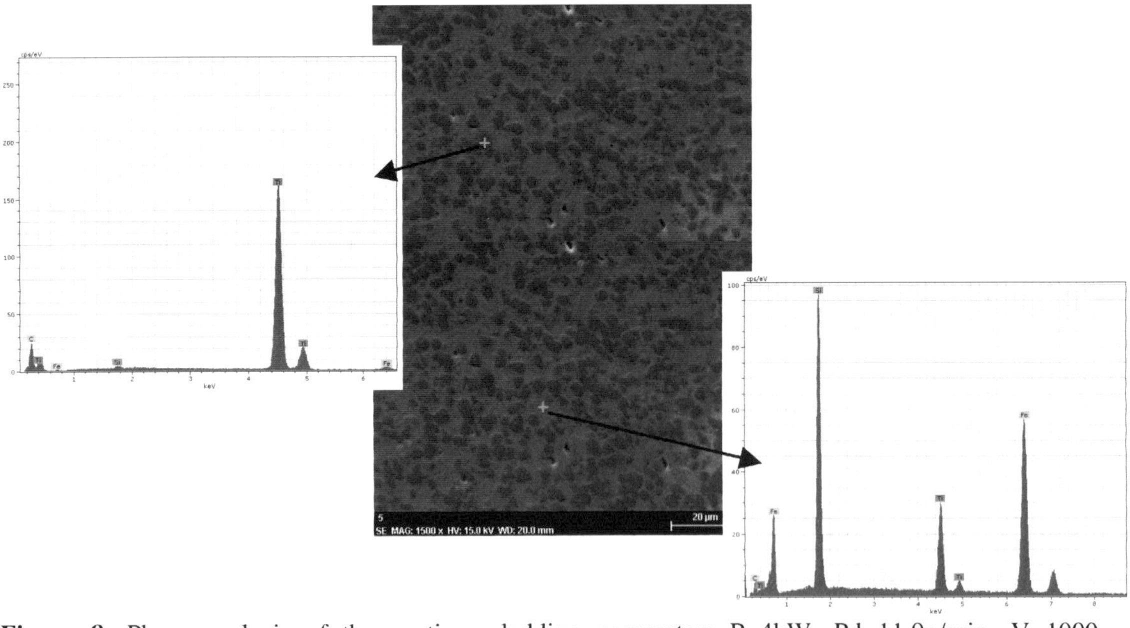

Figure 8. Phase analysis of the coating, cladding parameters P=4kW, Pd=11,9g/min, V=1000 mm/min.

3.5 Microstructure and microhardness profile of the clad layers

Microstructure of the clad layers is presented at figure 9. When the cladding velocity increase,(in case of both 4 and 5kW laser power,) more fine TiC precipitations are formed. In general, carbide inclusions are uniformly distributed in the coating.

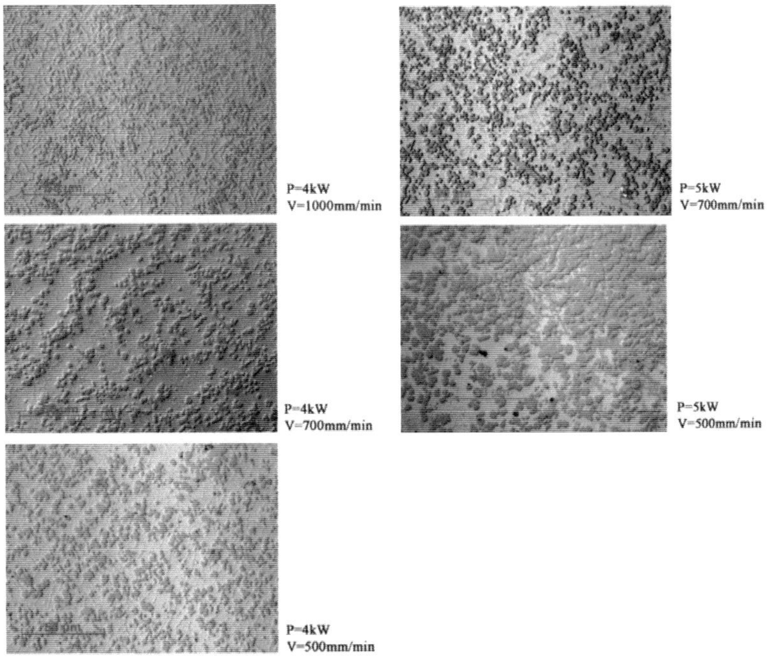

Figure 9. Microstructure of the clad layers, Pd=11,9g/min.

The microhardness profile across the clad samples is shown at figure 10, powder feeding rate Pd=11,9g/min. Microhardness values are notably higher in case of 11.9g/min powder feeding rate, than in 9.6g/min.

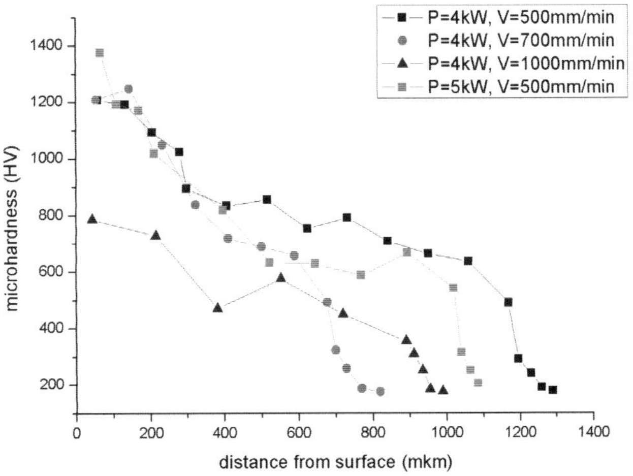

Figure 10. Microhardness profile across the clad samples, Pd=11,9g/min.

4. Conclusions

Research of the influence of lasers alloying modes on the elastic and plastic characteristics of the surface indicated that laser doping reduces microplasticity surface-treated steels. Structures formed by laser-hardening and consisting of martensite, residual austenite and a large quantity of carbide phase have the smallest Research microplasticity. Laser-hardening modes, leading to a large of carbide phase in the band structure of doping leads to increasing of heat resistance at 300oS compared with gas cementation. Doping of silicon increases heat resistance of hardened zones on 500S. Study of the electrochemical corrosion kinetics of the investigated steels by using potentiodynamic method in 3% solution of NaCl showed that Laser-hardening does not degrade the corrosion resistance compared with the traditional heat treatment, which is explained (as its shown by electron microprobe analysis results), by distribution of chromium in terms of area doping with high degree of homogeneity. Also, the possibility of hard and wear-resistant coatings formation from the Ti and SiC powders was demonstrated.

References

[1] Ocelík1 V, de Oliveira U, de Boer M and de Hosson J Th M *Thick Co and Fe based coatings by laser cladding - analysis of processing conditions, microstructure and properties, Beam technologies & Laser Application: Fifth International Conference (Saint-Petersburg, Russia, 23-28 September 2006)*

[2] Thivillon L, Pervushin D, Bertrand Ph and Smurov I *Industrial technology of laser assisted direct metal deposition* (Saint-Etienne, FR: Ecole Nationale d'Ingénieurs de Saint-Etienne (ENISE), DIPI Laboratory)

[3] Man H C, Zhang S, Cheng1 F T and Yue T M 2001 *Microstructure and formation mechanism of in situ synthesized TiC/Ti surface MMC on Ti-6Al-4V by laser cladding (Scripta Materialia 44(12))* pp 2801-2807

[4] Sen Yang, Wenjin Liu, Minlin Zhong and Zhanjie Wang 2004 *TiC reinforced composite coating produced by powder feeding laser cladding (Materials Letters 58)* pp 2958– 2962

[5] Cadenas M, Vijande R, Montes H T and Sierra JM 1997 *Wear* 212, 244

[6] Wang P, Yang Y, Ding G, Qu J and Shao H 1997 *Wear* 209, 96

Journal of Physics: Conference Series **416** (2013) 012007 doi:10.1088/1742-6596/416/1/012007

Characterization of Ti-Al surface alloy formed by pulsed electron-beam melting of film-substrate system

N Allain-Bonasso[1], V P Rotshtein[2], E Bouzy[1], L Germain[1], Yu F Ivanov[2], T Grosdidier[1]

[1]Laboratoire d'Etude des Texture et Applications aux Matériaux (LETAM-CNRS 3143), Université Paul Verlaine - Metz, Ile du Saulcy, 57045 Metz Cedex, France

[2]Institute of High-Current Electronics, 2/3, Akademichesky Ave., 634055, Tomsk, Russia

E-Mail: allain-b@univ-metz.fr

Abstract. This paper presents our first results of an ongoing research dedicated to the analysis of Ti-Al surface alloy obtained by a new duplex e-beam treatment. The crack-free Ti_3Al-based surface alloy was synthesized by cyclic pulsed melting/mixing of pre-deposited Al film (100nm) on Ti substrate with a low-energy (10–20keV) high-current electron beam (3μs, 3.5J/cm^2) followed by additional pulsed e-beam melting (100μs, 15keV, 15J/cm^2). The surface analysis has revealed the complexity of the local transformation paths associated with this synthesis mode. A fairly good surface chemical homogeneity was obtained but a duplex microstructures formed because of local changes in the primary phase under rapid solidification and the subsequent modifications of the solid state transformation paths.

1. Introduction

Bulk Ti_3Al/TiAl-based alloys have been studied for a long time owing to their attractive mechanical properties like high specific strength and stiffness at high temperatures as well as good oxidation resistance. These properties of Ti-Al intermetallics make them suitable for use as structural protective coatings. Unfortunately, they also possess certain disadvantages such as poor ductility and low toughness at room temperature. Several techniques have been used to fabricate Ti-Al surface layers or coatings: thin-film deposition/annealing [1], ion implantation [2], laser melting deposition [3]. The present work shows the first results on the surface characterisation of Ti-Al surface alloy formed by pulsed e-beam liquid-phase mixing of the film-substrate system [4]. This processing route consists of cyclic liquid-phase mixing of pre-deposited Al film with Ti-substrate by Low-Energy High-Current Electron Beam (LEHCEB) of microsecond duration followed by pulsed melting Ti-Al surface alloy by sub-millisecond (100μs) e-beam [5, 6] Low energy (10-20keV) high-current electron beam of microsecond duration is a fairly recently developed method for surface treatments [4, 5-8]. The pulsed electron irradiation induces (i) a rapid heating and cooling of the surface together with (ii) the formation of thermal stress and stress waves [4, 6-10]. As a result, improved surface properties of the material, often unattainable with conventional surface treatment techniques, can be obtained fairly easily. This is particularly true for tribological [6, 8, 11-13] and corrosion properties [4, 7, 12-16]. Generally present at the top surface is a zone that has been melted and subsequently solidified rapidly that is often a few μm in thickness [4, 7, 8, 17]. The repeated melting and rapid solidification sequences imparted to the material surface often leads to the dissolution of second phases precipitates (carbides, intermetallics) and the formation of a fairly homogeneous melted layer as was observed for the 316L [12, 18] and D2 [19] steels as well as NiTi shape memory alloys [20]. The so-formed homogeneous layers were demonstrated to improve the corrosion properties of these dual phase alloys. The treatment also often induces the formation of nano-scale structures formed by rapid solidification in the melted layer [19-25]. These ultra fine grains are not the only reason for the generally observed surface hardening and improved wear resistance. Surface strain hardening is also witnessed by the observation of slip traces [7, 26] or, in some cases depending on the stacking fault energy of the material, by the observation of deformation twins [27]. In addition, significant sub-surface hardening is generally observed and witnesses by the presence of two distinct hardness peaks [8, 10, 12, 28, 29]. The first hardness peak - extending generally over about 30/50μm – has been attributed essentially to the quasi-static thermal stress and the repeated action of the beam induced the formation of micro-deformation bands. The second peak -of lower intensity but extending over more than 100μm- was related to deeper hardening induced by the thermal stress wave [12].

As already mentioned, in addition to the previously listed advantages of the LEHCEB process for surface modification, it is also possible to use this electron beam technique for surface alloying by liquid-phase mixing of previously deposited layers [4, 7, 8, 24, 30]. In the present work, a Ti_3Al-based surface alloy has been synthesized by pulsed liquid-phase mixing of pre-deposited Al film on a Ti substrate. As it is also well established that the mechanical behavior of these TiAl-based alloys strongly depends on their microstructure and crystallographic texture [31], a special attention is paid here to the characterization of the microstructure that is expected to form from the melt at cooling rates of about 10^6K/s.

2. Experimental

An apparatus combining a dual stage device has been used for the preparation of the sample. In the first rig, an Al film has been deposited on a polycrystalline Ti substrate using a high-current pulsed vacuum arc evaporator [32]. The deposition rate was 0.8nm per pulse. The sample was then transferred in the second part of the chamber where the LEHCEB treatment was applied for melting and inter-mixing of the Al-Ti system. Further information on the LEHCEB source and it description can be found elsewhere [6, 7]. The working chamber was evacuated with a turbo-molecular pump to a pressure of 10^{-4}Pa. The irradiation of the sample was carried out in argon at partial pressure of 3×10^{-2}Pa. Prior to the Al film deposition, the Ti substrates were irradiated with LEHCEB (pulse duration of ~3μs, energy density of 5÷6J/cm^2, pulse number n=10) in order to avoid a local delamination of the thin film at pulsed melting because of potential surface cratering [4, 8, 19, 33]. The thickness of the deposited Al film in each synthesis cycle was 100 nm. After each film deposition the coated sample was subjected to pulsed melting with the same electron beam energy (3÷4J/cm^2). The number of cycles N of the Al film deposition followed by pulsed melting was 50; leading to a maximum total thickness of the deposited Al coating of ~5μm. The surface-alloyed sample was subjected to an additional pulsed multiple (n=9) e-beam melting under a stronger energy density level (10÷15J/cm^2) for a longer duration (100μs) to avoid surface cracking.

The surface morphology was examined using a JEOL 6500F FEG scanning electron microscope (SEM). The chemical composition of the surface layer was determined by EDS in the SEM. The SEM images were done under the secondary electron imaging (SEI) or backscattered electron imaging (BSE) conditions. The electron backscattering diffraction (EBSD) analysis was carried out with a step size of 0.05μm using the FEG-SEM equipped with the HKL-channel. The X-ray diffraction (XRD) measurements for the analysis of the present phases were carried out with a CPS 120 INEL diffractometer using the Cu-Kα radiation.

3. Results and discussion

3.1. Structure and chemical composition

As shown by the XRD diffractogramme obtained from the sample surface (figure1), the analyzed surface layer was essentially composed of α_2 (Ti_3Al) phase. EDS analysis have revealed that this layer contain ~ 34at.% Al rich (balance Ti) with traces of oxygen and carbon. As shown in figure 2, despite the presence of significantly different microstructure features, and figure 3, the repartition of Ti and Al over the entire surface was fairly homogeneous. This is illustrated by the SEM image in Figure 2a that shown two types of microstructure while the concentration maps given in figure 2(b) (Al) and 2(c) (Ti) are rather homogeneous. The slight changes, in particular in the Al map, are clearly not related to the differences in microstructure but more likely to the surface topology. A more detailed microstructure analysis is given in the following section, in particular in figure 3.

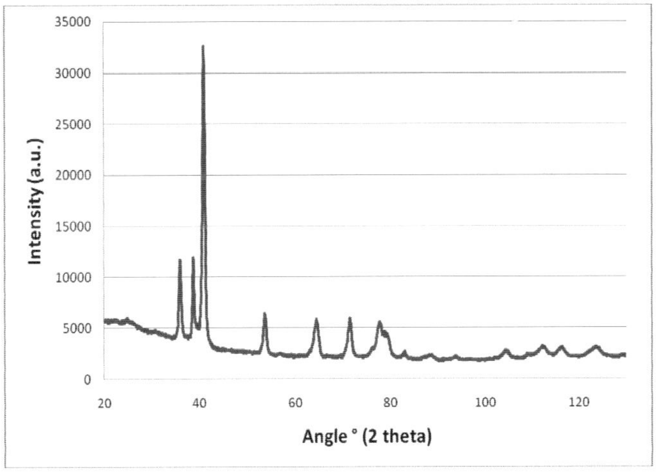

Figure 1. XRD diffractogramme obtained from the sample surface.

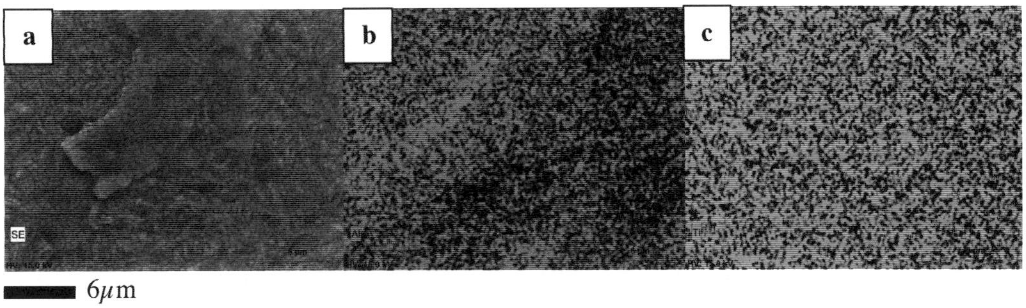

Figure 2. SEM images under secondary electron imaging condition (a) and corresponding composition maps of the same zone with Al (b) and Ti (c) repartitions.

3.2. Microstructure of the surface and local phase transformation paths.
The general surface topology is presented in figure 3(a) where grains having different height are clearly revealed. Higher magnification images of this zone are presented in figure 3(b) and 3(c). Two types of grains have been observed: grains containing nodular sub-micrometer domains (noted N) and larger grains containing lamellar structures (noted L). The N grains have sizes ranging from about 5 to 10μm while the L grains have sizes in the range 10 to 30μm. Quantitative manual image analysis using the intercept method indicated the fraction of L grains at the surface is about 85%. In order to gain more insights on the microstructure formation, a detailed EBSD analysis has been carried out and is detailed in the following section.

Figure 3. SEM images of the surface microstructure observed under SEI (a-c) and BSE imaging conditions (b).

3.3. Analysis of EBSD maps.

Figure 4 shows an example of EBSD map measured using the hexagonal α (α_2) indexing. About 46% of the pixels were indexed. The missing fraction corresponds to non-indexed points but no other phases were detected. The colour key of the map is given by the all euler orientation legend. The white domains in the map correspond to the non-indexed pixels. It is clear that the grains of type N have a fairly homogeneous colour indicating thereby that they present little internal misorientation. Comparatively, the lamellar domains in the grains of type L correspond to martensitic variants. The presence of martensitic variant is confirmed by the data given in table 1. In table 1 are presented the calculated misorientation obtained between each type of lamellar variant for the domain labelled L_1 in figure 4. As described in [34-35], the misorientations calculated from the inherited α_2 orientations correspond to the β□α transformations and these variants are issued from the same parent β phase. In many titanium and zirconium alloys, the high temperature (BCC) β phase transforms upon cooling into the (HCP) α phase with respect to the Burgers orientation relations [36]. The Burgers orientation was also obtained here (Table 1).

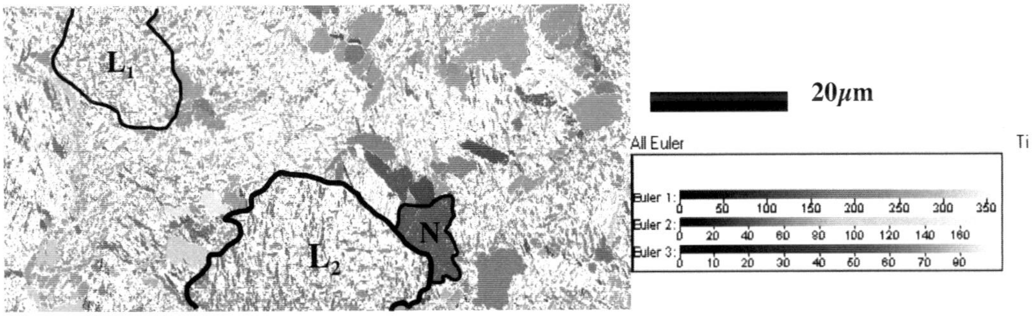

Figure 4. Measured orientation map of the α_2 phase

Thus, it is clear that the different microstructures obtained at the surface of the sample can be explained by different phase selection from the melt during the rapid solidification and the subsequent solid state transformation paths. The grains of N type have solidified directly as from the melt. Comparatively, the grains of type L come from the primary β phase that subsequently followed the β□α martensitic transformation.

Table1. Misorientation between two considered variants present in the L_1grain.

	Variant1	Variant 2	Variant3	Variant 4
Variant1	X	60° [1-210]	10°[-110-3]	90°[1-210]
Variant2	60° [1-210]	X	60°[-3210]	60°[-4-483]
Variant3	10°[-110-3]	60°[-3210]	X	90°[-12-10]
Variant4	90°[1-210]	60°[-4-483]	90°[-12-10]	X

4. Conclusion

A Ti_3Al –based surface alloy has been synthesized by cyclic LEHCEB melting and mixing of a pre-deposited Al film on a Ti substrate followed by pulsed melting with submilisecond e-beam to obtain microstructure without microcracks. For the investigated experimental conditions, the top surface chemistry was fairly homogeneous and contained about 34% Al. Despite the absence of significant chemical gradient at the top surface, two types of microstructure were revealed. This is the result of local changes in the transformation paths due to a local modification of the primary phase in solidification. While some grains (N type) have solidified directly as a hexagonal phase from the melt, other domains solidified as β and subsequently followed the β□α martensitic transformation.

The work was partially supported by RFBR under project № 10-08-91152-GFEN.

References

[1] Hampshire J, Kelly P J, Te D G 2002 *Thin Solid Films* **420-421** 386
[2] Tsyganov I, Wieser E, Matz W, Mücklich A, Reuther H, Pham M T, Richter E 2000 *Thin Solid Films* **376** 188
[3] Qu H P, Wang H M 2007 *Mater. Sci. Eng.* A **466** 187
[4] Rotshtein V P, Ivanov Yu F, Markov A 2006 *Review in book: Materials surface processing by directed energy techniques* ed Pauleau Y (Amsterdam: Elsevier) p 205-240

[5] Rotshtein V P *et al* 2010 *Proc. of 10th Intern. Conf. on Modification of Materials with Particle Beams and Plasma Flows (Tomsk, Russia)* p 416-419

[6] Ozur G E, Proskurovsky D I, Rotshtein V P, Markov A B 2003 *Laser Part.Beams* **21** 157

[7] Proskurovsky D I, Rotshtein V P, Ozur G E , Markov A B, Nazarov D I, Shulov S V, Ivanov Yu F, Buchheit R 1998 *J. Vac. Sci. Technol.* A **16** 2480

[8] Proskurovsky D I, Rotshtein V P, Ozur G E, Ivanov Yu F, Markov A B 2000 *Surf. Coat. Technol.* **125** 49-56

[9] Dong C *et al* 2003 *Surf. Coat. Technol.* **163-164** 620

[10] Qin Y, Dong C, Song Z F, Hao S Z, Me X X, Li J A, Wang X G, Grosdidier T 2009 *J. Vac. Sci. Technol.* A **27** 430

[11] Gao B, Hao S Z, Zou J X, Grosdidier T, Jiang L M, Zhou J Y, Dong C 2005 *J of Vac Sci Technol* A **23A** 1548

[12] Zou J X,. Zhang K M, Hao S Z, Dong C, Grosdidier T *in the press of Thin Solid Films*

[13] Fu F J, Tang G Z, Guo G W, Ma X X, Ozur G E 2010 *Nucl. Inst. Met. Phys. Res.* **268** 2395

[14] Zou J X, Zhang K M, Dong C, Quin Y, Hao S Z 2006 *Applied Physics Letters* **89** Art. N°041913

[15] Zhang K M, Zou J X, Grosdidier T, Dong C, Yang D Z 2006 *Surf. Coat. Technol.* **201** 1393

[16] Zhang K M, Zou J X, Grosdidier T, Dong C, Yang D Z 2006 *Surf. Coat. Technol.* **201** 3096

[17] Grosdidier T, Zou J X, Hao S Z, Dong C 2009 *Trans Ind. Inst. Metals* **62** 479

[18] Hao S Z, Wu P S, Zou J X, Grosdidier T, Dong C 2007 *Applied Surf. Sci.* **253** 5349-5354

[19] Zou J X, Grosdidier T, Zhang K M, Dong C 2006 *Acta Mat.* **54** 5409-5419

[20] Zhang K M, Zou J X, Grosdidier T, Dong C 2009 *J. Vac. Sci. Technol.* A **27** 1217

[21] Zhang K M, Zou J X, Grosdidier T, Gey N, Weber S, Yang D Z, Dong C 2007 *J Vac Sci Technol* A **25** 28

[22] Guan Q F, Yang P L, Zou H, Zou G T 2006 *J. Mat. Sci.***41** 479

[23] Guan Q F, Zou H, Zou G T, Wu A M, Hao S Z, Zou J X, Qin Y, Dong C, Zhang Q Y 2005 *Surf. Coat. Tech* **196** 145-149

[24] Rotshtein V P et al. 2006 *Surf. Coat. Technol.***200** 6378-6383

[25] Zou J X, Grosdidier T, Bolle B, Zhang K M, Dong C 2007 *Metal. Mater. Trans.* **38** 2061

[26] Samih Y, Bolle B, Allain-Bonasso N, Hao S Z, Dong C, Grosdidier T *in the press of Mater. Sci. Forum.*

[27] Zou J X, Zhang K M, Grosdidier T, Dong C, Qin Y, Hao S Z, Yang D Z 2008 *Mat. Sci. Eng.* A. **483-484** 302-305

[28] Hao S Z, Yao S, Guan J, Wu A M, Zhong P, Dong C 2001 *Current Appl. Phys.* **1** 203

[29] Qin Y, Zou J X, Dong C, Wang X G, Wu A M, Liu Y, Hao S Z, Guan Q F 2004 *Nucl. Inst. Met. Phys. Res.* B **225** 544

[30] Zhang K M, Zou J X, Grosdidier T, Dong C, Weber S 2008 *J. Vac. Sci. Technol.* A **26** 1407

[31] Sankaran A, Bouzy E, Fundenberger J J, Hazotte A 2009 *Intermetallics* **17** 1007–1016

[32] Popov S A, Proskurovsky D I, Pryadko E L, Batrakov A V, Markov A B, Ozur G E, Rotshtein V P 2009 *IEEE Trans. Plasma Sci.* **37** 1504

[33] Zhang K M, Zou J X, Grosdidier T, Dong C 2009 *J. Vac. Sci. Technol.* A **27** 1217-1226

[34] Gey N, Humbert M 2003 *J. Mater. Sci.* **38** 1289

[35] Germain L, Gey N, Humbert M 2007 *Ultramicroscopy* **107** 1129–1135

[36] Burgers W G 1934 *Physica* **1** 561

Role of the carbonate impurities on the surface state of pyrite and arsenopyrite under treatment by high power electromagnetic pulses (HPEMP): oxidation of 50-100 μm size particles.

I Filippova[1,3], V Chanturiya[2], L Filippov[1], M Ryazantseva[1,2], I Bunin[2]
[1]Université de Lorraine, Laboratoire Environnement et Minéralurgie, UMR 7569 CNRS, rue du Doyen marcel Roubault, 54501 Vandoeuvre-les-Nancy
[2] Research Institute of Comprehensive Exploitation of Mineral Recourses, Russian Academy of Science, 4, Krukovsky Tupik, Moscow, Russian Federation,
[3]National University of Science and Technology "MISIS", Leninsky prospekt 4, 119049 Moscow, Russia

E-mail: inna.filippova@ensg.inpl-nancy.fr

Abstract. Diffuse Reflectance Infrared Fourier Transform Spectroscopy (DRIFTS) and Transmission Electron Microscopy (TEM) have shown the variation of surface phase compositions of carbonate bearing pyrite and arsenopyrite as a result of the combined action of chemical oxidation and thermal processes after the treatment by high power electromagnetic pulses (HPEMP). The monitoring of the surface phase composition allowed to determine the correlation between the treatment conditions, the surface phase composition, and the flotation yield. Thus, HPEMP treatment may be regarded as a tool controlling the surface composition and the sorption ability of flotation collector onto minerals surface, and therefore, allowing to control the hydrophobic-hydrophilic surface balance. It was confirmed in this study that the flotation of pyrite with xanthate as a result of the influence HPEMP may vary depending on the presence of impurities such as calcite.

1. Introduction

Gold-bearing sulphides are significant contributors to overall world gold production. Flotation is the principal process for pre-concentration of the sulphides, for subsequent smelting, roasting or hydrometallurgical treatment [1]. It is often found that gold is associated mainly with arsenopyrite or pyrite in a sulphide ore: therefore, the pyrite/arsenopyrite separation would be of interest. Arsenopyrite has very similar properties to pyrite and the selective separation of these minerals using xanthates as collectors is problematic. When the arsenopyrite needs to be separated from pyrite, the former is usually depressed. It is noted that the boundary between flotation and depression of arsenopyrite is marked by the X^-/X_2 reversible potential [2]. The formation of ferric–hydroxide–arsenate species seems to inhibit the oxidation of xanthate ions into dixanthogen causing depression of arsenopyrite at Eh–pH conditions of dixanthogen stability[3-5].

Regulating the oxidation state of the pyrite and arsenopyrite by the addition of oxidants (such as H_2O_2, $KMnO_4$, MnO_2 and HClO) or reductants (sodium sulfide, sulfite and dithionite) is the key to selective separation using xanthate as collector at high pH. $KMnO_4$ is the preferred oxidant and when used as an arsenopyrite depressant is coupled with a control of the redox potential at between 400 and 500 mV [3, 6,7]). The precise control of the redox potential was found to be essential in order to prevent depressing the pyrite also.

Another way of modifying mineral properties of sulphides is to create "energetic impacts". A strong interest in this method was shown recently [8,9]. The method consists in using high power electromagnetic pulses (HPEMP). This was applied successfully for the selective disintegration of fine-grained mineral complexes for the break out of precious metals inclusions followed by their subsequent extraction in leaching solutions.

Also the possibility of applying this treatment method for selective separation of pyrite from arsenopyrite by flotation in neutral pH environment was observed [10-12]]. It was found that the flotation of pyrite with xanthate after HPEMP treatment may vary, depending on the presence of metal impurities (such as Co, As, Ni, Cu) that have a possible influence on the electrical properties of pyrite. Calcite is another mineral impurity that can modify sensitively the flotation behaviour of pyrite. From this follows the importance of studying the effect of impurities on the surface properties of pyrite. For this purpose, the X-ray Photoelectron Spectroscopy (XPS) and infrared spectroscopy were used as the most powerful tools for investigating the surface state of iron sulfides [13].

This work aims at further understanding of the arsenopyrite and pyrite flotation chemistry with xanthates, and of the surface oxidation of arsenopyrite and of pyrite in the presence of impurities such as calcite. In this work we examined the role of the carbonate impurities on the surface state of pyrite and arsenopyrite during oxidation under treatment by HPEMP. Solid characterizations (DRIFTS, XPS, TEM) were used to ensure a complete understanding of the reactional paths.

2. Materials

Two natural mineral samples used in the present study came from the Darasun ore field (Chita region, Russia). Both the minerals were characterised by X-ray diffraction, ICP-MS, electron microprobe, MEB and TEM analyses. The chemical compositions of pyrite and arsenopyrite used here are presented in Table 1. Minerals were found to contain about 9.3-955% impurities.

Table 1. Chemical analyses (weight %, ppm) of pyrite and arsenopyrite used in this study.

Sample	Fe%	S%	As%	Cu%	Sb%	Pb%	Zn%	Ca%	Al%	P%	Co_{ppm}	Bi_{ppm}
FeS_2	40.61	49.89	0.69	1.29	1.18	0.88	0.28	0.78	0.09	0.03	141.7	104.00
FeAsS	30.79	19.60	40.30	0.02	0.04	0.01	nd	0.18	0.53	0.02	0.01	40.20

Specimens of both the sulfides contain calcite $CaCO_3$, quartz SiO_2, kaolinite $Al_2Si_2O_5(OH)_4$. The electron microscope microanalysis of the pyrite specimen reveals Cu, Sb, and As bound with sulfide inclusions of tennantite-tetrahedrite group ($Cu_{12}As_4S_{13}$ – $Cu_{12}Sb_4S_{13}$) minerals with appreciably variable Cu, Sb and As ratio. Arsenic in FeS_2 specimens is also bound with arsenopyrite (FeAsS); Pb is bound with galena (PbS), pure or rich in impurities (Sb, Cu, As, Ag) and berthonite $CuPbSbS_3$ (pure or with As impurities). Zn is bound with sphalerite (ZnS) in association with grey copper ore and pyrite. In pyrite specimens Bi-Te mineral inclusions were found, specific for gold-ore deposits: bismuthinite Bi_2S_3 (inclusions up to 90 μm), tetradymite Bi_2Te_2S (up to 35 μ m) and tellurobismuthinite Bi_2Te_3 (up to 10 μm).

3. Experimental Procedure

Minerals were dry ground in a porcelain mortar, and a size fraction of 50-100 μm was selected by sieving. The samples were then wetted in distilled water before being treated by HPEMP and dried in a nitrogen atmosphere to minimize the natural oxidation of the minerals. The untreated wet samples that were dried under nitrogen will be referred to as initial samples in both the cases.

The treatment of the samples by electromagnetic pulses (HPEMP) was carried out in a laboratory unit [8] in air by using a series of pulses with a nanosecond front period between 5 and 30 ns. The electrical field component was 30 kV, whereas the frequency of pulses and pulses repetition were set at 100 Hz. The treatment doses varied from 0.05 to 1.5 kJ. The pulses parameters were kept constant meanwhile the treatment doses were controlled by changing the treatment duration.

DRIFT spectroscopy was used to characterize the surface species on the treated mineral particles. The spectra were taken using a BRUKER EQUINOX55 device.

XPS spectra were recorded at room temperature under a pressure of ~10^{-9} mbar using an X-ray photoelectron spectrometer KRATOS Axis Ultra equipped with a monochromatic Al Kα source (1462 eV). The instrument was calibrated using metallic copper (binding energy of 932.62 eV for the Cu3/2 line). The spectra were recorded around the S2p, C1s, O1s and Fe2p lines.

In addition to information obtained from IR and XPS was also used a high-resolution transmission electron microscope (TEM) Philips CM20 (200 kV) equipped with an Energy Dispersive Spectrometer (EDS).

4. Results and Discussion

Surfaces of HPEMP treated pyrite minerals have been studied by DRIFT spectroscopy, X-ray photoelectron spectroscopy (XPS) and transmission electron microscopy including microanalysis (TEM/EDS).

DRIFT analysis (figure 1) has shown that, before the treatment, the pyrite surface was represented by a mixture of Fe^{2+}/Fe^{3+} hydrated sulphates (nH_2O=7–9) dominated by melanterite ($Fe^{2+}SO_4 \cdot 7H_2O$). Sulphates, observed in spectra, are associated with ferrous (606 cm^{-1} ν4, 1093 cm^{-1} ν3), ferric iron (998 cm^{-1} ν1, 1043 and 1152 cm^{-1} ν3) and occasionally with both iron oxidation states (1020 cm^{-1}). The weak quantity of ferric oxyhydroxide, probably goethite FeOOH (the massif at 600-700 cm^{-1} and characteristic absorption band at 794 cm^{-1} can also be associated with Si-O of quartz), as well as of Ca carbonates (713, 874, 1437 cm^{-1}) and traces of calcium sulfates (1115 cm^{-1} with a shoulder at 1155 cm^{-1}) were detected.

After the HPEMP treatments (0.05÷1 kJ) of the sample, the mineral surface is strongly «gypsumed» because of FeS_2 oxidation and calcite transformation into gypsum as show the stretching (νOH) at 3548, 3491, 3408, 3244 cm^{-1} and the bending (δ_{OH}) at 1618, 1685 cm^{-1} vibration of crystal water and also stretching (ν_{SO4}) vibration of sulfate groups at 1137, 671, 606 cm^{-1} (Figure 1).

The chemical oxidation of FeS_2 under HPEMP treatments in the presence of water can provoke a local formation of sulphuric acid, which can react with calcium carbonate impurities to form gypsum according to the reaction:

$$FeS_2 + 15/4O_2 + 9/2H_2O + 2CaCO_3 \rightarrow FeO(OH) + 2CaSO_4.2H_2O + 2CO_2$$

With consideration for the local surface heating effect [8], it can be assumed that the increase in number of impulses can cause the thermal oxidation and decomposition of pyrite to form a series of final products, such as hematite (Fe_2O_3), magnetite (Fe_3O_4), iron (ferric or ferrous) sulphate ($Fe_2(SO_4)_3$, $FeSO_4$) and gaseous sulphur oxides (mainly SO_2) that react with $CaCO_3$ to form $CaSO_4$:

$$CaCO_3 + SO_2 + 1/2O_2 \rightarrow CaSO_4 + CO_2$$

Simultaneously, the absorption bands of carbonate group of calcite decreased considerably (1795, 1421, 874, 713 cm^{-1}).

An increased intensity of stretching and bending of OH-groups vibration of the sulphates crystal water is detected on the spectra of pyrite under a high treatment dose (≥1kJ). In addition, the vibration of free water OH groups appears in the spectra (3702 cm^{-1} and 3727 cm^{-1}) at high number of pulses.

In the case of arsenopyrite, no considerable variation of DRIFT spectra profile was detected with the increase in number of pulses. But similarly to the pyrite case, bending vibrations of free water molecules were observed. It is essential to mention that gypsum does not form at arsenopyrite surface and that calcite, available in a specimen, undergoes none transformations depicted by IR spectroscopy. TEM qualitative analysis for samples of pyrite, allows confirm in this study the formation of calcium sulphate under HPEMP treatment. As shown in figure 2, peaks corresponding to Ca, S and O are detected in the EDX spectra and it was confirmed that gypsum is formed in calcite-bearing pyrite samples during HPEMP treatment.

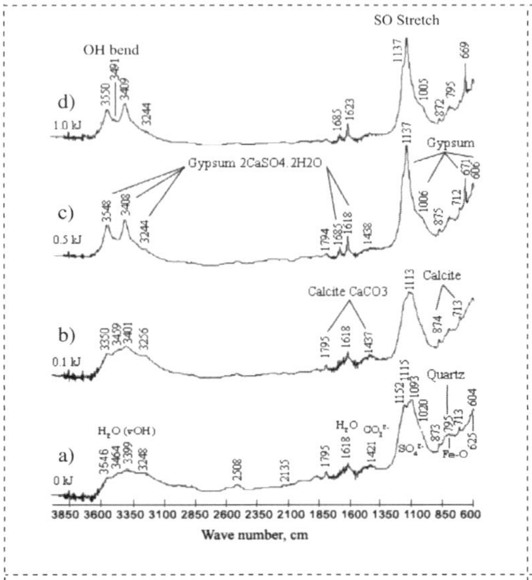

Figure 1. DRIFT spectra of the untreated and HPEMP treated pyrites samples. Treatment energy of samples: a) 0 kJ, b) 0.1 kJ, c) 0.5 kJ, d) 1.0 kJ.

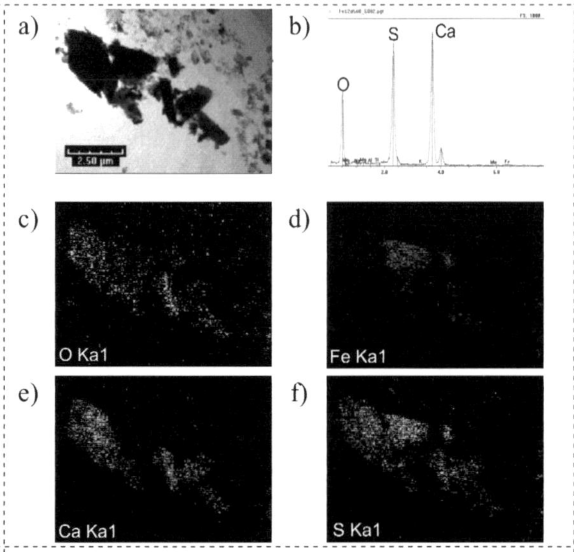

Figure 2. (a-b) TEM micrographs and corresponding EDS spectra of elements of gypsum in the pyrite sample after HPEMP treatment (0.05 kJ); (c-f) X-ray images which mapped the location of O, Fe, Ca and S.

In this study, the feasibility was explored for using transmission electron microscope (TEM) in conjunction with energy dispersive x-ray spectroscopy (EDS) to determine the oxygen directly and to quantify the sulfur content of the pyrite. The ability to perform chemical analysis of a sample during observation is an advantage to determine the degree of oxidation of pyrites and their chemical composition based on the detected elements after different treatment dose. In treated minerals, the TEM-EDS measurements show significant alterations in the respective surfaces when different HPEMP treatement were applied. In treated samples, the formation of Fe^{III} species containing oxygen were detected. The pyrite surface shows a layer of modified species with high oxygen content as a function of energy of HPEMP treatment (Figures 3 and 4). The samples of pyrite treated with a dose of 0.5 kJ produces less oxidized pyrite and iron oxide particles than in the samples treated with 0.05 kJ and 1 kJ (Figure 3). That is, there is no pattern of growth of oxide particles with the increase of the energy processing. This may indicate the presence of certain processes and the formation of compounds that protect the particles during the processing of pyrite oxidation [13].

Since we could not explain this result only from infrared spectroscopy or electron microscopy results, we used an additional X-ray photoelectron spectroscopy results presented in a previous paper [12]. As was reported by Chanturiya *et al.* [11,12] XPS results show considerable qualitative and quantitative changes of the surface layer composition of both the sulphides depending on the treatment conditions. After the treatment the increase of Fe oxidized ($FexOy-FeOOH.nH_2O$ - $Fex(SO_4)y.nH_2O$) was determined for all treated samples of pyrite from Fe2p and O1s lines.

At a treatment dose 0.05 kJ the intensive formation of iron oxide and hydroxide phase was observed in Fe2p and O1s lines. The surface concentration of iron oxides increased 2 times according to the fitting of O1s line. An increase of S° surface concentration from 12.95 % to 23.56 % was observed according to the S2p line [12].

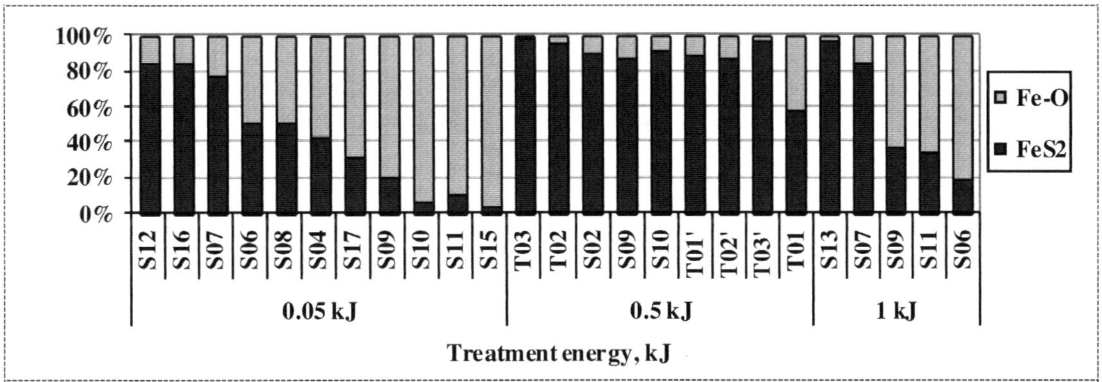

Figure 3. Schematic representation of the results of TEM-EDS analysis of change chemical composition of the FeS_2 samples as a function of energy of HPEMP treatments.

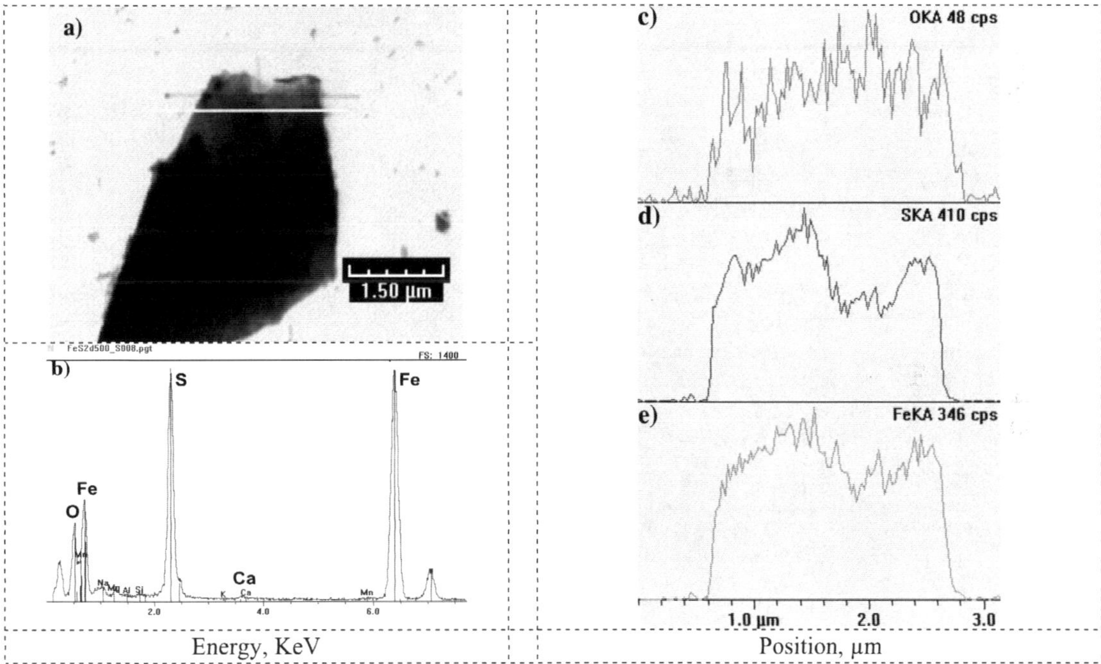

Figure 4. TEM observation of oxidized FeS_2 after HPEMP treatment (0.05 kJ): a-b) TEM micrographs and corresponding EDS spectra of elements Fe, S, O, Ca. c-e) The energy spectrum of the X-ray paterns emitted from the elements Fe, O and S.

A further increase of treatment dose (0.1 kJ) results in an increase of iron sulphates surface concentration. These results corroborate with the surface modification obtained in this paper and confirm the cyclic character of pyrite oxidation as a result of combined action of chemical oxidation and thermal processes after the treatment by high power electromagnetic pulses (HPEMP).

The treatment at 0.5 kJ results in a further increase of Fe sulphide, as well as a decrease of S° surface concentration (Figure 3). A 1 kJ treatment dose results in an increase of the surface concentration of the oxidation product. The process of elemental sulphur thermal removal was hypothetically discussed previously as an explanation of these results [12].

Arsenopyrite has shown an intensive oxidation at low treatment doses with formation of iron sulphates and As oxides. We have shown [10] that the surface hydration increased 3 times as compared with the initial sample. An increased sorption activity of HPEMP-processed pyrite at 0.5 kJ and 1.0 kJ in relation to HPEMP-unprocessed mineral was identified. Flotation tests have demonstrated that in the presence of ethyl xanthate, flotation activity of processed pyrite samples increased in average by 18.9% in comparison with the reference sample. On the contrary, the recovery of arsenopyrite decreased from 53.4% (0) to 23.7% (0.1 kJ). For 0.5 kJ and 1.5 kJ modes, arsenopyrite recovery was 36.1% and 29.07%, respectively [10].

5. Conclusions.

The phase composition of pyrite surface underwent significant changes during electric pulse treatment. The TEM study coupled to DRIFT analyses shows a cyclic oxidation way of pyrite surface as a result of combined actions of chemical and thermal action of high power electric pulse treatment. The pyrite activation at low-energy high-pulse treatment (<0.1 kJ) is ensured by the formation of elementary sulphur within a dose range of 0.1 kJ ÷ 1 kJ. The arsenopyrite depression is related to an increase in the general hydration of mineral surface during treatment, as well as a change in phase composition and electrophysical surface properties. The significant amounts of gypsum on the surface of the pyrite after HPEMP treatment is linked to the interaction of sulfates ions with the calcite present as impurity in the pyrite.

References

[1] Bulatovic S M 1997 *Minerals. Eng.* **10** 897
[2] Allison S A, Goold L A, Nicol M J and Granville A 1972 *Met. Trans. B: Process Met.* 2613
[3] Beattie M and Poling G W 1987 *Int. J. of Mineral. Processing* **20** 87
[4] Wang X H, Ahlberg E and Forssberg K S 1992 *J. Appl. Electrochem.* **22**, 1095
[5] Monte M B, Dutra A J, Albuquerque C R , Tondo L A and Lins F F 2002 Min. Eng. **15** 1113
[6] Draskic D, Grifing M M, Pavlica J 1982 *Proc XIV Int. Mineral Processing Congress*, Canadian Institute of Metallurgy, Toronto, Canada (1982) **IV** 5.1-5.13
[7] Li G, Zhang H 1989 The chemical principles of flotation, activation and depression of arsenopyrite *Int. Sym. on Processing Complex Ores*, eds Dobby G S and Rao S R (New York: Pergamon Press) pp. 61–70
[8] Chanturiya VA, Bunin I Z, Ryazantseva M V and Filippov L O 2011 *Mineral Processing and Extractive Metallurgy Review: An International Journal* **32** 105
[9] Can N M, Bayractar I 2007 *Minerals and metallurgical processing* **24** 185.
[10] Chanturiya V, Filippova I, Filippov L, and Ryazantseva M 2010 The effect of high power nanosecond electromagnetic pulses on the selectivity of pyrite flotation from arsenopyrite. *XXV Int Mineral Processing Congr, Brisbane Australia, 6-10 Septembre 2010* pp1743-1752
[11] Chanturiya V, Filippova I, Filippov L, Ryazantseva M, and Mullet M 2008 Influence of High Power Electromagnetic Pulses on Surface State and Floatability of Carbonate-Bearing Pyrite and Arsenopyrite. *Proc of 3rd France-Russia Conf. "New achievements in Materials and Environmental Sciences", 7-9 November 2007, Metz, France* (EDP Sciences) pp 111-114.
[12] Chanturiya V A, Filippova I V, Filippov L O, Ryazantseva M V and Bunin I 2008 *J. of Mining Sci.* **44** 518
[13] Descostes M, Mercier F, Thromat N, Beaucaire C and Gautier-Soyer M 2000 App. Surface Sci. **165** 288.

Synthesis of Li_5AlO_4 involving a repulpation mixing in a Li_2CO_3 saturated solution.

Ol'ga N Krivolapova and Vadim P Tarasov
Interbranch Educational Centre of Science of Recycling of Chemical Sources of a Current, *National University of Science and Technology "MISiS"*, 4 Leninsky Prospect Moscow, Russia 119049

E-mail: onk@misis.ru

Abstract. Two methods of the pentalithium aluminate synthesis (Li_5AlO_4) are investigated in the study: by sintering dry mixture of lithium carbonate and hydroxodialuminate and mixture obtained as a result of repulpation mixing in Li_2CO_3 saturated solution. The pentalithium aluminate synthesis was studied using thermogravimetry and infrared spectroscopy. The researches on the Li_5AlO_4 synthesis kinetics were carried out using gas chromatography and the pycnometric density, bulk density and tap density were also determined.

1. Introduction

At present time metallic lithium is produced by electrolysis [1]. This method has significant disadvantages. The use of extremely corrosive and hygroscopic salt: lithium chloride, and also the necessity of chloride trapping and neutralizing, the utilization of exhausted electrolyte consisting of eutectic mixture of potassium and lithium chlorides should be referred to them. To obtain a high purity lithium, both initial lithium chloride salt and also electrolyte salts must be additionally purified.

Most of these disadvantages are avoided in the vacuum metal-thermal method of the lithium production. The analysis of studies for the metal-thermal lithium production [2], [3] showed that the use of lithium aluminates, in the place of lithium monoaluminate $LiAlO_2$, and pentalithium aluminate Li_5AlO_4 as initial materials was outstandingly the most preferable.

It must be pointed out that the method of metal-thermal reduction of lithium monoaluminate has a significant drawback along with the obvious merits: a low lithium content in this compound– 10.45%. From the point of view of the improvement of technical and economic indicators of the reduction process the method of the pentalithium aluminate reduction in which the lithium content amounts to 27.78% is more preferable.

Lithium carbonate sintering with lithium hydroxodialuminate (GODAL) [4] according to the reaction:

$$9Li_2CO_3 + Li_2O \cdot 2Al_2O_3 \cdot 11H_2O \xrightarrow{T} = 2Li_5AlO_4 + 11H_2O \uparrow + 9CO_2 \uparrow \quad (1)$$

should be recognized as the most efficient method to produce pentalithium aluminate.

The application of this method revealed along with the positive side: time decrease of the pentalithium aluminate synthesis to 2 h instead of 16-18 h when using aluminum oxide [5], but also some disadvantages. Original particles sharply differ in particle sizes: Li_2CO_3 – (1.2 – 2.4) – (1.6 – 3.2)$\cdot 10^{-5}$ m; GODAL – (0.3 – 0.06) – (0.4 – 0.02)$\cdot 10^{-5}$ m. In this connection the considerable dispersity of fine-dispersed particles of GODAL within the entire capacity of the mixer is observed. Obtaining an homogeneous mixture is then complicated, what results in mechanical losses, the dust loading in the working place requires to use covered gears and a powerful ventilation. Considerable time and high energy consumption are required to reach a proper mixing process. Synthesized pentalithium aluminate has a low bulk density that reduces the efficiency of the vacuum thermal plants upon its aluminothermal reduction.

2. Experimental

Two methods of Li_5AlO_4 synthesis were investigated in this study: on one hand by sintering a dry mixture of lithium carbonate and hydroxodialuminate and on another hand mixture obtained as a result of repulpation mixing in the Li_2CO_3 saturated solution. Repulpation mixing in lithium carbonate–saturated solution prevents solid Li_2CO_3 from passing from the mixture into solution and the solubility of lithium hydroxodialuminate does not exceed 0.005 g/l. Besides, mixing in the saturated lithium-carbonate solution runs considerably more intensively. According to [6], this is mainly attributed to the fact that the liquid injection into the mixture favors the decrease of the electric attraction forces among particles and, consequently, their better dispersion in the mixture. In addition, under the action of capillary forces an increased pressure is formed in thin particles fissures that assists in disengaging particles associated into aggregates and results in component post-grinding. Upon mixing in saturated solution the particle mobility increases that favors, in its turn, a more intensive mixing.

It must also be pointed out that as a consequence of repulpation mixing of original products, their subsequent filtration and drying, every particle of lithium hydroxodialuminate (GODAL) is coated with a lithium carbonate film, i.e. a close contact is formed at the boundaries between powder particles. In this connection the reaction surface sharply increases.

2.1. Synthesis methods

Lithium carbonate of the especially pure grade and "GODAL" of the following composition: Li_2O = 6.9 mass %; Al_2O_3 = 47.5 mass %; H_2O = 45.6 mass % that complies with $Li_2O \cdot 2Al_2O_3 \cdot 10.95H_2O$ compound were used as initial materials. GODAL was produced according to the method defined in [7], [8].

The amount of initial materials was calculated in compliance with the synthesis reaction equation (see equation 1).

Dry mixing was performed in the mixer of the «drum tumbler» type repulpation one, in lithium carbonate-saturated solution using an agitator. Mixing in saturated solution was performed at a S:L ratio of 1:2; 1:3; 1:4; 1:5. Burden material mixing at a S:L ratio of > 1:4 results in increasing the pulp viscosity and as a consequence, in mixing deterioration. Decreasing the S:L ratio below 1:4 leads to an excessive carbonate solution consumption. Mixing results were controlled by the chemical sample analysis. Samples were treated with water. In this case lithium carbonate having a solubility of 13.3 g/l passes into the solution and GODAL remains in precipitate. The phase relation in samples selected from prepared mixtures was determined by the subsequent solution analysis for lithium and drying up to the GODAL mass consistency and calculating lithium therein. In case of the application of the repulpation mixing technology the burden material homogeneity was achieved in 0.5 h and in case of dry mixing in 5-6 h.

2.2. Analysis

2.2.1. Thermogravimetric analysis.

The thermogravimetric analysis of the mixture of GODAL and lithium carbonate at a molar ratio of 1:1 was carried out in the "DU PONT-1090" plant at the atmospheric pressure. The heating rate was 10 degrees/s. Platinum was used as crucible material. Sensitivity: DTA – 1/5; DTG – 1/15; TG – 200 mg. In the process of linearly programmed heating of the original mixture a series of successive solid-phase chemical transformations takes place which are recorded in curves of the differential-thermal analysis. Within the temperature interval of 100-190°C a low endothermal effect is noted in dry and repulpation mixture that is accompanied by a sample mass loss corresponding to three water molecules and a partial disintegration of the GODAL structure. The second endothermal effect within the interval of 210-295°C is accompanied by the further sample mass loss and corresponds to the complete water removal from the GODAL structure. That well accords with the results of the GODAL differential-thermal analysis performed in study [8]. Within the temperature interval of 310 up to 470°C and of 470 up to 700°C two inadequate exothermal effects are noted in the differential-thermal curve of dry mixture heating. The first exothermal effect with a minimum at 420°C is attributed, obviously, to the recrystallization of GODAL X-ray amorphous components obtaining Li and Al oxides mixture from GODAL decomposition products. The second exothermal effect is accompanied by a sample mass decrease corresponding to the CO_2 content in the original mixture that evidences the interaction of lithium carbonate with GODAL thermal decomposition products with the formation of pentalithium aluminate. The process of the α-Li_5AlO_4 synthesis out of the mixture terminates at 700°C. The further temperature rise leads to an additional exothermal effect in the DTA curve within the temperature interval of 980 – 1,000°C. This effect is not accompanied by any mass decrease and can be attributed to the polymorphous transformation of α-Li_5AlO_4 to its high-temperature form β-Li_5AlO_4. When comparing the chemical interaction in the dry mixture with the interaction in the repulpation mixture it must be pointed out that the process of repulpation mixing ensures itself the delivery and securing of lithium carbonate particles on the GODAL surface. That assists in activating the solid-phase interaction.

A conclusion can be made on the ground of the analysis results that the synthesis of pentalithium aluminate from dry mixture runs within the temperature interval of 470 up to 700°C, and from repulpation one within the temperature interval of 310 - 635°C. Thus, repulpation mixing of initial products (lithium carbonate and GODAL) activates to a significant extent the solid-phase chemical interaction: the reaction rate increases, the pentalithium aluminate synthesis temperature decreases.

2.2.2. Infrared (IR) spectroscopy.
For the IR spectroscopy the reaction mixture was sintered at temperatures of 500, 550, 600, 650, 700, 750, 800, 850 and 900°C. The sintering time at every temperature was 1 h. Infrared spectra of synthesis products were recorded in the IR spectrometer SPECORD-75JR within the band of wave numbers of 1,600 to 400 cm^{-1} (6.5 – 25 μm). Samples for taking IR spectra were prepared by mixing weighed specimens with potassium bromide. A pellet of pure potassium bromide was used as a reference standard.

Synthesis products at 500 – 550°C within the band of 1,480 – 1,410 cm^{-1} and 1,080 – 1,010 cm^{-1} have absorption bands characteristic for carbonates [9]. That confirms the data of the X-ray phase analysis that synthesis products at 500 and 550°C contain Li_2CO_3 residues.

2.2.3. Research of the Li_5AlO_4 synthesis kinetics applying the gas chromatography method.
The interaction reaction of lithium carbonate with GODAL dehydrated as a result of the thermal GODAL decomposition is accompanied by the release of gaseous carbon dioxide. Therefore the rate of the pentalithium aluminate synthesis can be judged by the variation of the CO_2 concentration in the gas phase. The comparative analysis of the pentalithium aluminate synthesis kinetics was performed using gas chromatography.

Inertial argon-carrier gas was let pass through the quartz reactor inside which there is an alumina fiber boat with burden material (4 g) at a given rate.

The reactor was heated in the tube resistance furnace. The temperature was controlled and regulated by means of a chromel-alumel thermocouple. After the set temperature had been achieved, the furnace was moved along the quartz reactor to the area of the boat location and sampling was started. Sampling was carried out in three minutes. The gas mixture analyzed was investigated after drying for the CO_2 content. A strictly controlled sample of 20 cm^3 was subjected to analysis. The results of the chromatographic analysis for CO_2 in the form of a D.C. signal were plotted in the data recorder.

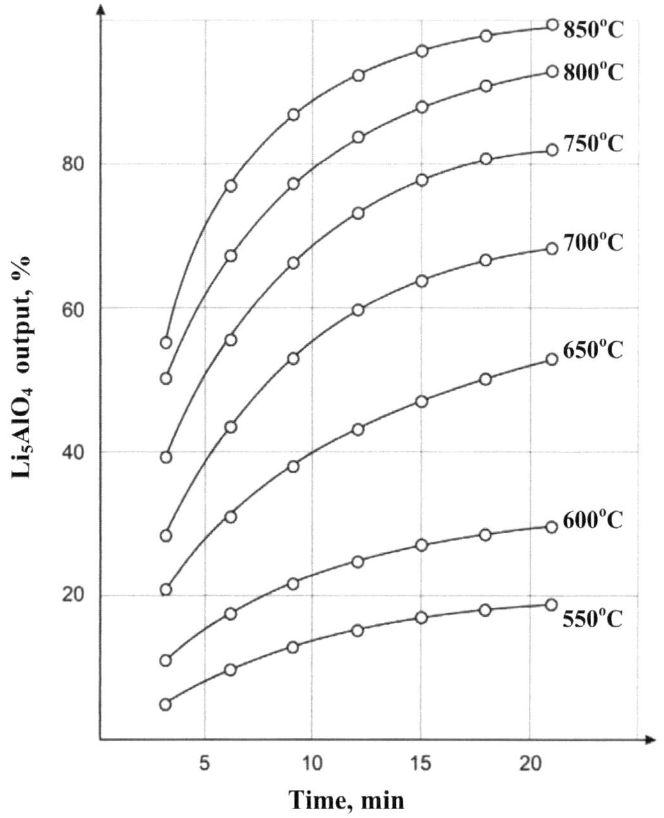

Figure 1. Li_5AlO_4 synthesis kinetics (dry mixture)

2.2.4. Research of physical and technological properties.

When using pentalithium aluminate as initial a raw material for vacuum thermal production of lithium, its physical and technological properties are of a major importance. First of all that applies to the vacuum thermal gear efficiency. To that end a comparative analysis of physical and technological properties of Li_5AlO_4 powders synthesized from dry and repulpation mixtures was performed at various temperatures.

The pycnometric density was determined in kerosene. The measured pycnometric density of Li_5AlO_4 synthesized from the repulpation mixture confirms the data of the X-ray phase analysis that α-Li_5AlO, is formed at 635°C and its polymorphous transformation to β-Li_5AlO_4 terminates at 700°C. Upon the Li_5AlO_4 synthesis from the dry mixture the crystallization of the α-phase of pentalithium aluminate completes at 700°C and within the temperature interval of 750 - 900°C the process of the polymorphous transformation of α-Li_5AlO_4 to β-Li_5AlO_4 takes place. Since lithium carbonate sintering with GODAL both in dry and also in repulpation mixtures was performed for 2 h at every temperature this phenomenon confirms that the repulpation mixing of initial products accelerates the Li_5AlO_4 synthesis.

3. Results

The relation of the tap density and bulk density of pentalithium aluminate as a function of the synthesis temperature was studied here. The obtained results are presented in Figure 1. The both indicators in the samples obtained from the repulpation mixture increase with the temperature rise although the pycnometric density remains unchanged after 700°C. The research of sizes of pentalithium aluminate particles demonstrated that with the temperature rise the particle size increased from $(0.8 - 1.2) \cdot 10^{-5}$ m at 700°C to $(2.0 - 2.4) \cdot 10^{-5}$ m at 900°C. Upon the phase transition the β-Li_5AlO_4 «bunches» collective crystallization occurs up to sizes of $(6.8 - 8.0) \cdot 10^{-5}$ m. Besides, particles in the β-phase vary their form changing-over to the isometric rounded.

4. Conclusions

Thus, having conducted the comparative analysis of the pentalithium aluminate synthesis using dry and repulpation mixtures it became obvious that the repulpation mixing of initial products favors a more uniform mixing, provides the delivery and securing of lithium carbonate particles on the GODAL surface that activates to a significant extent the solid-phase chemical interaction:

- the rate of the chemical reaction run increases;
- the temperature of the pentalithium aluminate synthesis decreases;
- the bulk density of pentalithium aluminate increases.

Acknowledgments

This research was carried out with financial support of the National University of Science and Technology "MISiS" Creation and Development Program.

References

[1] Kulifeyev V K, Miklushevsky V V, Vatulin I I 2006 *Lithium* (Moscow: MISiS) pp 172-76

[2] Kotsupalo N P, Riabtsev A D 2008 *Chemistry and technology of producing lithium compounds from lithium-bearing hydromineral raw* (Novosibirsk: Academic Publishing House Geo)

[3] Pantiukhina M I, Pletnev E E, Batalov N N 2007 Synthesis and research of electrophysical characteristics of lithium-conducting phases for high-temperature current sources *Urgent problems of the chemical production, assessment of technical risks. Physico-chemical principles of the rational use of natural and technogeneous resources. Chemical aspects of the present-day power engineering. Petrochemistry. Oil processing and catalysis: Proc. 18 Mendeleyev congress on general and applied chemistry* (Moscow, September 23-28, 2007) (Moscow: Granitsa) vol 3 p 212.

[4] Tarasov V P 1990 A method to produce pentalithium aluminate *Bulletin of Author's Certificate* (USSR: Nauka) 10

[5] Ginestra L A, Jacono M L , Porta P 1972 *J. Therm. Anal.* **4** 7

[6] Kotsupalo N P, Guseva I V, Yevteyeva O G 1967 *Properties of lithium hydroaluminate* (Novosibirsk: Rare alkali elements) pp 92-99

[7] Miklushevsky V V, Vatulin I I, Tarasov V P 2003 Research of kinetic mechanisms and determination of technological parameters of the alumothermal reduction process of pentalithium aluminate in vacuum *J. Non-ferrous metals* **3** 55-58

[8] Tarasov V P 2002 Physico-chemical bases and technology of the utilization of lithium chemical current sources *Doctor thesis* (Moscow: MISiS)

[9] Louson K 1964 *Infrared spectra of inorganic substance absorption* (Moscow: MISiS)

Tailoring of analytical performances of urea biosensors using nanomaterials

W Nouira[1,2*], H Barhoumi[1], A Maaref[1], N Jaffrézic Renault[2*], M Siadat[3]

[1] Laboratoire de Physique et Chimie des Interfaces, Faculté des Sciences de Monastir, Avenue de l'Environnement, 5019 Monastir, Tunisia
[2] Laboratoire des Sciences Analytiques, Université Claude Bernard, Lyon 1, 43 Boulevard du 11 novembre 1918, 69622 Villeurbanne, France
[3] Laboratoire d'Automatique et de Sciences Comportementales, Université Paul Verlaine, 7 Rue Marconi, 57070 Metz, France

E-mail: widednouira@yahoo.fr ; nicole.jaffrezic@univ-lyon1.fr

Abstract. This paper is a contribution to the study of enzymatic sensors based on nanoparticles of iron oxide (FeNPs). Urease enzyme was immobilized on FeNPs using layer-by-layer (LbL) deposition method. FeNPs were first coated with polyelectrolytes (PE): Poly (allylamine hydrochloride), PAH and Poly (sodium 4-styrenesulfonate), PSS for enzyme immobilization and then with enzyme. It has been confirmed through zeta potential measurements of FeNPs that the enzyme is immobilized on the surface. We evaluated the sensitivity of biosensors for urea by potentiometric and capacitive measurements on silicon / silica / FeNP-LBL-urease structures. The recorded capacity-potential curves (C-V) show a significant shift of flat band potential towards negative potentials in the presence of urea, the observed values of sensitivity vary between 30 and 40 mV/p[urea]. It has been shown that the proposed method for the immobilization of urease can increase the dynamic range of urea detection (10^{-4}M to 10^{-1}M) compared to the immobilization of urease without FeNP ($10^{-3.5}$ M to $10^{-2.5}$ M). When the number of PAH-PSS layers was increased the sensitivity of detection was modified. This effect is due to partial inhibition of the enzyme in presence of FeNPs, which was shown by measurements in homogeneous phase.

Keywords: Biosensors; polyelectrolyte; nanoparticles; silica; urease.

1. Introduction

Magnetic particles were traditionally used for concentration, separation, purification and identification of molecules and specific cells. Inorganic nanoparticles (NPs) with specific size and morphology have attracted increasing interest due to the influence of their shape and size on their electronic, optical, magnetic and catalytic properties [1]. Inorganic NPs play most important roles in the development of the nanoelectrochemistry, because inorganic NPs exhibit higher surface area to volume ratios than their bulk counterparts, so inorganic NP modified electrochemical interfaces will provide larger electrochemically active areas and therefore probably lead to higher detection sensitivity for target molecules; then, some novel NPs, particularly metal NPs, can easily act as enhancing agents for effective acceleration of electron transfer between electrode and detected molecules, so leading to more rapid current response for target molecules; further inorganic NPs can act as a supramolecular assembling unit with advanced functional properties for constructing a variety of architectures on the surface of electrodes and further tailoring of an electrochemical-sensing interface [2]. A review from N. Jaffrezic-Renault and al presents the use of magnetic nanoparticles for biosensor applications. They covered three biosensing systems: affinity biosensors, enzymatic biosensors and bio-bar codes [3]. In recent years, the magnetic properties of some NPs have also been used as labels in biosensing [4, 5]. So, Cheng proposed a sensitive electrochemical biosensor for DNA sequence mutation detection based on the separation and enrichment of magnetic nanobeads. [6]. It can be seen that the application of nano-materials greatly improves the detecting sensitivity. Electrostatic layer-by-layer (LbL) self-assembly provides a convenient and versatile method to build micro/nanostructures on two- or three-dimensional substrates at the molecular level. [7, 8], this technique was developed first by Decher et al. [9, 10]. The construction of LbL shells involves consecutive polyelectrolyte adsorption on a colloid-template followed by decomposition of the template core. Different materials, including linear polyelectrolytes (synthetic and natural), lipids, proteins, and inorganic nanoparticles can be used in nanoassembly processes [11]. Through the LbL method, Elaissari et al. used polyelectrolytes to encapsulate oil in water magnetic emulsion and obtained magnetic particles [12]. Further, Souiri et al used the LbL technique involving the immobilization of a bioactive protein (or peptide) by trapping it onto a previously "conditioned surface" obtained by alternatively adsorbing oppositely charged polyelectrolyte layers [13]. In this study we propose to work with the enzymatic biosensors based on iron NPs using the LbL technique.

The aim of this work is to develop a potentiometric and capacitive biosensors for urea detection. The enzymatic reaction catalyzed by urease is the following one:

$$CO(NH_2)_2 + H^+ + 2H_2O \xrightarrow{Urease} 2NH_4^+ + HCO_3^-$$

The reaction can be monitored either by measuring the local production of ammonium ions or by measuring the local variation of pH. The enzymatic reaction was monitored in homogeneous medium (solution) using a potentiometric ammonium electrode. For an enzymatic biosensor, the local variation of pH, when urease enzyme was immobilized, was transduced by an insulator-semiconductor structure, SiO2/Si3N4/SiO2/Si. The electrochemical response of the developed urea biosensor was evaluated using the capacitance potential measurements [14]. Variation of the flat-band potential of the Electrolyte-Insulator-Semiconductor structure varies when concentration of urea in solution varies. The effect of the conditions of immobilization of urease on the insulator surface (without and with FeNPs) where studied and the analytical features of the obtained biosensor were determined.

2. Materials and Methods

2.1. Materials

Magnetic ferric oxide nanoparticles (FeNPs) were obtained from Sigma - Aldrich. The nanoparticle size was less than 50 nm. These nanoparticles were suspended in ultrapure water. Cationic polyelectrolyte used in this study, Poly(allylamine Hydrochloride) (PAH) and anionic polyelectrolyte is Poly(sodium 4-styrenesulfonate) (PSS) were purchased from Sigma - Aldrich (Figure 1). Bovine serum albumin (BSA), glutaraldehyde (GA) (grade II, 25% aqueous solution), NaH2PO4 (> 99%) and Na2HPO4.12H2O (> 99%) were obtained from Sigma - Aldrich. Acetone, trichloroethylene and isopropanol were purchase from Fluka. Urease (Type IX from Jack Beans, 0.28G solid 70400 U/g), Stock concentrated solutions were prepared in 10 mM phosphate buffer solution (pH 7.6) and stored at 4°C. All aqueous solutions were prepared using 18 MΩ cm^{-1} ultrapure water produced by a Millipore Mili-Q system.

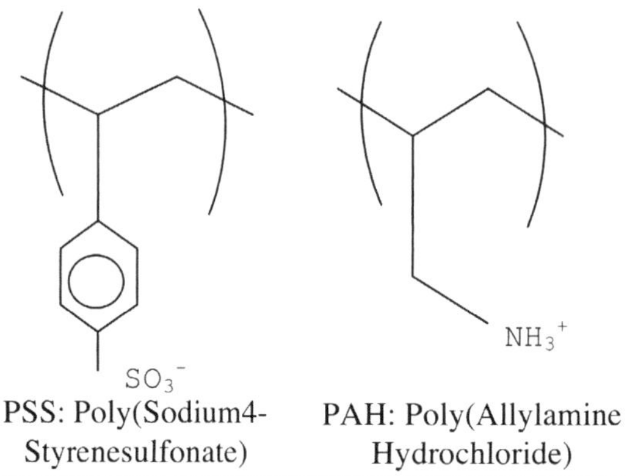

Figure 1. Polyelectrolytes.

2.2. Potentiometric measurements of ammonium ions produced by the enzymatic reaction
The ammonium electrodes with polyvinyl polymeric chloride membrane (PVC) and the reference Ag/AgCl used in this study were provided by the Nico2000 Company. The potential difference between both electrodes was measured by a potentiometer (ELITS 8051) provided Nico2000 company.

2.3. Determination of electrophoretic mobility of coated FeNPs
To control the final surface composition of LbL NPs, we used polycation (PAH) or polyanions (PSS) as the outermost layer for the shells NPs. The zeta potential of polyelectrolyte NPs with different outermost layers were measured using dynamic light scattering (DLS) by a Zetasizer Nano-ZS (Malvern Instruments). Measurements on very dilute samples were performed at ionic strength (10^{-3}M NaCl). The average value of at least five measurements was taken at a given condition.

2.4. Preparation of the urea biosensor

2.4.1. EIS substrate and capacitance-voltage (C-V) measurements. The Electrolyte-Insulator-Semiconductor (EIS) were purchased from the Institute of Microtechnology of the University of Neuchatel (Switzerland). The EIS capacitors were fabricated by thermally growing 80 nm SiO2 on <100> p-type silicon wafer with 3-5 Ω.cm resistivity and 400μm thickness. Silicon oxynitride of 100 nm thickness was then deposited by LPCVD. The nitride surface was then oxidized for obtaining a 50 nm thick oxide layer. The ohmic contact was obtained by deposition of a gallium–indium mixture on the back side of the Si/SiO2/Si3N4 structures (Figure 2). Before the deposition of the biomembrane, the IS structures were treated using an optimized cleaning procedure: surface transducers were cleaned in successive baths of trichloroethylene, acetone, isopropanol and sulphochromic mixture (sulphuric acid and chromic acid), washed with ultrapure water, dried under nitrogen atmosphere at room temperature and placed at 70 °C for 10 min. The EIS structures were characterized by capacitance-voltage (C-V) method. This latter were performed using an impedance analyzer (Voltalab 40, Hach Lange) at a frequency of 10 kHz and with a signal amplitude of 10mV, the potential being varied from -0.5 V to +2.5 V. The solution used for testing the urea sensitivity was 10 mM phosphate buffer at a fixed pH value of 7.6.

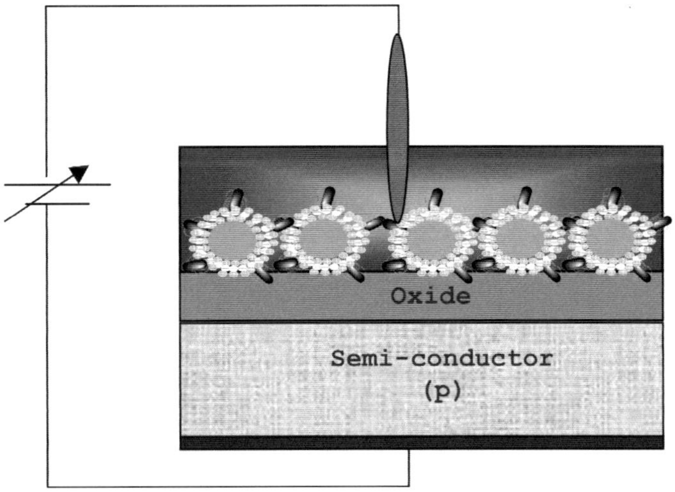

Figure 2. Scheme of the EIS structure with urease functionalized FeNPs.

2.4.2. Coating of ferric oxide nanoparticles. Nanoparticles (FeNPs) were first homogenized under mechanical stirring for 15 min in ultrapure water (Fe3O4:1%w/w). The nanoparticles were then coated with an initial "preconditioning" layer of PAH (positive charge), then a layer of PSS of opposite charge was deposited to form the first PE bilayer. Each PE coating step was performed in aqueous solutions of filtered ultrapure water at 5mg/mL of PE, under similar stirring conditions during 15 min. The nanobeads were then recovered by applying a mild magnetic field (using a small magnetic stir bar). The PE rich supernatant phase was then eliminated and the beads were rinsed twice in filtered ultrapure water under mechanical stirring during 15 min, and separated again under a mild magnetic field (Figure 3).

2.4.3. Urease immobilization. The globally negatively charged enzyme (purchased at Sigma–Aldrich) was immobilized on the nanoparticles coated with three layers of polyelectrolytes (FeNP-(PAH-PSS-PAH)) by electrostatic interaction in a 1mg/mL urease/phosphate buffer at pH 7.6. This step was performed by separating the particles from the urease solution, rinsing in phosphate buffer, separating them via magnetic field application (Figure 3).
In this study, two procedures were followed for the biosensor preparation:

- without FeNPs: 10 µL of a 10 mM phosphate buffer pH 7.6 containing 6% (m/v) BSA, 10% (m/v) glycerol with 4% (w/w) of urease was deposited on the surface of the IS structure. The sensors were then placed in saturated Glutaraldehyde (GA) vapours for 30 min. After incubation, the membranes were dried in free air for 15–30 min.
- with FeNPs: 10 µL of a 10 mM phosphate buffer pH 7.6 containing 6%(m/v) BSA, 10% (m/v) glycerol with 1% (w/w) of urease/PE/NPs was deposited on the surface of the IS structure. The sensors were then placed in saturated Glutaraldehyde (GA) vapours for 30 min. After incubation, the membranes were dried in free air for 15–30 min.

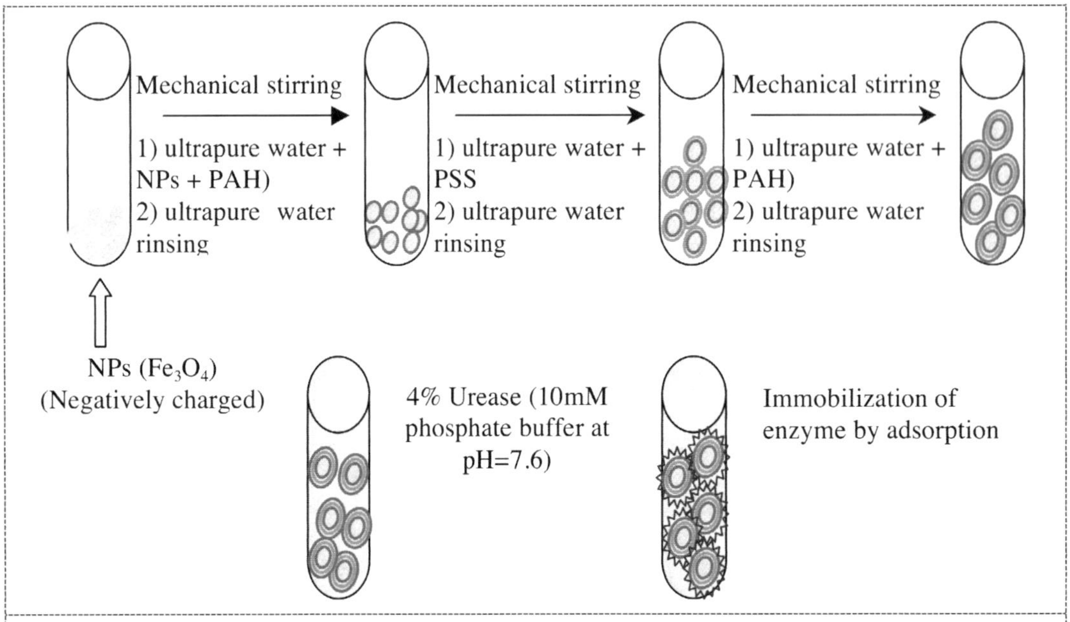

Figure 3. LbL technique and urease immobilization.

3. Results and discussion

3.1. Characterization of coated FeNPs through electrophoretic mobility of coated FeNPs

The measured values of zeta potential of the uncoated and coated FeNPs are presented in Table 1. When the outermost layer is a polycation (PAH), a positive surface potential is observed whereas when the outmost layer is a polyanion (PSS) a negative surface potential is observed. When urease is the outermost layer, a negative potential is observed showing the successful immobilization of urease.

Table 1. Surface potentials for uncoated and coated iron oxide nanoparticles.

Name	**Temperature (°C)**	**Zeta Potential (mV)**
Uncoated FeNP	25	-12.1 ± 0.3
FeNP + PAH	25	25.5 ± 0.7
FeNP + PAH + PSS	25	-38.9 ± 0.2
FeNP + PAH + PSS + PAH	25	12.4 ± 0.7
FeNP + PAH + PSS + PAH + Enzyme	25	-24.7 ± 0,7

3.2. Variation of the enzyme activity in presence of iron oxide nanoparticles

Heavy metals inhibit urease activity. The mechanism of the inhibition of urease is based on interactions between the heavy metal ions and the groups of thiol/méthylthiol of cystéine/méthionine presented at the active centre of the enzyme. Consequently, several urease biosensors are reported in the literature for the detection of heavy metal ions through enzyme inhibition [15-21]. Activity of urease is monitored in homogeneous phase by measuring produced ammonium ions (cf. Figure 4). When 1% of FeNPs is added, the concentration of produced ammonium ions decreases for more than 50 %. The green curve shows that the presence of copper ions (Cu^{2+}) leads to a complete inhibition of enzyme activity by reducing the signal to zero. These results confirm the inhibition of urease in presence of FeNPs.

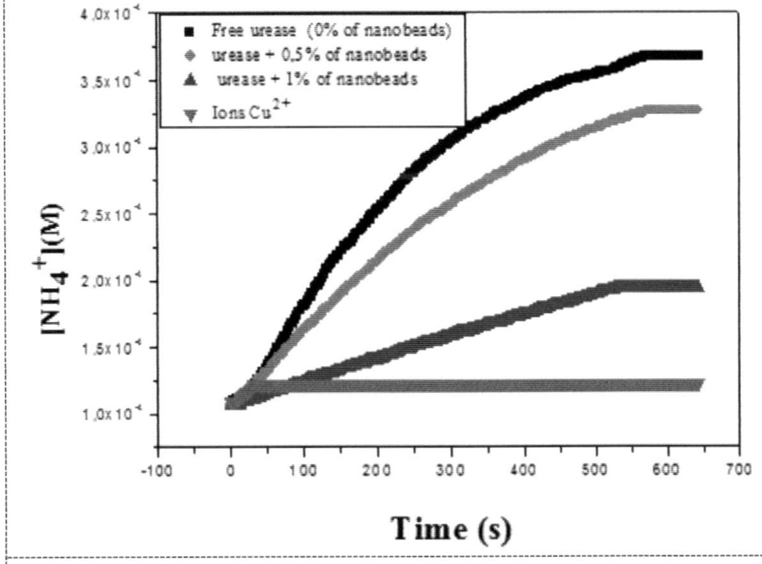

Figure 4. Variation of NH^+_4 concentrations versus time when urea (0.5 μmole) is hydrolysed in presence of 0.3 mg of urease.

3.3. Analytical features of the EIS biosensor

3.3.1. Urease immobilized without nanoparticles. Figure 5 shows the potentiometric response of EIS-urease structure for different urea concentration. The modified electrode shows a linear response to urea addition from $10^{-3.5}$M to $10^{-2.5}$M with a sensitivity of 26 mV/pUrea.

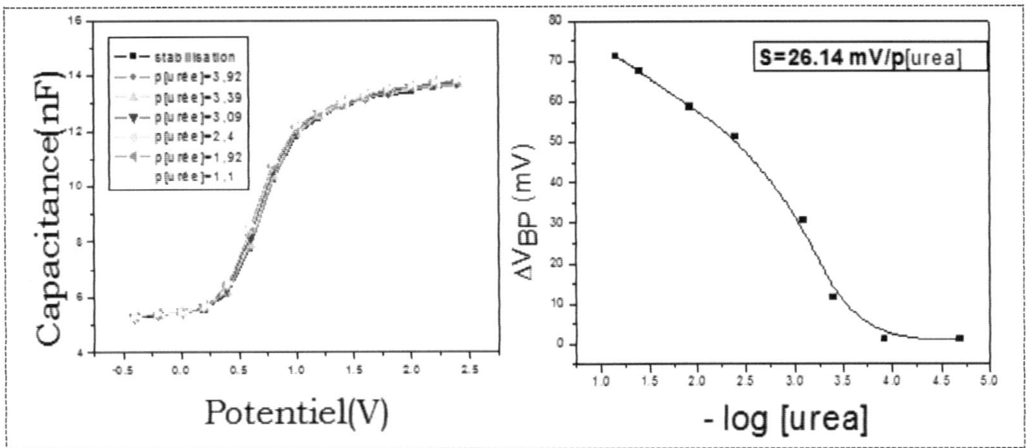

Figure 5. Capacitance response with urease immobilized without NPs.

3.3.2. Urease immobilized with nanoparticles. When urease is immobilized on FeNP-PAH-PSS-PAH, and deposited on the surface of the EIS, the effect of urea concentration is presented in Figure 6. It is shown that the proposed method for the immobilization of urease can increase the dynamic range of urea detection (10^{-4}M to 10^{-1}M) compared to the immobilization of urease without FeNP ($10^{-3.5}$M to

$10^{-2.5}$M). This effect is due to partial inhibition of the enzyme by these particles, it was shown by measurements in homogeneous phase (cf. Figure 4).

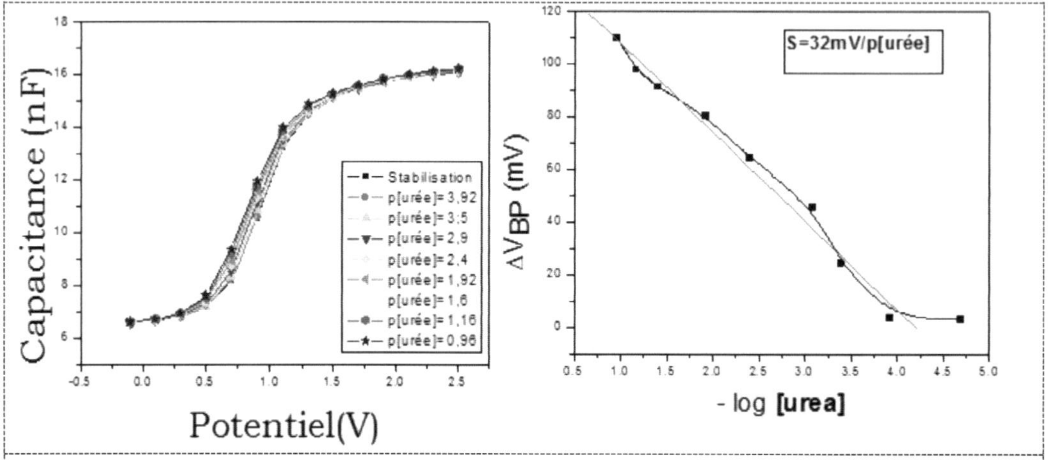

Figure 6. Capacitance response with urease immobilized on FeNP/PAH/PSS/PAH.

3.3.3. Effect of the number polyelectrolyte layer on FeNP on the response of biofunctionalized EIS. The effect of the number of PAH-PSS layers deposited on FeNP on the biosensors response was investigated. The obtained results are gathered in Table 2. An increase of the urea biosensor sensitivity is observed when the number of PAH-PSS layers increases. This effect can be explained by a higher protection of urease enzyme against inhibition by iron ions.

Table 2. Effect of the immobilisation procedure of urease on EIS surface, on the analytical characteristics of the urea biosensor.

Composition of the biofilm on EIS structure	Sensitivity (mV/p[Urea])	Domain of linearity	Limit of detection
Urease without nanoparticles	26	$10^{-3.5}$M – $10^{-2.5}$M	$10^{-3.7}$M
FeNP/PAH/PSS/PAH/Urease	32	10^{-4}M – 10^{-1}M	10^{-4}M
FeNP/PAH/PSS/PAH/PSS/PAH/Urease	33.4	$10^{-3.5}$M – $10^{-1.7}$M	$10^{-3.4}$M
FeNP/PAH/PSS/PAH/PSS/PAH/PSS/PAH/Urease	34	$10^{-3.7}$M – $10^{-2.5}$M	$10^{-3.8}$M

4. Conclusion

In this paper, layer-by-layer technique has been used for the immobilisation of urease on FeNPs for the fabrication of an EIS urea sensor. This procedure allows increase of the sensitivity of detection (26 mV/pUrea □ 33 mV/pUrea) and tailoring of dynamic range of the biosensor. It is expected that such a promising concept of inhibitor-based biosensors will be of interest for different applications of this urea sensor.

5. Acknowledgements

The authors thank EGIDE for support through UTIQUE programs N° 09G1128 and N° 10G1103 and EC for support through NANOBIOSENS Project (PIRSES-GA-2008 2300802). W. Nouira thanks Ministry of Education in Tunisia for mobility grant.

6. References

[1] El-Sayed M A 2001 Acc. Chem. Res. **34** 257.
[2] Tao L, Shaojun D and Erkang W 2009 Analytical Chemistry.**81** 2144.
[3] Jaffrezic-Renault N, Martelet C, Chevolot Y and Cloarec J P 2007 Sensors **7** 589.
[4] Shen W, Schrag B D, Carter M J and Xiao G 2008 Appl. Phys. **93** 33903.
[5] Lowery T J, Palazzolo R, Wong S M, Prado P J and Taktak S 2008 Anal.Chem. **80** 1118.
[6] Cheng G F, Huang C H, Zhao J, Tan X L, H E P G and Fang Y Z 2009 Chin. J. Anal. Chem. **37** 169.
[7] Ai H, Jones S A and Lvov Y M 2003 Cell Biochem Biophys **39** 23.
[8] Lvov Y 2000 In: Lvov Y, Mohwald H editors "Protein architecture:interfacial molecular assembly and immobilization biotechnology" (New York, Dekker), 125–167.
[9] Wasinger V C, Cordwell S J, Cerpa-Poljak A, Yan J X, Gooley A A, Wilkins M R, Duncan M W, Harris R, Williams K L and Humphery I 1995 Electrophoresis **16** 1090.
[10] Fleischmann R D, Adams M D, White O, Clayton R A, Kirkness E F, Kerlavage A R, Bult C J, Tomb J F, Dougherty B A and Merrick M 1995 J. Science **269** 496.
[11] Hua A, Pink J J, Shuai X, David A. Boothman J G 2005 J.Biomedical. Materials Research **73** 303.
[12] Veyret R, Delair T, Elaissari A and Magn Magn J 2005 Mater **293** 171.
[13] Souiri M, Mora-Ponsonnet L, Glinel K, Othmane A, Jouenne T and Duncan A C 2009 Colloids Surf B Biointerfaces **68** 125.
[14] Jaffrezic-Renault N, De A, Clechet P and Maaref A 1989 Colloids and Surfaces **36** 59.
[15] De Castro M D J and Herrera M C 2003 Biosens. Bioelectron **18** 279.
[16] Andres R T and Narayanaswamy R 1995 Analyst. **120** 1549.
[17] Tsai H C, Doong R A, Chiang T C and Chen K T 2003 Anal. Chim. Acta **481** 75.
[18] Zhylyak G A, Dzyadevich S V, Korpan Y I, Soldatkin A P and El'skaya AV 1995 Sens.Actuators B **145** 24.
[19] Kukla A L, Kanjuk N I, Starodub N F and Shirshov Y.M 1999 Sens. Actuators B **57** 213.
[20] Starodub N F, Kanjuk N I, Kukla A L and Shirshov Y M 1999 Anal. Chim. Acta **385** 461.
[21] Senillou A, Jaffrezic-Renault N, Martelet C and Cosnier S 1999 Talanta **50** 219.

Morphology of carbonates particles precipitated from saline waste solution: Influence of magnesium

L O Filippov, M Grandjean, I V Filippova, M Pelletier[1]

LEM, Laboratoire Environnement et Minéralurgie, UMR 7569 CNRS-Nancy Université, rue du Doyen M.Roubault, BP 40, Vandoeuvre-lès-Nancy, France

E-mail: lev.filippov@ensg.inpl-nancy.fr

Abstract. The role of a very low concentration of Mg on the nature, morphology and surface of carbonate particles during soda-ash residual brine carbonation has been studied. The Mg concentration of 200 mg/kg in brine slows down the kinetic of carbonation, modifies the shape of precipitated particles and new carbonated phases are precipitated. The existence of aragonite and (Ca, Mg) hydrated phases is supposed for Ca:Mg ratio equivalent to 24:1 in solid fraction.

1. Introduction

The ammonia soda process is used to produce sodium carbonate and bicarbonate from two raw materials, lime and NaCl brines. The by-products are residual brines, they are now landfilled and the surnageant is sent to the nearest river or the sea. Efforts have to be done on brine treatments to decrease their environmental impact. In this study, by-products are carbonated by bubbling to avoid unsuitable phase formation during landscape storage. This method has two environmentally advantages: the precipitation of stable carbonates and the carbon dioxide sequestration. During carbonation of brines, initial chemical composition and contained mineral phases affect precipitated carbonates nature and reactions kinetics. Availability of calcium and magnesium [1]seems to be significant factors influencing carbonation with pH, temperature and CO_2 pressure [2].The main components of brine are chlorine, calcium, potassium and sodium. Four carbonatable phases are identified: $Ca(OH)_2$, $Mg(OH)_2$, CaOHCl and CSH [3,4,5]. Competition between ions present in solution can influence the nature of precipitated carbonates: a high concentration in Ca^{2+} favors calcite formation [6]. Contrary to Ca, the ions of Mg stabilize amorphous, unstable and hydrated phases [1,7] and so decrease the quantity of well crystallized carbonates. Magnesium incorporation also inhibits carbonate growth and seems to change their morphology, from rhombohedral to spindle-like shape [8,9,1]. Magnesium influence is well known in sea water or in brine with a Mg:Ca ratio superior to 1:1.

The Mg:Ca ratio is generally inferior or equivalent to 1:3000 in liquid fraction of residual brines from soda-ash process. Thus, the aim of this study is to understand the role of a very low concentration of Mg on brine carbonation.

Infrared spectroscopy and electronic microscopy are applied to study the role of Mg on the nature, morphology and surface carbonates. The final objective is to establish a relation between low Mg concentration in brine and properties of carbonated phases.

2. Experimental procedure

To study the role of Mg on carbonation, brines from ammonia-ash process were filtrated (<10μm) to eliminate solid fraction (>100μm). Indeed, solubilization of solid phase could add new ions in liquids and so perturb the initial equilibrium. Resulting liquids are then carbonated by bubbling (15%CO_2 - 85% air, total flowrate: 67 L/h) in a 30 L reactor at 55°C. The same experiment is done with different concentrations of added soluble magnesium ($MgCl_2.6H_2O$, Merk). Four Mg concentrations have been studied: 0, 25, 100 and 200 mg/kg of brine. During brine carbonation, pH decreases from 10.3 to 5.7. Thus, monitoring the pH is used to control the carbonation progress. The pH value is measured continuously during experiments. Samples are taken at three stable pH values: 9.5, 8 and 7.2. Samples are then filtrated at 0.45 μm in a Teflon Millipore compressed air system (2 bar). Liquids are acidified for analyses. Solids are frozen and lyophilised to eliminate free water. Chemical compositions of solid and liquid phases have been determined by inductively coupled plasma mass spectroscopy (ICP-MS) excepted for CO_3^{2-} done by ACA and Ca. Solid samples have been studied by diffuse reflectance infrared Fourier transform spectroscopy. Infrared spectra were recorded with a Fourier Transform Infrared (FTIR) spectrometer (BRUKER IFS 55) equipped with a large band (5500–600 cm^{-1}) mercury–cadmium telluride (MCT) detector cooled at 77K. Solids were investigated morphologically by scanning electron microscopy (SEM), with a Hitachi S-4800. Dispersed solid powder was deposited on a carbon ribbon and examined at 5 kV. For particles analyses, a high-resolution transmission electron microscope (TEM), a CM 20 Philips was used at 200 kV with energy dispersion spectroscopy (EDS). In TEM, the powder was dispersed in alcool and deposited on a copper grid.

3. Results and discussion

3.1. Chemical analysis

Chemical composition of solid fraction shows significant evolution of carbonate composition in function of Mg content. Initial CO_3 concentration at pH 10.3 is considered near to 0 mg/kg. The carbonate concentration in filtrated brines without added Mg (Figure 1) is around 380 mg/kg and does not evolve between pH 9.5 and pH 7.2. It signifies that most of the carbonation occurs between initial pH 10.3 and pH 9.5. When a weak quantity of Mg (25 mg/Kg) is added, carbonation rate is not affected. On the contrary, CO_3 concentration continuously raises (125 to 350 mg/kg) from pH 10.3 to 7.2 for 200 mg/kg of Mg.

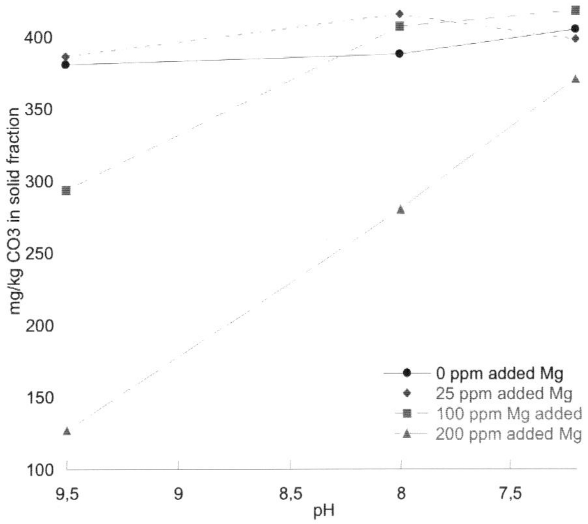

Figure 1. Evolution of CO_3 content in the solid fraction during carbonation of filtrated brine in the presence of different concentrations of additional Mg.

According to the literature [8,9,1], Mg adsorbed on carbonates inhibits their growth and so decreases carbonation rate. The new results of this study concern the non-linear influence of Mg on carbonation rate. Indeed low Mg concentrations (25 mg/kg) do not influence carbonate precipitation rate whereas higher Mg quantities (100-200 mg/kg) decrease significantly the carbonation rate. The Mg concentration can reach 120 mg/kg in the non filtrated brines, the influence of Mg ions on carbonate rate can not be considered as negligible in these conditions.

The Ca:Mg ratios have been calculated in liquid and solid fractions. Figure 2 presents the evolution of Ca:Mg ratio during carbonation in function of Mg quantity. At alkaline pH (9.5), the additional Mg is precipitated immediately in a non soluble stable $Mg(OH)_2$ phase. Between pH 9.5 and 8, Mg is solubilized and the ratio Ca:Mg is decreasing in liquids. For the three first samples (0, 25 and 100 ppm Mg) a stabilization is observed for pH values inferior to 8 (Figure 2a). The solubilization rate of Mg seems to be comparable to the rate of Mg incorporation in carbonates, the Ca:Mg ratio is then stabilized. For higher Mg concentration (200 ppm) is magnesium solubilized in solution faster than its incorporation in carbonates.

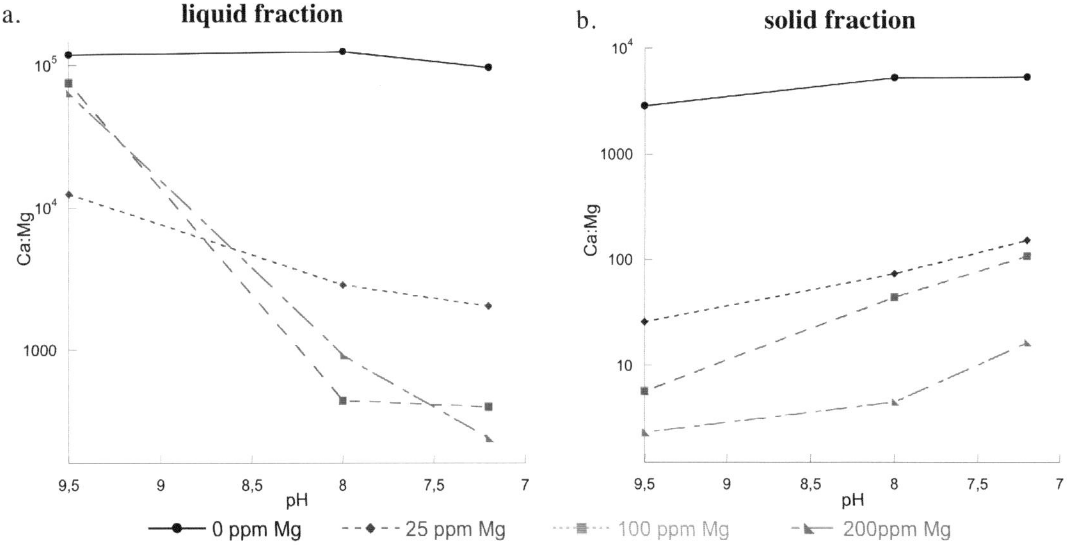

Figure 2. Evolution of Ca:Mg ratio in liquid (a) and solid (b) fraction during carbonation of filtrated brine in the presence of different concentration of Mg.

Figure 2b shows that the Ca:Mg ratio in solid fraction increases during carbonation. Its means that Ca is more carbonated than Mg in all samples. Considering the low Mg concentrations and the slow Mg solubilization rate, this result is not surprising. At stable neutral pH (7.2), $Mg(OH)_2$ should be completely solubilized, then Mg present in solid fraction can be associated exclusively to carbonates. Without additional Mg, less than 0.1 wth% of Mg is present in carbonates. For 25 and 100 ppm of added Mg, the capture of Mg by carbonates is estimated at 1%. These values are coherent with adsorption of Mg ions on calcite lattice for advanced carbonation (pH 7.2). In the last solid sample (200 ppm Mg), Ca:Mg ratio is equal to 24:1 for neutral pH values. Are all Mg ions (5 wth%) adsorbed at the surface of calcite or are others carbonated phases precipitated ?

3.2. FTIR spectral analysis

The all diffuse reflectance FTIR spectra of carbonated brine samples are almost similar at pH 9.54 (Figure 3a). Main peaks 875, 1462 and 713cm^{-1} correspond respectively to v_2, v_3 and v_4 of calcite CO_3 vibrations [10]. Peaks at 2513, 2875 and 2987 cm^{-1} are attributed to calcite harmonics. Gypsum is identified by 1134-1162 cm^{-1} doublet and 1637 cm^{-1} corresponding respectively to $v_3(SO_4)$, and

$\nu_2(H_2O)$.A detailed observation of spectra between 3600 and 3800 cm^{-1} (Figure 3b) indicates a peak at 3698 cm^{-1} assigned to the infrared active hydroxyl stretching vibration of $Mg(OH)_2$. At pH 9.5, this peak is present in the three samples with added Mg. The $Mg(OH)_2$ band disappears at pH 8 for 25mg/kg of Mg. When carbonation is advanced (pH 7.2), this band is only visible for higher Mg concentrations (200 mg/kg of added Mg). Thus, more brine is carbonated, less $Mg(OH)_2$ band is visible. Others absorption features at 3639 and 3656 cm^{-1} are observed at pH 9.5 and 200 mg/kg of added Mg (Fig.3b). The first peak corresponds to OH stretching vibration of $Ca(OH)_2$ and is visible in this conditions only. A pure sample of $Ca(OH)_2$ (Merk) presents a peak in diffuse reflectance at 3645cm^{-1}, the 6 cm^{-1} of difference in the studied Mg-sample is attributed to the substitution of Ca by Mg in portlandite lattice. The other band (3656 cm^{-1}) may be attributed to OH stretching vibration of a hydrated magnesium carbonate, Hydromagnesite $Mg_5(CO_3)_4(OH)_2•4(H_2O)$ [10].

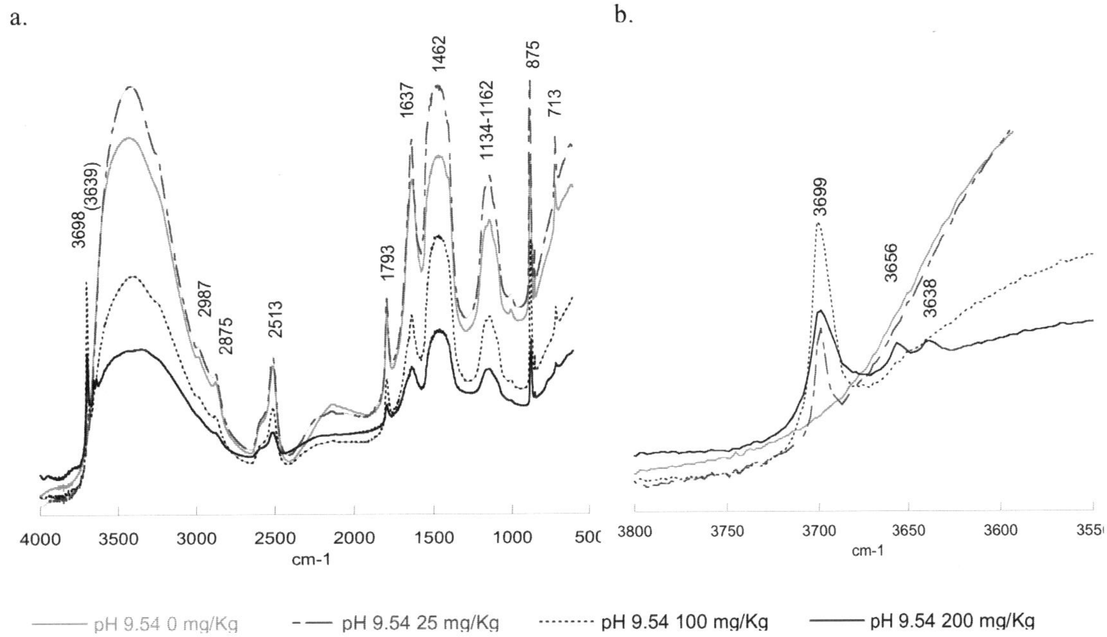

Figure 3. FTIR spectra of carbonated brine (pH 9.54) with different Mg concentrations (a) and (b) 3500-3800 cm^{-1} detailed spectra.

IR-spectra of carbonated brines at pH 7.2 are almost identical to the precedent spectra at pH 9.5 excepted little features. Detailed spectrum of 800-950 cm^{-1} vibrations (Figure 4) shows three peaks. The major peaks at 871 and 712 cm^{-1} correspond to $\nu_2(CO_3)$ and $\nu_4(CO_3)$ of calcite. A third feature is observable at 854 cm^{-1} for 200 mg/kg of Mg and could be assigned to $\nu_2(CO_3)$ of aragonite or to an hydrated Mg carbonate, Nesquehonite ($MgCO_3.3H_2O$) [10]. The hypothesis of aragonite presence is reinforced by an anomaly at 1082 cm^{-1} only present in the sample carbonated in presence of 200 mg/kg of Mg.

A model of phases evolution during brine carbonation in presence of a low but non negligible Mg concentration can be proposed. Calcium hydroxides present initially in brine disappear at pH 9.5 excepted for higher Mg concentration. Without additional Mg, $Ca(OH)_2$ is carbonated at the beginning of bubbling, between pH 10.3 and 9.5. Magnesium presence changes the equilibrium, Mg^{2+} competes with Ca^{2+} and so inhibits calcite precipitation [11,12]. Contrary to $Ca(OH)_2$, $Mg(OH)_2$ seems to be solubilized and carbonated continuously from pH 9.5 to pH 7.2.

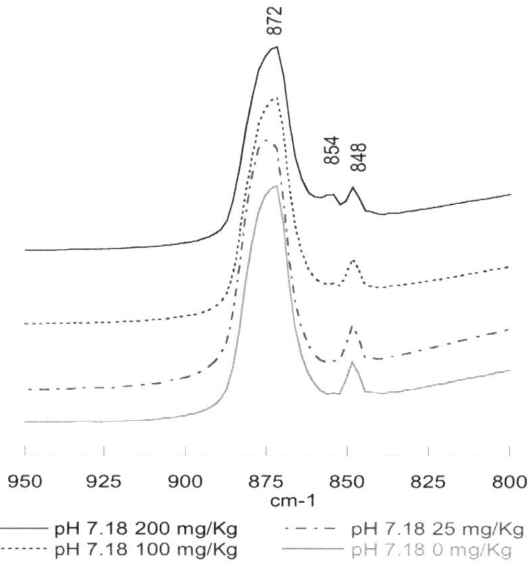

Figure 4. Detailed FTIR spectra (3500-3800 cm^{-1}) of carbonated brine (pH 7.18) with different added Mg concentrations.

In non filtrated brines, Mg is slowly liberated by Mg-minerals (silicates, hydroxides...) and its concentration in liquids can exceed 100 mg/kg for neutral pH. The proportion of Mg liberated in solution increases as pH decreases. IR results indicate that hydrated Mg hydroxy-carbonates can be precipitated at the beginning of carbonation. And then, in neutral pH, Mg adsorded at the carbonate surface seems to stabilize unstable calcium carbonates (aragonite) as confirmed by Figure 3 that corroborates with result of Folk [8].

3.3. Microscopic observations

Photos presented in Figure 4 show the consequences of Mg addition on the morphology of carbonate particles. With low Mg concentration, the size of calcite particles (Figure 4a) can reach 7μm with rhombohedral and sharped extremity. Many ramifications are observed. Magnesium presence on carbonate surface leads to decrease particles size (3-4μm maximum for 200 mg/kg of Mg) and particles extremities become round. The number of particle ramifications falls significantly. In theory this morphologic changes are explained by the difference of Ca and Mg ions diameters.

a.

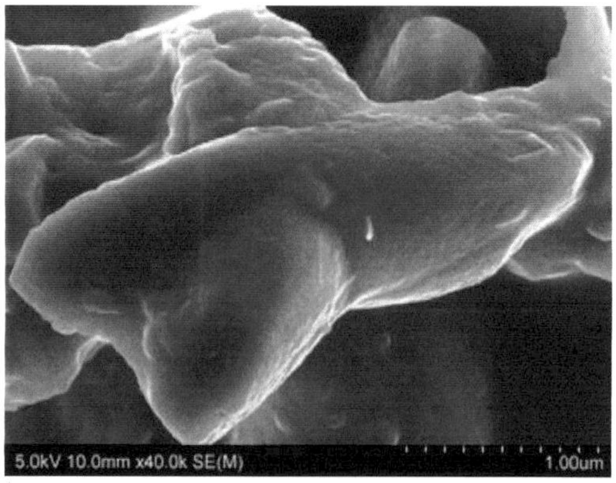

b.

Figure 4. Particle photos (SEM) of filtrated brines solid fraction carbonated until pH 7.2 without added Mg (a) and with 200 mg/kg of added Mg (b).

SEM photo (Figure 4b)also reveal the presence of a film at the surface of carbonate particles for highest Mg concentrations. It can correspond to the precipitation of amorphous phase by a dissolution-reprecipitation process at the interface carbonate-brine solution and later the interface air-carbonate during liophylisation[1]. Analyses with TEM-EDS indicate that only 2-3% of Mg are incorporated near particle extremities, it is sufficient to radically change calcite morphologies.

4. Conclusion

This study demonstrates that low Mg concentrations (200 mg/kg) have an influence on brine carbonation.

- In residual brines, $Mg(OH)_2$ is solubilized for pH values inferior to 9.5. The incorporation of Mg in portlandite particles stabilizes hydroxide phases and decreases the availability of Ca at basic pH (9.5).
- Mg has an influence on carbonation of calcite. Magnesium slows down the kinetic of carbonation and modifies precipitated particles shape.
- At 200 mg/kg of additional Mg, new carbonated phases have been precipitated. Hydrated magnesium carbonates and an intermediate polymorph of $CaCO_3$ (aragonite) have been detected by infrared spectroscopy. Amorphous carbonated phases are observed on calcite surface in presence of Mg

This study reveals the importance of Mg parameter to control the carbonation of low Mg:Ca brines. Higher Mg concentrations have to be tested in further works as well as higher CO_2 flow rates to confirm findings results and envisage a better control of precipitated phases.

References

[1] Loste E, Wilson R M , Seshadri R. and Meldrum FC 2003 The role of magnesium in stabilising amorphous calcium carbonate and controlling calcite morphologies, *J. Cryst. Growth* **254** 206

[2] Han Y S, Hadiko G, Fuji M and Takahashi M 2005 Effect of flow rate and CO_2 content on the phase and morphology of $CaCO_3$ prepared by bubbling method, *J. Cryst. Growth* **276** 541

[3] Filippov L O, Filippova IV and Yvon J 2007 *Rapport ANR–DECALCO Part II Caractérisation des échantillons solides de SR et SRC.*

[4] Bodenan F, Lassin A, Hottier M, Filippov L, Durance M V and Piantone P 2008 Piégeage et valorisation de déchets alcalins par passivation au CO_2 industriel. Final report *ANR–DECALCO*, BRGM/RP-**56361**

[5] Filippova I V, Filippov L O, Piriou P and Grandjean M 2011 Carbonation of residual brines produced by ammonia-soda prtocess, *this issue*

[6] Chen P C, Tai C Y and Lee K C 1997 Morphology and growth rate of calcium carbonate crystals in a gas-liquid-solid reactive crystallizer. *Chem. Eng. Sc.* **52** (21-22) 417

[7] Brecevic L and Kralj D 2007 On Calcium Carbonates: from Fundamental Research to Application. *Croatia Chem. Acta* CCACAA **80** (3-4) 467-484

[8] Folk R L 1974 The natural history of crystalline calcium carbonate: effect of magnesium content and salinity. *J. Sed. Petrology* **44** (1) 40

[9] Lahann R W 1978 A chemical model for calcite crystal growth and morphology control. *J. Sed. Petrology* **48** (1) 337.

[10] Farmer V C 1974 *The Infrared Spectra of Minerals.* Monograph 4 (London : Mineralogical Society)

[11] Davis K J, Dove P M and De Yoreo J J 2000 The role of Mg^{2+} as an impurity in calcite growth. *Science* **290** 5494 1134

[12] Davis K J, Dove P M, Wasylenki L E and De Yoreo J J 2004 Morphological consequences of differential Mg^{2+} incorporation at structurally distinct steps on calcite. *Am. Mineralogist* **89** 714.

Journal of Physics: Conference Series **416** (2013) 012012 doi:10.1088/1742-6596/416/1/012012

Development of vinylic and acetylenic functionalized structures based on high permeable glassy polymers as membrane materials for gas mixtures separation

D.Roizard[1], Y.Kiryukhina[2], A. Masalev[2], V. Khotimskiy[2], V.Teplyakov[2], D. Barth[1]

[1]Laboratoire Réactions et Génie des Procédés (CNRS3349, Nancy-Université), ENSIC, 1 rue Grandville, 54000 Nancy, France

[2]A.V.Topchiev Institute of Petrochemical Synthesis Russian Academy of Sciences, 119991 Moscow, Leninsky pr., 29, Russia

E-mail: Denis.Roizard@ensic.inpl-nancy.fr

Abstract. There are several challenging separation problems in industries which can be solved with the help of membrane technologies. It is the case for instance of the purification of gas energy carriers (i.e. H_2, CH_4) from CO_2 as well as the CO_2 recovery from flue gas. Glassy polymers containing trimethylsilyl residues like poly(1-trimethylsilyl-1-propyne) [PTMSP] and polyvinyltrimethylsilane [PVTMS] are known to exhibit good membrane properties for gas separation. This paper reports two ways of improving their performances based on the controlled introduction of selective groups – alkyl imidazomium salts (C_4I) and polyethyleneglycol (M-PEG)- able to enhance CO_2 selectivity. CO_2 Isotherm sorption data and permeability measurements have shown that the membrane performances could be significantly improved when C4I and M-PEG were introduced as residues covalently bounded to the main polymer chain. Moreover the introduced bromine reactive centres could also be used to induce chemical crosslinking giving rise to more resistant and stable membranes to organic vapours. With the C4I groups, the CO_2 sorption could be enhanced by a factor 4.4.

1. Introduction

Polymeric membranes have attracted the attention of many scientists since more than 30 years in the field of gas separation because membranes are able to induce the molecular separation of species having relatively similar physical properties under rather simple conditions compared to conventional methods. For instance, the separation of the two major constituents of air can be achieved by cryogenic distillation; but applying a simple pressure difference between the two sides of a well chosen membrane can also lead to the production of nitrogen of high purity [1] as commercialized by Mosanto in the 80s with the Prism gas module. The separation principle through dense membranes is explained by the solution-diffusion mechanism early evidenced by Graham in 1861 [2] and Long [3]: in the 1st step, the molecules are sorbed into the polymer network according to their affinity with the polymer structure, i.e. it corresponds to a thermodynamic equilibrium; in the 2nd step these molecules,

[1] To whom any correspondence should be addressed.

called permeant, diffuse through the membrane with respect to their activity gradient maintained between the two sides of the membrane: this is the kinetic step which is also the limiting one.

In the field of gas permeation, the interactions of the molecules with glassy polymers such as PVTMS or PTMSP are generally small with gas having a low critical temperature like H2, O2, ... Therefore the membrane selectivity is ruled by a size sieving effect, the smallest molecules being the fastest permeating ones. For a given gas **i**, the driving force of the permeation through a membrane is its pressure difference between the two sides; under steady state conditions, the mass transfer is known to follow the first Fick law [4]:

$$J_i = -D_i \frac{dC_i}{dx} \tag{1}$$

with **Di**, the diffusion coefficient of the gas **i**, and **dCi/dx** the thermodynamic gradient of **i** within the membrane thickness. For a gas under low pressure, the concentration Ci can be related to the fugacity fi as:

$$C_i = S_i.f_i \approx S_i.Pr_i \quad (Nm^3/m^3.Pa) \tag{2}$$

where $\mathbf{S_i}$ is the sorption coefficient of the gas **i** for the considered membrane, and $\mathbf{Pr_i}$ is the gas partial pressure.

Therefore the mass transfer **Ji** can be written as:

$$J_i = D_i.S_i \frac{\Delta Pr_i}{x} = P_i \frac{\Delta Pr_i}{x} \quad (Nm^3/m^3.s) \tag{3}$$

where $\mathbf{P_i = D_i.S_i}$ is the permeability coefficient, and **x** the membrane thickness.

Hence for a binary mixture, the selectivity $\alpha_{i/j}$ is equals to the ratio of the permeability coefficients:

$$\alpha_{i/j} = \frac{P_i}{P_j} \quad \text{i.e.} \quad \alpha_{i/j} = \frac{D_i}{D_j}.\frac{S_i}{S_j} = \alpha_{Di/j}.\alpha_{Si/j} \tag{4}$$

The usual $\mathbf{P_i}$ unit is the Barrer: 1 Barrer=$10^{-10}cm^3$(STP).cm/cm^2.s.cm(Hg),i.e. 0.75.10^{-15} m^3.m/m².s.Pa.

PVTMS [5] and PTMSP [6-7] can lead to very interesting gas separation membranes for H_2 and CH_4 purification because they are endowed with very high permeation fluxes; but these membranes are too sensitive to hydrocarbons vapours and lack also of selectivity for CO_2 [8-9]. This paper reports two ways of improving the membrane performances by the controlled introduction of selective groups in the polymer structure able to enhance the membrane stability and the CO_2 selectivity.

2. Materials and methods

Bromination of polymers. The halogenation was carried out with Br_2 or N-bromosuccimide (NBS) on polymer membranes, but it was shown that the substitution degree was better controlled with NBS. The substitution was confirmed by FTIR and quantified by elemental analysis. In the case of PTMSP the maximum Br content in the modified polymer was 27% wt., but only 60% wt. in the case of PVTMS [10].

Membrane crosslinking. 1,12-diaminododecane was used to react with the bromine-containing films to generate crosslinks through the polymer network. Thickness of films: 50-200 μm, T = 60 ^{0}C, [diamine] / [Methanol] =0.5 mol/L, Duration of treatment: 60 min.

Introduction of CO_2 selective groups in PTMSP. The Bromine was used as reactive center to incorporate polyethyleneglycol groups (M-PEG) and ionic-liquid –like groups (imidazolium C_4I).

Introduction of ethylenoxide groups into polymer structure of PTMSP. Reaction of bromine-containing polymer with low molecular poly(ethyleneglycol) monomethyl ether was carried out in presence of sodium under argon atmosphere at 25°C; THF was chosen as reaction solvent and the reaction duration was 24 h. According to elemental analysis data, it was established that the yield reached 90% of the brominated centers. Taking into account the degree of bromination (maximum 27% wt. for PTMSP) it was possible to introduce up to 60% wt. of PEG in the polymer structure.

Introduction of ionic liquid like groups. The modification of the brominated polymers required two steps. The first one was the quaternization of the bromine-containing polymers (PTMSP and PVTMS) with N-butylimidazol. Reaction was carried out under argon atmosphere at 25°C, CCl_4 was chosen as solvent with reaction duration of 48 h. The maximum reaction yield reached was about 40% in the case of PVTMS and 30% in the case of PTMSP, that are respectively 28% wt. and 21.5% wt. of N-butylimidazolium bromide.
The second stage is anion-exchange reaction with KPF6. The reaction was carried out under argon atmosphere at 25°C, in a THF-acetone solvent mixture (ratio 4:1) with reaction duration of 48 h. The reaction yield was almost quantitative according to the elemental analysis data.

Preparation of blends of PTMSP and PVTMS with low molecular model compounds. Two low molecular model compounds, i.e. M-PEG and C_4I, were synthesized from chloromethyl trimethylsilane (CMTMS) to get molecular models analogous to the PTMSP and PVTMS structures in order to induce better compatibility when blending with polymers.
The schemes of synthesis of models M-PEG is represented on Figure 1. Reaction was carried out under argon atmosphere at 25°C in THF with reaction duration 24h. The reaction yield was up to 90% (as calculated using elemental analysis data).

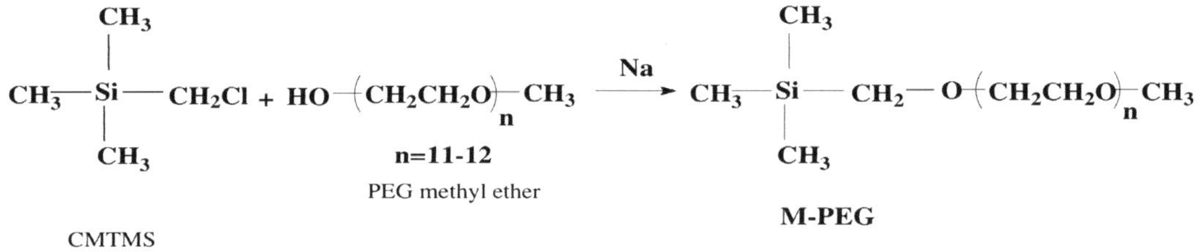

Figure 1. Scheme of synthesis of the model compound M-PEG used to prepare polymer blends

The scheme of synthesis of models C_4I is represented on Figure 2. The first step of reaction was carried out under argon atmosphere and at 45°C in THF with a reaction duration of 12h; the yield of the halogen substitution was about 90% (elemental analysis). In the second step at 25°C, acetone was chosen as a solvent with reaction duration of 48h; the yield of second stage was almost quantitative.

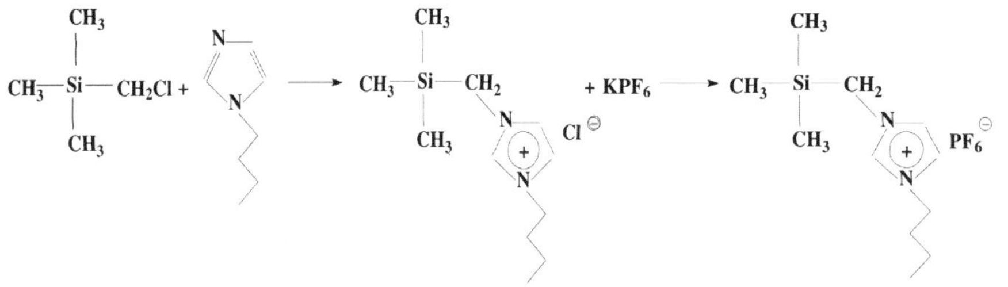

Figure 2. Scheme of the two steps synthesis of the imidazolium model C_4I used for polymer blend

Blend films. They were prepared from mechanical stirring of polymer solutions with various weight amounts of the low-molecular models. The films were then obtained by casting from chlorobenzene or CCl_4 polymers solutions.

Gas permeation experiments. The well-known barometric time lag method was used to determine the permeability coefficients of O_2 and N_2 as reported elsewhere [11].

CO_2 sorption measurements. They were carried out with a magnetic suspension microbalance (Rubotherm, precision 0.1μg) at 35°C, in a pressure range of 3 to 10 bar.

3. Results and discussion

Creation of the reaction centres is necessary on PVTMS and PTMSP (Fig. 3) for the introduction of groups which may provide high separation selectivity of CO_2-containing gas mixtures and also to improve polymer stability towards hydrocarbons by chemical crosslinking. Bromine atoms can be one of such reactive centres. It was established earlier that bromination of PTMSP and PVTMS takes place at one of the silicon methyl groups, with formation of the following structure $(CH_3)_2$- Si-$CH_{x-1}Br_x$, where x=1-3. The introduced bromine atoms were further used to induce polymer crosslinks to improve membrane chemical stability and to generate CO_2 selective groups.

Figure 3. Chemical structure of PVTMS (left hand side) and of PTMSP (right hand side)

3.1. Crosslinking of bromine containing PVTMS

Polymer cross-linking is one of methods which can be used to obtain more stable membranes; the substitution of bromine with a di-functional reagent generates indeed covalent bonding between the polymer chains and thus reduces the swelling ability of the network towards chemical solvent. After reaction with the α,ω-diamino alkyl reagent, the modified polymers were found to be insoluble in organic solvents. Density of the cross-linking was regulated by bromine content in initial polymer; higher bromine content led to higher crosslinking density.
Data on gas permeability for cross-linked PVTMS films are given in table 1. As one can see, permeability coefficients decrease as cross-linking density increasing whereas ideal selectivity does not change (e.g., $\alpha_{PO2/N2}$). Obtained crosslinked PVTMS samples combine high gas separation selectivity parameters with the resistance to organic solvents of different types.

Table 1. Gas permeability of Br-containing PVTMS upon treatment by 1,12-diaminododecane.

Br content in polymer, % wt	P_{N2}, barrer		P_{H2}, barrer		P_{O2}, barrer		P_{CO2}, barrer		Selectivity α P_{O2}/P_{N2}	
	before	after	before	after	before	after	Before	after	before	after
0	10	10	193	193	40	40	160	160	4	4
4	9	17	180	216	36	44	144	145	4	3
17	4.2	6	61	149	15.3	26	50	99	2.5	4
44	2	0,3	44	14	6	1,3	25	4	2.3	4
61	0.4	-	12	-	0.7	-	5.9	-	2.6	-

3.2. Introduction CO_2 selective groups.

Two different groups were used to enhance PVTMS and PTMSP CO_2 affinity: ethylene oxide groups [12] and ionic liquids like groups [13], on the basis of quaternized N-butylimidazol. In order to establish the influence of the nature and quantity of functional groups needed, as well as the best method of introduction of these groups to get reversible interaction with CO_2, two ways of membrane materials preparation were investigated: a) the chemical modification of brominated PTMSP and PVTMS with CO_2 selective groups; b) the preparation of blends of PTMSP and PVTMS, with the low molecular compounds having structures close to chemically modified polymers.

3.3. *Investigation of CO_2 sorption*

Sorption was studied on the two types of PVTMS and PTMSP modified films, i.e. grafted and blend ones; the recorded measurements are presented on Figures 4-5 and compared to the initial PVTMS and PTMSP sorption data. The data shown correspond to measurements recorded at 35°C under increasing pressure of CO_2 up to 10 bar.

The difference of sorption capacity between the two initial polymers can be explained by the differences between the available free volumes of the two polymer networks; indeed it is well-known that PTMSP has a much higher free volume than PVTMS which induces in turn a much higher sorption capacity for gas [14]. With CO_2, the recorded data indicate that the sorption capacity is at least two times the PVTMS capacity under the set experimental conditions used.

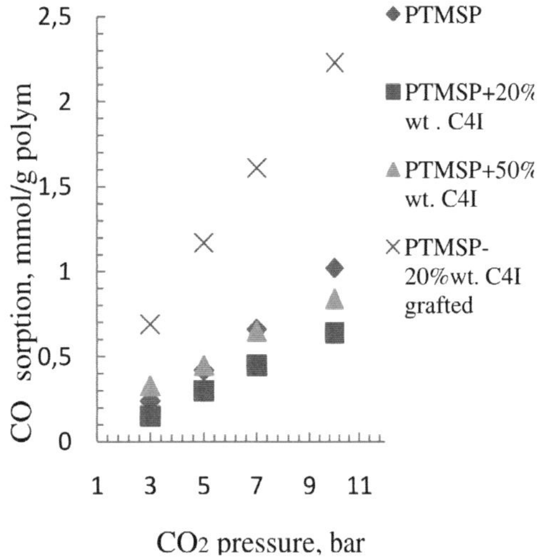

Figure 4. CO_2 sorption in PTMSP with ionic liquid-like groups (grafted polymers and blend at T=35 C

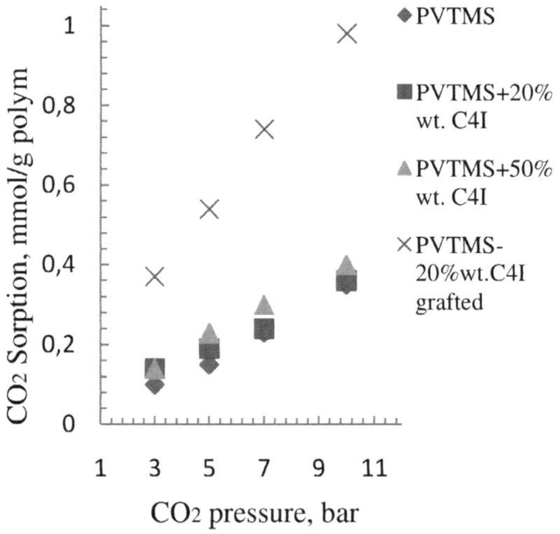

Figure 5. CO_2 sorption in PVTMS with ionic liquid-like groups (grafted polymers and blends) at T=35 C

For blends systems, either with PVTMS or with PTMSP, the sorption capacity was increased with the CO_2 pressure like with the native polymers, as expected. The blends having the higher amount of additive (here C_4I in the Figures) led to the higher sorption values either PVTMS or PTMSP, but C_4I gave slightly higher values than M-PEG. However, surprisingly, the CO_2 sorption was not increased at all. The same trend was also obtained with M-PEG blends (data not shown here).

For grafted polymers, the sorption levels obtained with the two C_4I grafted polymer series were higher than the ones reached with M-PEG grafted polymers (data not given here). C_4I grafted polymer exhibited a very different trend than blends; indeed, CO_2 sorption was markedly increased compared to the initial polymers, both with PVTMS and with PTMSP: in each case, the sorption was multiplied by a factor 2.5 to 3. Nevertheless, like for the case of native polymers, the C_4I grafted PTMSP gave better results than C_4I grafted PVTMS.

This last set of results seems to indicate that the high free volume characteristic of the PTMSP network was also saved in the grafted PTMSP structure. Obviously, the lower sorption values recorded with the blend series indicate that it was not the case with the addition of low molecular weight models in PTMSP.

Conclusion

It was shown that adequate modifications of PVTMS and PTMSP could give gas membranes more stable to organic vapors. With the two polymer series, CO_2 selective groups were successfully grafted on the trimethylsilyl substituent of the polymer backbone.
It was established that the level of sorption of the modified materials essentially depends on the method used for the introducing of the CO_2 selective moieties. Indeed the data showed that the CO_2 sorption in chemically modified polymers was higher than that in the blends having similar mass composition in the CO_2 selective moieties, i.e. either polyethyleneoxide groups or n-butylimidazolium groups.
Moreover with the grafted PVTMS and PTMSP polymers, it was possible to surpass the initial level of CO_2 sorption of the native polymer. For each type of grafted polymers, ionic liquids-like groups induced a much greater influence on CO_2 sorption properties than polyethylene oxide groups.
Therefore the two grafted polymer series can be considered as promising new materials for the purification of permanent gas from CO_2 rich mixtures.

Acknowledgments. The authors are grateful to the financial supports granted by the French Embassy of Moscow, ARCUS-Région Lorraine, LIA "Plate" CNRS and Russian Foundation for Basic Research (project 08-03-92497-НЦНИЛ_a).

References

[1] Clarks D et al., 1988 *US Patent* 4,772 392
[2] Graham T. 1995 *J Membr Sci.* **100** (1) 27-31
[3] Long L. *J Chem Educ.* **21**(3) 139-41
[4] Kamaruddin HD, Koros WJ. 1997 *J Membr Sci* 135(2) 147-59
[5] Le Roux JD, Teplyakov VV, Paul DR. 1994 *J Membr Sci* **90**(1-2) 55-68
[6] Syrtsova DA, Borisevich OB, Shkrebko OA, Teplyakov VV, Grinshpan DD, Khotimskii VS, Roizard D. 2007 *Sep. and Purif. Techn.* **57**(3) 435-9
[7] Yampolskii YP, Korikov AP, Shantarovich VP, Nagai K, Freeman BD, Masuda T, Teraguchi M, Kwak G. 2001 *Macromolecules* **34**(6) 1788-96
[8] Plate NA, Durgaryan SG, Khotimsky VS, Teplyakov VV, Yampolskii Y. 1990 *J Membr Sci* **52** 289
[9] Kelman SD, Rowe BW, Bielawski CW, Pas SJ, Hill AJ, Paul DR, Freeman BD. 2008 *J Membr Sci* **320**(1-2) 123-34.
[10] Masalev A A, Khotimskii VS, Bondarenko GN. 2008 *Polym. Sci. Ser.A* **50** 47–53
[11] Morlière N, Corriou JP, Favre E, Roizard D. 2002 *Desalination*;144(1-3)109-13
[12] Bondar VI, Freeman BD,Pinnau I. 1999 *J. Polym. Sci., Part B: Polym. Phys.* **37**, 2463–2475
[13] Yoo S, Won J, Kang SW, Kang YS, Nagase S. 2010 *J Membr Sci* **363**(1-2) 72-9
[14] Hofmann, D, Heuchel, M, Yampolskii, Y. 2003 *Materials Research Society Symposium – Proceedings* **752** 149-154

Synthesis of potassium ferrate using residual ferrous sulfate as iron bearing material

N Kanari,[1] L Filippov,[1] F Diot,[1] J Mochón,[2] I Ruiz-Bustinza,[2] E Allain[1] and J Yvon[1]

[1]Laboratoire Environnement et Minéralurgie, UMR 7569 CNRS, Nancy-University, ENSG, BP 40, 54501 Vandoeuvre-lès-Nancy Cedex, France

[2]CENIM - CSIC, Av. Gregorio del Amo 8. Madrid 28040, Spain

E-mail: ndue.kanari@ensg.inpl-nancy.fr

Abstract. This paper summarizes the results obtained during potassium ferrate ($K_2Fe^{VI}O_4$) synthesis which is a high added value material. This compound that contains iron in the rare hexavalent state is becoming a substance of growing importance for the water and effluent treatment industries. This is due to its multi-functional nature (oxidation, flocculation, elimination of heavy metals, decomposition of organic matter, etc.). The most well known synthesis methods for potassium ferrate synthesis are those involving the chemical and/or electrochemical oxidation of iron (II) and (III) from aqueous solutions having a high alkali concentration. These methods are generally characterized by a low Fe^{VI} efficiency due to the reaction of the potassium ferrate with water, leading to the reduction of Fe^{VI} into Fe^{III}. Concerning the work pertinent to this paper, the synthesis of $K_2Fe^{VI}O_4$ was achieved by a simultaneous reaction of two solids (iron sulfate and KOH) and one gaseous oxidant (chlorine). The synthesis process is performed in a rotary reactor at room temperature and the global synthesis reaction is exothermic. The effects of different experimental parameters on the potassium ferrate synthesis are investigated to determine the optimal conditions for the process.

1. Introduction

The denomination "ferrate" is generally attributed to compounds containing iron at oxidation state higher than Fe^{III}. Though the ferrates (IV, V and VI) are cited in the literature, the ferrate (VI) seems to be the best known and the most studied.

The importance of the ferrates' synthesis is related to their considerable and innovating applications in different fields. In aqueous solution, the ferrate ion (FeO_4^{2-}) is reduced, generating both ferric hydroxide precipitate and nascent oxygen, as shown by equation (1). Ferrate is used for water treatment due to its powerful oxidizing capacity (oxidation of organic and mineral materials, biocide agent) and flocculating property of the resulting ferric hydroxide. The ferrate can replace the chlorine in the pre-oxidation stage of water and partly the iron and aluminum salts [$FeCl_3$ and $Al_2(SO_4)_3$] that are used as coagulating and flocculating agents. Furthermore, the decomposition of the ferrate according to equation (1) generates alkaline media favorable to the precipitation of heavy metals into hydroxides. These properties exhibited together by one single compound make the ferrate an ongoing material particularly interesting and valuable for the treatment of water.

The most attractive property of the ferrate ion is its strong oxidizing ability. The redox potentials of the Fe^{VI}/Fe^{III} couples [$E°(FeO_4^{2-}/Fe^{3+})$ and $E°(FeO_4^{2-}/Fe(OH)_3)$] are 2.20 and 0.72 V respectively at pH values equal to 1 and 14. These values are significantly higher than those corresponding to the Mn^{VII}/Mn^{IV} [$E°(MnO_4^-/MnO_2)$ = 1.679 and 0.588 V at pH 1 and 14, respectively] or Cr^{VI}/Cr^{III} couples [$E°(Cr_2O_7^{2-}/Cr^{3+})$ = 1.33 V in acidic medium, $E°(CrO_4^{2-}/Cr(OH)_3)$ = -0.12 V in alkaline medium]. The use of ferrate as oxidant instead of chromate and/or permanganate is a promising choice from the environmental point of view. One may underline that the oxidants based on chromium and manganese are corrosive, and they are irritant for the skin and for sensitive body organs such as the eyes. The use of chromium compounds as oxidant is also considered toxic for mankind and the environment [1]. As a contrast, ferrate reduction products are not harmful for humans and the environment as they are essentially composed of iron oxy/hydroxides.

$$4\,FeO_4^{2-} + 10\,H_2O \rightarrow 4\,Fe(OH)_3 + 8\,OH^- + 3\,O_2 \qquad (1)$$

However, the goal of this study is not to focus on the development of alkali ferrates for different applications but to summarize a process developed for $K_2Fe^{VI}O_4$ synthesis by solid-gas reactions.

2. Materials and Methods

The iron bearing material for the synthesis of potassium ferrate is chosen to be iron sulfate heptahydrate ($FeSO_4.7H_2O$). The industries of TiO_2 and surface treatment of steel are the main producers of ferrous sulfate and it is considered as a waste for these industries. Titanium dioxide is produced from its raw materials (ilmenite, rutile, anatase, slags, etc.) using "sulfate" and "chloride" process. Ilmenite and titanium slag are used for TiO_2 production using the "sulfate" process. Equations (2) and (3) represent the principal reactions occurring during the treatment of ilmenite to produce TiO_2 in using the sulfate route. Thus, ilmenite is reacting with sulfuric acid generating a solution containing titanyl sulfate and iron sulfate. The solution is treated with metallic iron to reduce ferric ions to ferrous ones, hence avoiding the precipitation of ferric hydroxide. Ferrous sulfate is then crystallized as $FeSO_4{\cdot}7H_2O$ and is separated from the liquor. Finally, $TiOSO_4$ is treated to obtain TiO_2.

The major environmental disadvantage of the sulfate process is the large quantities of spent acid and ferrous sulfate generated. The amount of iron sulfate produced depends on the quality of the raw materials used for the titanium dioxide production and can reach up to 6 tons of $FeSO_4{\cdot}7H_2O$ generated per ton of TiO_2 produced. Since the production of TiO_2 in Europe is essentially made through the "sulfate" method, about 5 million tons of ferrous sulfate heptahydrate are annually generated.

Potassium hydroxide pellets used for the ferrate synthesis had a diameter of about 5 mm. Note that potassium hydroxide contains about 15 % H_2O (by weight). The gaseous reagents (Cl_2, N_2, O_2) used in this research had a purity higher than 99 %.

$$FeTiO_3 + 5\,H_2O + 2\,H_2SO_4 \rightarrow TiOSO_4 + FeSO_4{\cdot}7H_2O \qquad (2)$$

$$TiOSO_4 + 2\,H_2O \rightarrow TiO(OH)_2 + H_2SO_4 \qquad (3)$$

The main apparatus and experimental procedure for the ferrate synthesis were described previously [2]. The solid sample ($FeSO_4·xH_2O$+KOH) was placed in the reactor before a given gas mixture containing chlorine was circulated allowing the ferrate synthesis by the contact of the two solid phases with the gas mixtures (air+Cl_2 or N_2+O_2+Cl_2). Though the synthesis was achieved in room conditions, a thermocouple was placed into the sample to measure the temperature increase due to the exothermic nature of the synthesis reactions. The synthesis products, after drying at 65 °C, were subjected to Mössbauer spectrometry to determine the Fe^{VI} synthesis efficiency of the process.

3. Results

The potassium ferrate synthesis using $FeSO_4·xH_2O$ and chlorine as oxidant is described by equation (4) producing (x+4) moles of water (x being the amount of structural solvation water moles in the ferrous sulfate). However, potassium hydroxide used for the ferrate synthesis contains about 15 % H_2O. In this condition, the real reaction of ferrate synthesis could be represented by equation (5). This suggests that the amount of water released during the synthesis is high and consequently, it can react with the product and decrease the Fe^{VI} yield. For these reasons, it was suggested to use a partially dehydrated ferrous sulfate for the synthesis of potassium ferrate.

$$FeSO_4·xH_2O + 2\ Cl_2 + 8\ KOH \rightarrow 2\ K_2(Fe_{0.5}, S_{0.5})O_4 + 4\ KCl + (x+4)\ H_2O \quad (4)$$

$$FeSO_4·xH_2O + 2\ Cl_2 + 8\ KOH.0.55H_2O \rightarrow 2\ K_2(Fe_{0.5}, S_{0.5})O_4 + 4\ KCl + (x+8.4)\ H_2O \quad (5)$$

The effects of a good number of parameters (as shown below) on the potassium ferrate synthesis were checked in order to find the best conditions for a high Fe^{VI} efficiency of the process.

- Reactor shape and volume (from 0.2 to 2.5 L),
- Flow rate of gases (from 40 to 700 L/h),
- Type of diluting gas (air and/or N_2),
- Chlorine content in the gas mixtures (2 to 7 % Cl_2),
- Type of iron salt ($FeSO_4·H_2O$, $FeSO_4·OH$),
- K/Fe ratio of solids,
- Loading of solids as the reaction progresses,
- Residence time (30 to 120 minutes).

After an optimization of the above-mentioned parameters, a flow-sheet [3] of the potassium ferrate synthesis was proposed (figure 1). According to this flow-sheet, the $FeSO_4·7H_2O$ has to be dehydrated at temperatures of about 125 °C allowing a control of the water quantity during the potassium ferrate synthesis, and an increase of the reactivity of ferrous sulfate.

The success of the whole process depends on the reactor design and experimental parameters to create the following conditions: (i) a good contact of three substances ($FeSO_4·xH_2O$, KOH and Cl_2); (ii) a preferential diffusion of KOH towards the reaction zone (iii) a rapid evacuation of the water released from the synthesis reaction; and (iv) a high and uniform heat transfer as the overall synthesis reaction is exothermic. Although the potassium ferrate synthesis was achieved without external heat supply, the reactor temperature increases due to the exothermic nature of the reactions. An increase of the reactor temperature between 45 °C to 175 °C was observed, depending on the experimental parameters. Good results of potassium ferrate synthesis were obtained for temperatures lower than 100 °C with a Fe^{VI} yield as higher as 60 %.

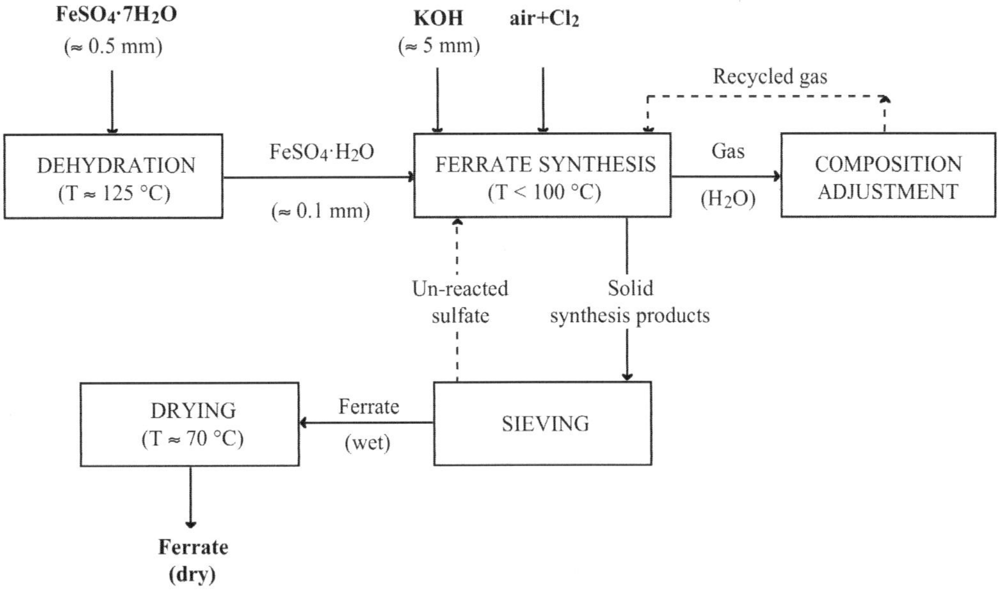

Figure 1. Suggested flow-sheet for the synthesis of potassium ferrate.

The last step of the proposed process consists in the separation of the synthesis products from non reacted ferrous sulfate and drying of the obtained potassium ferrate. The separation of synthesis products from the non reacted ferrous sulfate was facilitated by the important difference of particle size between the ferrate (≈ 5 mm) and that of the ferrous sulfate (≈ 0.1 mm). Thus, a simple sieving of the synthesis product will allow their separation. Ferrate could be dried at temperatures lower than 70 °C. However, in order to not compromise its stability, the drying temperature has to be the lowest possible and never exceed 100 °C.

4. Conclusions

Residual iron sulfate is an appropriate raw material for the preparation of potassium ferrate. The process was achieved in room conditions and the overall synthesis reaction was exothermic. The conversion efficiency of Fe^{II} to Fe^{VI} (Fe^{VI} yield) during potassium ferrate synthesis was higher than or equal to 60 %. This work shows the possibility to transform a waste (iron sulfate) into a useful value added product (potassium ferrate).

Acknowledgments

This work was performed in the frame of contract N° BRPR-CT97-0392. Thanks to the financial support of the European Union (DG-XII).

References

[1] Delaude L and Laszo P 1996 *J. Org. Chem.* **61** 6360
[2] Kanari N, Ostrosi E, Ninane N, Neveux N and Evrard O 2005 *JOM* **57** 39
[3] Kanari N, Evrard O, Neveux N and Ninane L 2001 *JOM* **53** 32

Journal of Physics: Conference Series **416** (2013) 012014 doi:10.1088/1742-6596/416/1/012014

Carbonation of residual brines produced by ammonia-soda process

I V Filippova, P Piriou, L O Filippov, J Yvon, M Grandjean.

Laboratoire Environnement Minéralurgie, UMR 7569 CNRS,-Nancy-Université, France

inna.filippova@univ-lorraine.fr

Abstract. This work deals with the carbonation of residual brines produced during the manufacture of soda ash to avoid the unsuitable phase transformation during the land storage. The study resulted in a demonstration pilot, which showed the feasibility of such an approach and the possibility of his extension to an industrial scale. Carbonation of the residual brines is a promising process as it entirely transforms $Ca(OH)_2$, "CaOHCl" and CSH into calcite, avoids the further phase evolution, allows to obtain a neutral pH which considerably reduce the land storage impact on environment and shorten by around 10 % the global CO_2 emission of the ammonia-soda process.

1. Introduction

Nowadays accelerated carbonation begins to be used for the treatment of wastes and contaminated soils and for the sequestration of CO_2 [1, 2], the most important greenhouse gas whose atmospheric concentration has been raising from 280 ppm to 380 ppm since the beginning of the industrial era. This kind of mineral sequestration of CO_2 is adapted to modern carbon management, whose main ideas are the capture and sequestration of CO_2, the increase of efficiency of energy conversion and the use of low carbon or carbon free energy sources [3]. Thus carbonation of the residues of the ammonia-soda process, which makes them passive and avoids further phases transformation, is a modern and promising way of dealing with such hazardous for environment materials, mainly through chloride leaching [4].

Previous studies have already shown the possibility of using these residual materials to produce calcium-magnesium phosphates, cosmetic chalk and/or gypsum of standard value. Some authors also suggested the use of these residues for the neutralization of combustion gases [5-6]. Same authors focussed on the optimal parameters for the precipitation of calcium sulphate and performed a simple economic study of their process to show the feasibility of such an approach [7]. Recently CO_2 sequestration in different brine solutions has been studied. For example carbonation of brine solutions containing fly ashes from coal combustion have been proved to have a great potential for CO_2 sequestration [8-9].

The objectives of this study are to determine the phases present in the residual brines which are sensitive to carbonation, to obtain a magnitude order of carbonation of theses phases, to assess the decrease of CO_2 emissions of the overall ammonia-soda process and to highlight the limits of this process.

2. Material and methods

Carbonation reactions were carried out in a Pyrex reactor with a 1900 mL volume capacity at a constant temperature of 55°C which corresponds to the temperature of the residual brines getting out of the ammonia-soda process factory. Gas injected by bubbling contained 15 % of CO_2 and 85 % of air corresponding to the composition of the gas at the outlet of the boiler. The mixture was continuously stirred for a better bubble dispersion and to prevent sedimentation of particles, however it must be noted that, in industrial conditions, the possible use of a counter–current bubbling column avoids using a stirring device. Indeed, in a continuous reactor the large amount of gas injected is sufficient to ensure an efficient mixing of the brines. When pH is considered as stable, 250 mL of the mixture are sampled and then filtrated at 0.45 μm in a Teflon Millipore compressed air system. The liquid phases are acidified, whereas the solid phases are frozen and freeze-dried to withdraw the free water. Crystalline phases present in the solids are determined by X-Ray Diffraction (XRD) using a Bruker D8 Advance diffractometer equipped with a Co-Kα_1 computed monochromatic source (1,789Å) at 35 kV and 45 mA with a linear Lynx-Eye detector at 0.035° by 2 theta step from 3° to 64° and a 3 seconds counting time. Presence of sulphate, hydrated, silicate or carbonate phases are determined by Diffuse Reflectance Infrared Fourier Transform Spectroscopy (DRIFTS) using a Bruker IFS 55 equipped with a large band (600-5500 cm^{-1}) mercury-cadmium telluride (MCT) detector cooled to 77 K. Fine chemical analysis of Cd, Cr, Mn, Ni and Pb concentrations in the liquid phases are performed using Inductively Coupled Plasma Mass Spectroscopy (ICP-MS VARIAN 820-MS).

3. Results and discussion.

3.1. Order of carbonation

After filtration and freeze-drying of a sample the composition of residual brine was deduced from associating XRD and DRIFTS results. XRD shows the presence of portlandite and calcium hydroxychloride CaOHCl. The vibrations at 3644 cm^{-1}, 3698 cm^{-1} and 3570 cm^{-1} in the DRIFTS spectra are assigned to the hydroxyl stretching vibration in $Ca(OH)_2$, $Mg(OH)_2$ and CaOHCl respectively (figure 1). The large band at 966 cm^{-1} corresponds to the vibration of Si-O in the C-S-H, and the large band centred on 1430 cm^{-1}, to the C-O elongations in carbonates. Other bands are diverse overtones, deformations or combinations of the same groups.

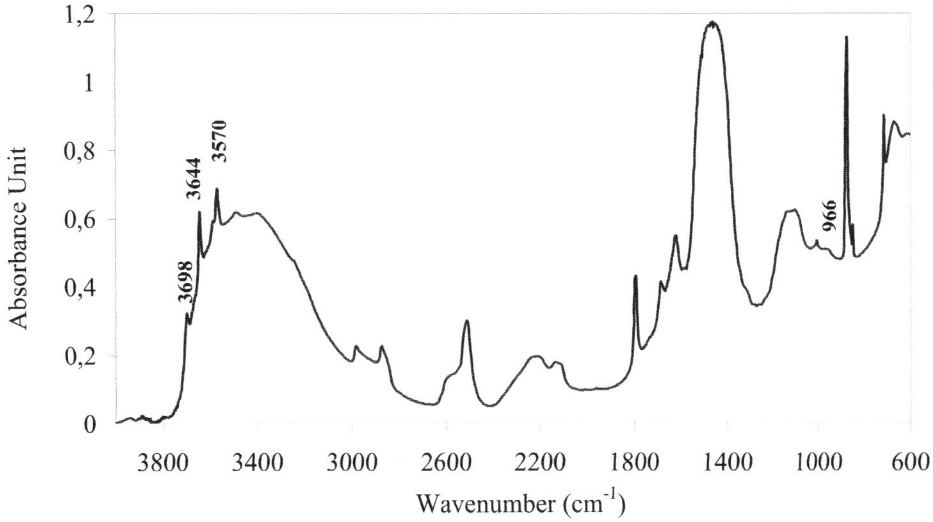

Figure 1 Diffuse reflectance FT-IR spectrum of solid phase from initial brines

At carbonation pH (close to 6) all peaks characterising the presence of these minerals have disappeared (figure 2). However these mineral phases are not directly carbonated. In fact CO_2 must first be solvated, hydrated and ionised giving carbonate ion CO_3^{2-} (reaction 1) which reacts with free Ca^{2+} according to reaction (2).

$CO_{2\,(g)} + H_2O = CO_3^{2-}{}_{(aq)} + 2\,H^+$ (1)

$Ca^{2+}{}_{(aq)} + CO_3^{2-}{}_{(aq)} = CaCO_{3\,(s)}$ (2)

These two reactions displace the solid-solution equilibria of $Ca(OH)_2$, $Mg(OH)_2$, CaOHCl and CSH (reactions (3) – (5)), due to the consumption of free Ca^{2+} and to the slow disappearance of OH^- ions when pH decreases. Indeed the appearance of CO_3^{2-} in water comes along with the appearance of two H^+ according to reaction (1). This reaction explains why the pH tends to decrease when CO_2 is injected in the reactor.

$Ca(OH)_2 = Ca^{2+} + 2\,OH^-$ (3)

$CaOHCl = Ca^{2+} + OH^- + Cl^-$ (4)

$Mg(OH)_2 = Mg^{2+} + 2\,OH^-$ (5)

$x\,CaO - y\,SiO_2 - z\,H_2O = x\,CaO + y\,Si(OH)_4 + (z\text{-}2y)\,H_2O$ (6)

The solubilization of the C-S-H (reaction (6)) is a two-step reaction. Indeed CO_3^{2-} reacts with free Ca^{2+} (reaction (2)) modifying the equilibrium between sequestered CaO and solubilized CaO, that is $Ca(OH)_2$ (reaction (3)). It cannot be accepted that C-S-H are directly carbonated. They are dissolved due to an equilibrium between CaO contained in C-S-H and free Ca^{2+} continuously consumed by carbonation.

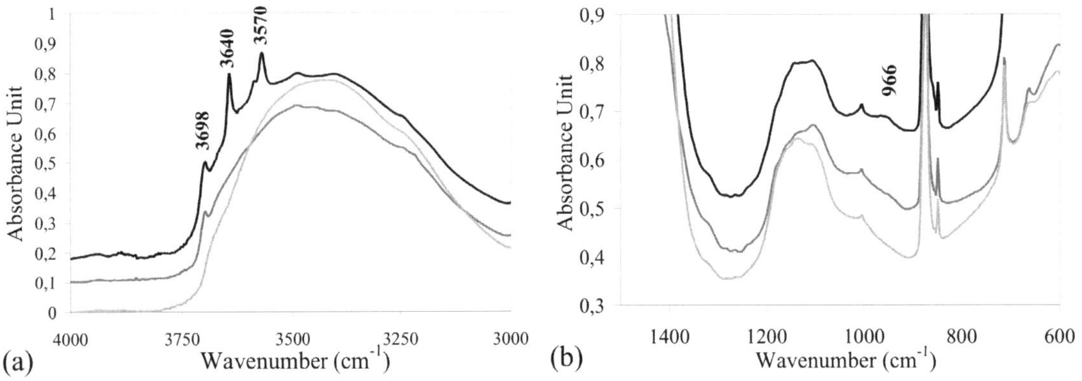

Figure 2 Evolution of diffuse reflectance infrared spectra of residual brines with pH

Due to kinetics and solubility products an order of carbonation of phases was obtained. CaOHCl which crystallizes [1a] at the surface of $Ca(OH)_2$ first disappears followed by $Ca(OH)_2$ whose dissolution is relatively quick due to the facts that both Ca^{2+} and OH^- are consumed by carbonation. C-S-H then disappear after its slow but complete dissolution, releasing colloidal silica in the brine (reaction (6)), whose dissolution is favoured by the alkaline pH. $Mg(OH)_2$ finally disappears from the mixture when pH is low enough to ensure the total solubilization of brucite [4].

3.2. Less CO_2 emission for the ammonia-soda process

Using the BAT (Best Available Techniques) document of the IPPC (Integrated Pollution Prevention and Control) directive [10] the ammonia-soda process has been schematized in 12 steps with the software BILCO ® [1]. This model has been intentionally designed for the production of one ton of sodium carbonate Na_2CO_3 so that the values of the flow-rates of the wastes (CO_2 and brines) are now given per ton of Na_2CO_3 produced. Thus, the production of one ton of Na_2CO_3 comes along with the emission of 351 kg of CO_2 [1]. The CO_2 consumption for the carbonation of the residual brines depends on the value of the pH at the end of the carbonation. For a final value of pH 7 the consumption of CO_2 is around 35 kg per ton of produced Na_2CO_3, then with a 10 % decrease of the CO_2 emission from the ammonia-soda process. For a worldwide annual production of 60 Mt of Na_2CO_3 this process would allow the sequestration of 2.1 Mt of CO_2.

3.3. *Further phase transformation avoided.*

Residual brines contain mix layered double hydroxides (LDH) of Ca^{2+} and Al^{3+} such as hydrocalumite $[Ca_2Al(OH)_6][Cl_{1-x}OH_x].3H_2O$. During landfield storage these LDH lead to the formation of expansible phases such as ettringite which shorten the designed life-time of tailing dumps [12]. Furthermore these expansible phases are also responsible for the creation of a lens in the middle of the tailing dumps which prevents them from the complete leaching of chloride [12].

The vertical zonality model of phase evolution in dump proposed in [12] reveals the essential role played by Al (and Mg) in the blocking of the maturation of the solids resulting from the effluent decantation and contact with the air. In the case of a waste disposal processed by a simple decantation without carbonation, since the kinetics of the natural carbonation, controlled by the diffusion of CO_2, is not fast enough, the solids precipitate and free water preserve an alkaline pH buffered by the portlandite. This alkaline pH will cause, on one hand, to continue the dissolution of residual silico-aluminous minerals contained in the brine (mullite) and, on the other hand, to support the "alkaline" mineral formation (layered double hydroxides and cancrinites). These neoformed phases by their structuring capacity (swelling and retention of water) favour the relative impermeability and will delay the carbonation of the level where they occur.

Decreasing the pH causes the sequestration of aluminium in the solid phase as predicted by Pourbaix' diagram of Al. It can be noted that carbonation decreases the mobility of aluminium because of pH decreasing and avoids formation of layered double hydroxides.

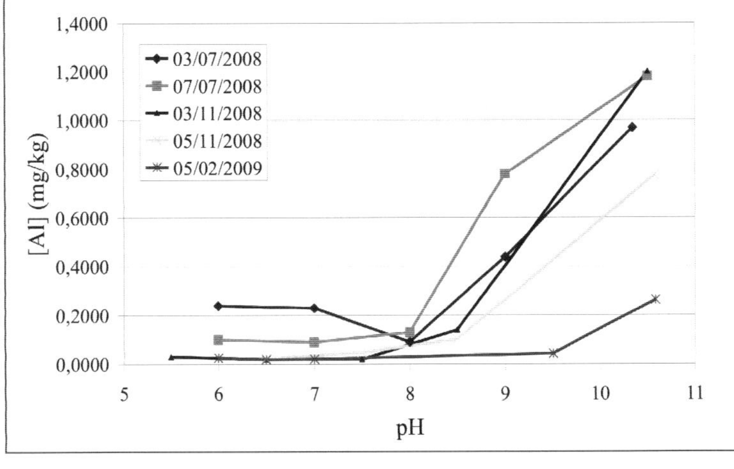

Figure 3 Evolution of aluminium concentration in the liquid phase with pH. (Evolution is given for five different dates)

3.4. *Limits of the carbonation process.*

The CO_2 sequestration method by carbonation of industrial brines has limits. Indeed when decreasing the pH, two limits clearly appear which lead to the necessary incomplete carbonation of the residual brines of the ammonia-soda process.

3.4.1. Heavy metals solubilization.

Heavy metals such as Cr, Pb, Mn, Ni or Zn are known for their behaviours, which strongly depend on the pH value. Most of these heavy metals are more solubilized for acid pH than for alkaline pH. Due to the decrease of the pH value, it can be assumed that the carbonation leads to increase the heavy metals concentrations in the liquid phase.

For the reason that liquid phase of the residual brines produced by the ammonia-soda process is then rejected in rivers, oceans or lakes, proceeding with the entire carbonation of the brines has to be avoided. A pH value between 7.0 and 7.5 can be considered as a good choice to limit the heavy metals mobilization.

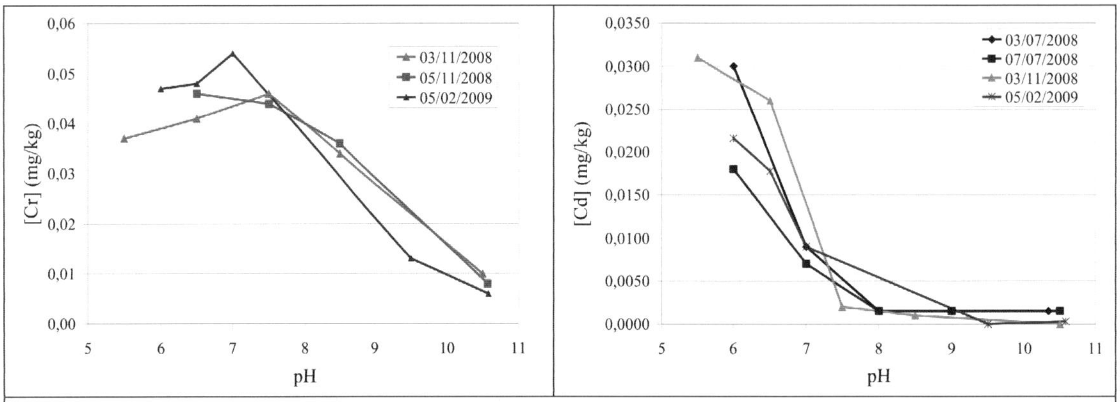

Figure 4 Evolution with pH of Cr and Cd concentration in the liquid phase at different dates

3.4.2. Calcite solubilization for acid pH

In order to understand the calcite dissolution when the acidity growths, the acido-basic reactions of carbon dioxide and the dissociation reaction of calcite are shown below :

$H_2O, CO_2 = H^+ + HCO_3^-$ ($pKa_1 = 6.37$)

$HCO_3^- = H^+ + CO_3^{2-}$ ($pKa_2 = 10.32$)

$CaCO_3 = Ca^{2+} + CO_3^{2-}$ ($pK_S = 8.72$ at 55°C)

Thus, decreasing the pH value leads to the decrease of CO_3^{2-} concentration which may lead to the solubilization of the calcite if the product of the concentration of Ca^{2+} by the concentration of CO_3^{2-} is lower han the product of solubility of calcite at 55°C calculated with an experimentally obtained expression [11]. Moreover, for acid pH, HCO_3^- ions attack $CaCO_3$ releasing free Ca^{2+} in brines as $Ca(HCO_3^{2-})$ is not a solid compound, modifying the equilibrium between free Ca^{2+} and "sequestered" Ca.

4. Conclusion.

The analysis of the results leads to the three main conclusions:

First, phases presents in the residual brines sensitive to carbonation are identified and an magnitude order of carbonation of them is obtained by coupling infrared spectroscopy and X-ray diffraction methods. Thus, the first phase which disappears is CaOHCl, followed by the disappearance of $Ca(OH)_2$ at pH around 10, then C-S-H disappear for pH around 8.5 and finally $Mg(OH)_2$ at pH around 7.5.

It must be noted that carbonation is not only interesting because of the precipitation of new mineral phases (calcium carbonate essentially) from different mineral phases ($Ca(OH)_2$, CaOHCl, C-S-H). Indeed it also deeply modifies the global equilibrium of the system, leading to the decrease of mobility of elements such as aluminium, which is responsible of the appearance of unsuitable phases during land-storage.

However this process has two main negative aspects when pH reaches acid values. This means that CO_2 sequestration cannot be achieved to the maximum. Indeed, when pH is too low, some hazardous heavy metals such as chromium and cadmium are solubilized. Moreover calcite is also solubilized due to the acido-basic equilibrium of carbonate, which leads to the disappearance of CO_3^{2-} and the appearance of HCO_3^-, which attacks calcite.

References

[1] Bodénan F, Lassin A., Hottier M, Filippov L, Durance M V and Piantone P 2008 Piégeage et valorisation de déchets alcalins par passivation au CO_2 industriel : *Rapport final ANR-Decalco, BRGM/RP-56361, 151 p.*

[2] Fernández Bertos M, Simons S J R, Hills C D, Carey P J 2004 *J. Hazardous Materials* **B112** 193

[3] Soong Y, Fauth D L, Howard B H, Jones J R, Harrison D K, Goodman A L, Gray M L and Frommell E A 2006 *Energy Conversion and Management* **47** 1676

[4] Bodénan F , Hottier M , Lassin A , Filippov L, and Piantone P 2008 Reducing CO_2 emissions from the Solvay process by entrapment in saline waste, *Proceed. of 2nd Int.Conference on Engineering for Waste Valorisation, WASTEENG'08, Patras, Greece Juin 2-5, 2008,* ISBN 978-960-530-101-9. P171, 7 pp.

[5] Kasikowski T, Buczkowski R, Dejewska B, Peszynska-Bialczyk K, Lemanowska E and Iglinski B 2004 *J. of Cleaner Production* **12** 759

[6] Kasikowski T, Buczkowki R and Lemanowska E 2004 *J. of Environmental Management* **73** 339

[7] Kasikowski T, Buczkowski R and Cichosz M, 2008 *Int. J. Production Economics* **112** 971

[8] Soong Y, Goodman A L, McCarthy-Jones J R, Baltrus J P 2004 *Energy Conversion and Management* **45** 1845-1859

[9] Nyambura M G, Mugera G W, Felicia P L and Gathura N P 2011 *J. Environmental Management* **92** 655

[10] CEFIC 2004 IPPC BAT reference document – Large volume solid inorganic chemicals family – Process BREF for soda ash, issue n°3 - http://www.cefic.be/files/Publications/ESAPA_Soda_Ash_Process_BREF3.pdf

[11] Dorange G, Marchand A, Le Guyader M 1990 *Revue des sciences de l'eau* **3** 261.

[12] Filippov L O , Yvon J, Filippova I V, Godon B, Perrin P and Piantone P, 2008 Characterisation and descriptive model of saline waste evolution. *Proceed. of 2nd Int.Conference on Engineering for Waste Valorisation, WASTEENG'08, Patras, Greece Juin 2-5, 2008*, ISBN 978-960-530-101-9, P 143 2pp.

Journal of Physics: Conference Series **416** (2013) 012015 doi:10.1088/1742-6596/416/1/012015

New approaches to the estimation of the geosystem properties transformation in technogenesis

G Sarapulova, K Fedotov
National research Irkutsk state technical university, 664074, Russia, Irkutsk, Lermontov St 83

E-mail: sara131@mail.ru

Abstract. A new approach to the estimation of environmental situation of the urbanized territories of a large city is offered. The approach is based on a complex of physical and chemical parameters adequately describing transformation of soil properties, with the use of landscape-geochemical method and GIS technologies. The pollution of soil horizons by heavy metals (HM) and mineral oil (MO) in a zone of influence of dangerous industrial objects exceeds by orders the maximum permissible concentration (MPC). Sharp deterioration of a nitric and carbon mode, increase of the alkalinity and the decrease in buffer activity of ground were revealed. The dynamics of technogenic streams and aureole of the contaminants migration in soils can result not only in the further transformation of their properties, but in closing the migration cycles of MO and HM. As a rule this is accompanied by the formation of secondary local sites of toxic substances accumulation and abnormal geochemical fields - *laterally technogenic module.* The revision of approaches to the analysis of soils under technogenesis and the development of new system of ecological monitoring represent a challenging goal.

1. Introduction

Industrially overloaded regions of Siberia and boundary territories cause the problems of nature protection, the decision of which is complicated by the absence of the state program of environmental safety [1]. In the Baikal Natural territory, which is being under UNESCO protection, multiple technogenic influence on the environment characterized by synergistic effect is observed. The scale of soils and the water-currents degradation, the out-of-date schemes of sampling, the simplified techniques of contaminants determination (dated back to the beginning of the last century), and the absence of modern system of monitoring do not promote to the development of new approaches to correct estimation of environmental situation of the territories.

The urgency of the investigations conducted is due to the necessity of obtaining the reliable criteria of transformation of natural geosystems under the influence of technogenesis. New approaches are based on revealing the correct characteristic describing modern and predicted situation of technogenic-changed soil and water geosystems involving a complex of physical and chemical parameters and models [2].

In the urbanized territories, technogenic streams of contaminants can essentially differ from those of in natural environment. However, the migration of pollutants in geosystems and their transformations under the action of various environmental factors in the course of technogenesis and the formation of new geochemical fields are not practically studied.

2. The purposes

The investigation is aimed at the revealing of regularities of pollutants migration and accumulation in soil water to understand the mechanism of their transformations during technogenesis. This is connected with necessity of the revision of out-of-date notions on the dynamics and normalization of contaminants in geosystems, especially in the urbanized territories and in a zone of influence of large industrial complexes. The new approach to ecological diagnostics of technogenically modified soils and water-currents are necessary. This will allow the modern quality monitoring under changing conditions of the environment to be developed. The necessity of the development of new system of ecological monitoring was underlined at session of the State Advice of the Russian Federation on May 27th, 2010.

We carry out systematic geoecological research according to the complex program:

- studying the transformation of properties of soil and water ecosystems to reveal migration streams of the contaminants in the urbanized territories of large cities;
- studying the transformation of soil and water-currents properties in a zone of technogenic influence during the extraction and processing of mineral raw material under mountain conditions.

3. Methods and objects of research

Soil is the main buffer, which takes the whole complex of long-term influence from industrial production and anthropogenic sources. Therefore, objects of the research are soils and grounds ecologically vulnerable areas of Siberia and the urbanized territories of cities in a zone of influence of dangerous objects, which according to the forecast of mineral resources development will be intensively developed up to 2030 [3].

The present work deals with the results of the investigation of technogenically changed soils (urbicsoils) in a zone of dangerous industrial objects of a large city with the population of 1.5 million. The objects of research are gasoline stations (GS), petroleum depots, dams of hydraulic ash-disposal areas of thermal power stations located on territory of a city. The pollution was estimated under the content of mineral oil (MO) and heavy metals (HM) - Cu, Cd, Ni, Pb, Zn, Hg, as well as As.

Methodological methods of investigations:

- landscape and geochemical in combination with cartographical are most appropriate for ecological estimation of territories;
- determination of physical and chemical and geochemical parameters of soils, which adequately reflect the content and forms of contaminants and their properties;
- establishing the spatial and time dynamics of the contaminants distribution in soil horizons;
- compiling the monoelement maps using GIS-technologies to forecast the conditions of geosystems;
- modelling of the pollutants migration in soil ecosystem under the influence of environmental factors.

The following methods were used to study the soil samples: physical and chemical, including atomic absorption spectroscopy with the inductive connected plasma, gel-penetrating chromatography, electronic absorption spectra, microelement analysis, wet and dry combustion, DTA, and also the correlation analysis, computer modeling and GIS-technologies and so on.

4. Discussion of results

The fertile layer of northern territories does not exceed 5-10 cm. Therefore, long-term intensive accumulation of diverse pollutants inevitably leads to the significant changes of morphology and structure of soils, transforming them into lands, technical soils and urbicsoils.

The urban territory is a specific geoecological system having original structure and composition of components [1, 3]. Under conditions of urbanization, ecosystems, as a rule, are transformed: the relief and lithogenous basis, hydro-geological and climatic conditions are changed, geochemical processes in soils being broken. As a result, new natural-technogenic soil complexes with new properties are formed. There are various geochemical anomalies and geopathogenic zones, which features are not practically investigated. The analysis of microelement structure of soils on background and technogenically transformed sites allows to estimate a level of landscapes pollution. Therefore, the characteristic of territorial and intraprofile distribution of microelements and substances in soils, definition of features of their migration and accumulation constitute a basis for an estimation of geosystems stability under the conditions of technogenesis [4-6].

To analyze the soils, the most widespread contaminants, MO and HM, included in the priority list of contaminants according to Russian specification SanPiN 2.1.7.1287-03, have been chosen. These are the most widespread toxicants in a zone of negative influence of gasoline stations, petroleum depots, the mining enterprises, etc. They belong to 1st and 2nd danger classes since they possess high affinity to physiologically important organic compounds and are capable to inactivate them in the living processes. In the state ecological control, the number of samples for one controllable object does not exceed 1-2. It does not allow to reveal fairly a real picture of the soil pollution and to estimate the ecological safety of territory.

On the basis of the developed and tested model of sampling, the number of soil samples was increased up to 30 for every studied object. The coverage of the territory with asphalt and its landscape inclination has been taken into account. The soils were investigated in the depth of 30-40 cm i.e. up to a zone of roots of grass and bushes.

In the result of investigations, the contents of HM in soil horizons of territories, exceeding by orders maximum permissible concentration (MPC) (table 1), have been found. This has caused a specific accumulative type of distribution of both HM and MO into depth of 40 cm. Calculated coefficients of concentration of chemical elements *Kc* and the obtained associative series, for example, $Cu_{4.5}$-$Ni_{2.7}$-$Zn_{2.4}$-$As_{2.2}$ are indicative of the emergence of geochemical anomaly in the city centre due to technogenic pollution (*Kc* should not exceed 1).

However, the organization of soil-chemical monitoring should account for not only the composition of soil, its property, presence of the contaminants, but also the factors influencing the mobility of chemical elements. The following parameters exert the most significant action on the migratory streams of substances and elements: acidity and alkalinity of soils, an oxidation-reductive mode, the content of humus, the presence of readily soluble salts. Acid properties of soil, pH values and the content of carbon and nitrogen, were chosen by us as one of key parameters of its resistance to technogenic influence.

Table 1. The content of heavy metals[a] and As (mg/kg) in soils of gas station site.

№ point of sample splitting	Cu	Cd	Ni	Pb	Zn	Hg	As
1	15.04	0.071	11.54	20.36	58.61	1.53	5.13
2	16.42	0.090	10.36	18.23	58.30	1.38	4.92
3	15.10	0.061	10.36	16.78	62.30	1.42	4.95
4	18.86	0.070	6.55	16.66	64.30	1.29	5.26
5	18.52	0.050	8.66	19.36	61.54	1.81	4.95
6	18.00	0.050	8.78	18.77	54.35	1.83	4.65
7	17.68	0.050	9.23	18.69	64.87	1.79	5.62
MPC	5	0.5	10	32	55	2.1	2

[a] Total forms

The experiment has shown a weak alkaline character of soil in a zone of the studied industrial objects. So, for soil samples around of the gas station and petroleum depots, the pH values equal 7.50-7.64. For natural pure soils, the pH values range 4.5-6.5. This acidity provides self-clearing properties of soil and decreases the negative effect of toxic metals in the case of soil pollution. The registered value of pH is attributable to the precipitation of urban alkaline dust containing Ca and Mg carbonates. The alkaline dust comes from a highway and building objects using solutions with lime carbonate and cement. Such substrates are easily blown away, realizing to atmosphere Ca and Mg, the latter precipitate on soil. Therefore the value of pH for urbicsoils can reach 8-9.

The dynamics of distribution of metals and As in the analyzed soils in dependence on the pH value, determined by correlation analysis (software package Statistica in Excel), does not obey to the normal law (see, for example figures 1 and 2). This testifies to the breaking of exchange geochemical and migratory processes in the urbanized ground.

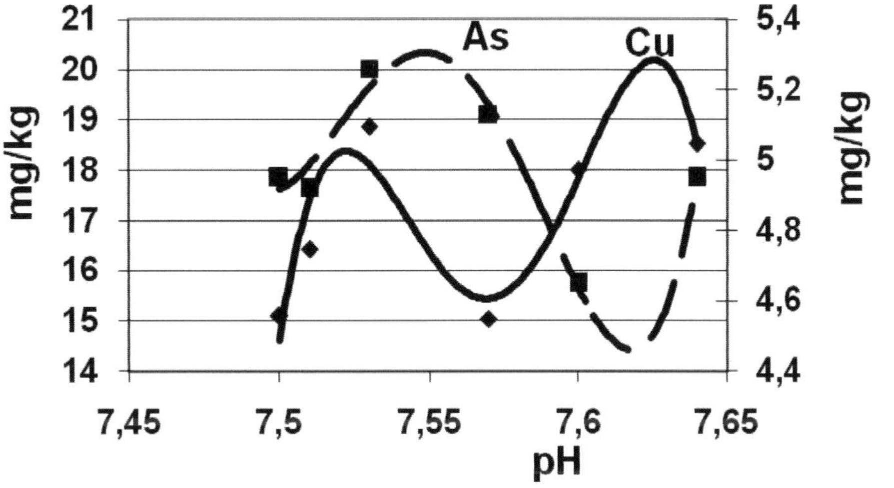

Figure 1. Dynamics of As and Cu content depending on pH value.

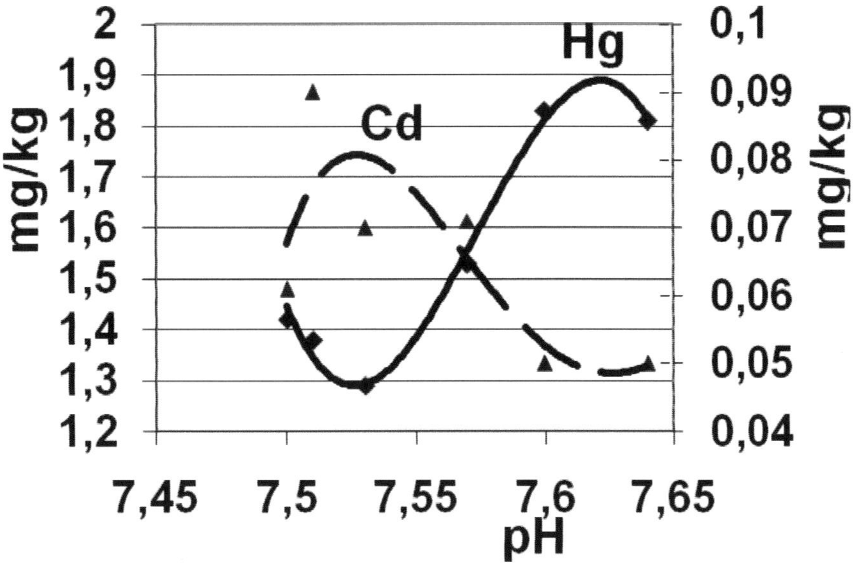

Figure 2. Dynamics of Cd and Hg content depending on pH value.

Buffer properties of soils are caused by the content of humus substances, which basic components are carbon and nitrogen. The method of wet combustion (Tyurin's method) was employed to determine the N content (%) in soil samples (table 2), content of HM was found by atomic absorption spectroscopy on a Perkin Elmer-500 instrument, and urbicsoils mineral oil content was defined by spectrophotometric analysis [5].

Table 2. The content of the N_{total}, mineral oil in soil and value of pH.

Sample	Depth (cm)	N_{total} (%)	pH_{salt}	Mineral oil (mg/kg)
1	0-5	0.065	7.57	668
4	0-5	0.098	7.53	490
7	0-5	0.090	7.61	570
8	10	0.074	8.09	600
9	20	0.115	8.09	400
10	0-5	0.063	6.98	820
11	10	0.098	7.35	610
12	20	0.061	7.86	590

From the table 2 follows that the soil is extremely impoverished by nitrogen, and MO content exceeds maximum concentration limit by 2-3 times (maximum concentration limit ranges 100 to 300 mg/kg according to different techniques). The deterioration of a nitric content of soil along with essential increase of pH values (9 to 7.6) was observed. Strong decrease of organic carbon content in soils was registered. Together with a drop in the content of nitrogen, it leads to drastic deterioration of buffer properties of urbicsoils and increases environmental danger of urban territory. The key ratio of C/N is 20-25, while this value for pure soils is 10.

Monoelement maps of MO and HM distribution (for example, figures 3 and 4) have been compiled using the GIS-technologies (Surfer 8, MapInfo). It has been found, that HM are not accumulated in humus-deficient soils of the polluted territory and migrate to landscapes.

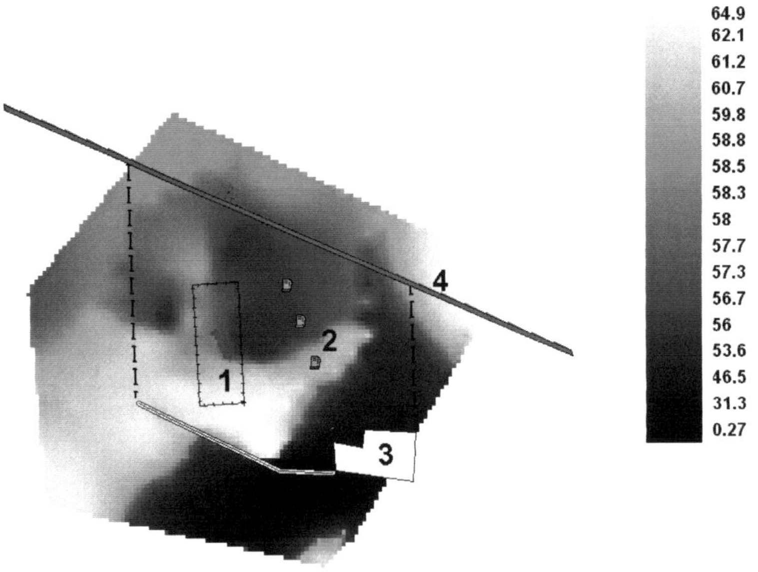

Figure 3. Zn distribution map in a zone of influence of the enterprise: 1, 2 - sources of pollution, 3 - office, 4 - a motorway.

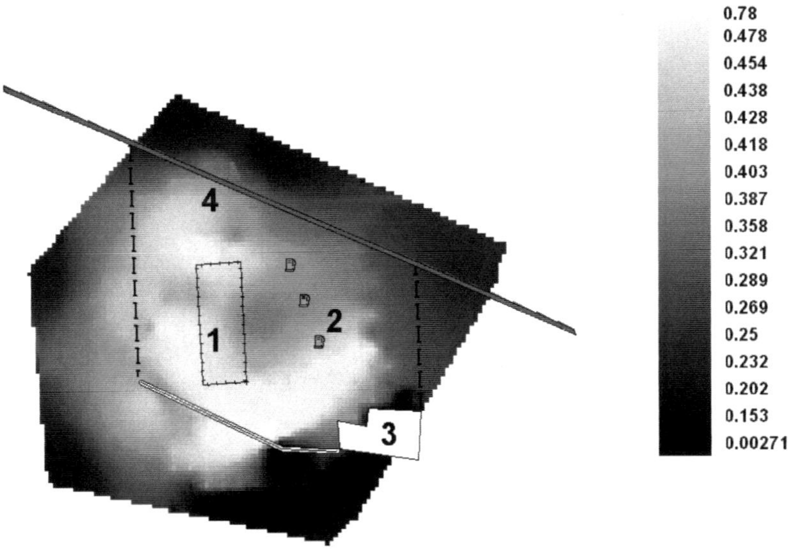

Figure 4. Mineral oil distribution map in a zone of influence of the enterprise: 1, 2-sources of pollution, 3 - office, 4 - a motorway.

Aureoles of their distribution and content confirm the formation of abnormal geochemical zones and accumulation of chemical compounds and elements both in radial and lateral directions. Figure 5 show the particular character of dependence of MO content on concentration of nitrogen in soils of the area.

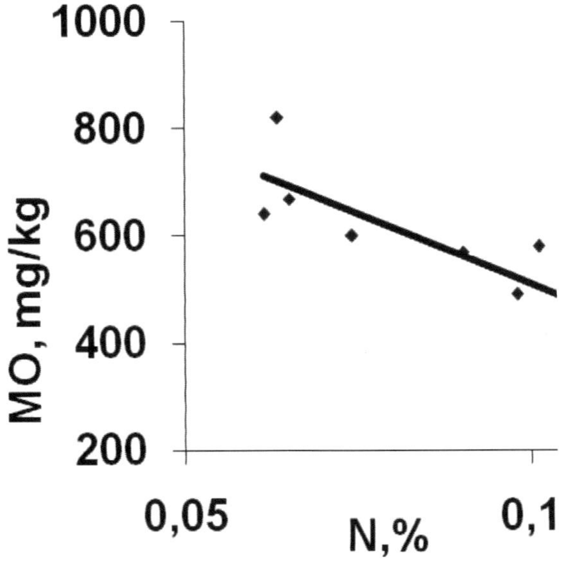

Figure 5. Dependence of mineral oil content on concentration of nitrogen in soils.

In technogenically-changed soils with the broken natural morphological attributes, the HM and MO are not detected on territory of pollution object. Migratory streams and accumulation of contaminants are found far outside the contamination source in a radius of 100-200 m and more [6].

Thus new zones of the enhanced ecological danger, technogenic geochemical modules, (*laterally technogenic module*) are formed. This fact convincingly proves the necessity of change of approaches to the analysis of soils during monitoring of ecological conditions in urbanized territories. Ecological geochemical and the cartographical analysis of urbanized soils has found their substantial pollution with HM and MO, sharp deterioration of nitric and carbon content, increase of alkalinity, decrease in buffer activity of soil. It is explained by signs of transformation of soil properties and makes it important to develop new approaches to the estimation of its ecological conditions [7, 8].

Conclusions

Physical and chemical parameters and the geochemical characteristics which adequately reflect the change of soil ecological systems have been determined.

Dynamics of technogenic streams in urbicsoils which can result not only in transformation of its properties, but in closing the cycles of MO and HM migration has been found. As a rule, it is accompanied by the formation of secondary local sites of contaminants accumulation, *laterally technogenic module.*

Monoelement maps of the distribution of heavy metals and mineral oil in a zone of influence of dangerous objects have been compiled. The data obtained prove the necessity of change of the approaches to monitoring of soils. The results of research constitute the basis for the development of modern system of ecological monitoring under technogenesis. Integration of methods of modern ecological diagnostics will promote to the acceptance of optimal decisions in nature management and to operative revealing the zones of ecological danger.

The investigations are of significance not only for ecological diagnostics of geosystems. The developed approach will allow to discover new geochemical regularities as well as to find the associations of elements in technogenic landscapes for practical purposes in the course of the development of technogenic deposits of minerals.

For optimization of an estimation of technogenically modified soils it is possible to recommend:

- development of the state program of soils monitoring in a zone of influence of industrial objects;
- development of the reasonable schemes, methods and choice of places of approbation of the polluted territory;
- revision of contaminants normalization taking into account their transformations and migration;
- creation of banks of the geochemical and cartographical data for aureoles of pollutants distribution in a zone of dangerous objects and in the urbanized territories.

References

[1] Kochurov V I 1998 *Geoecology: ecodiagnostics and ecological economic balance of territory* (SGU Smolensk)
[2] Kolomyets Y, Rosenberg G and Glebov O 2000 *Natural complex of the big city: Landscape and ecological analysis* (the Science Moscow)
[3] Ladonin D V 2003 Heavy metal compounds in soils - problems and methods of studying *Soil science* **6** 682
[4] Sarapulova G I, Gantomor S and Batsukh N 2009 *Actual problems of oil-and-gas branch of Mongolia and a way of their decision* (Ulan Bator) **1/15** 262
[5] Gantomor S 2009 Heavy metals pollution of the urbanized territory of Ulan Bator *The Post-graduate student* **6** 96
[6] Sarapulova G I, Gantomor S and Altantsetseg M 2010 *Natural and engineering science* (Moscow) **2** 250
[7] Dobrovolsky V V 2009 *Geochemistry of soils and landscapes* (Nauchnyi Mir Moscow)
[8] Arzhanova V S and Elpatievskii P V 1990 *Geochemistry of landscapes and technogenesis* (Nauchnyi Mir Moscow)

Effect of ultrasound on flotation kinetics in the reactor-separator

L O Filippov[1,2], A S Matinin[1,2], V D Samiguin[1] and I V Filippova[1,2]

[1]National University of Science and Technology «MISIS», Moscow, Russia

[2]Laboratoire Environnement et Minéralurgie, UMR 7569 CNRS-Nancy Université, Vandoeuvre-lès-Nancy, France

Email : levfil@gmail.com

Abstract. Effect of the ultrasound on flotation kinetics in reactor-separator has been studied for chalcopyrite/quartz mix mineral system. Under ultrasound treatment, recovery of chalcopyrite into bulk concentrate is higher than that at reagent-only treatment. It can be explained by increased of flotation rate for slow fraction as defined by Kelsall model. The slow fraction flotation rate increase multiplied by 6 vs. ultrasound treatment. Additional effect of the ultrasound treatment has been noticed under conditions when gangue minerals detachment from bubbles can be controlled. Reactor-separator has advantages over other types of flotation cells for this purpose providing a special zone for the ultrasound treatment that can be easily designed in this impeller less machine. The ultrasound influence on particles collision probability is able to explain of chalcopyrite recovery increase in the concentrate and activation chalcopyrite particles flotation.

1. Introduction

The use of the ultrasound conditioning in flotation has been expanding in the recent years. Application of the ultrasound at 20 kHz frequency and approximately 0.3 W/cm^2 power causes cavitation in water or slurry. The cavities collapse obtained under ultrasound treatment results in very high energy densities [1]. Ultrasound treatment at these characteristics has been used for pre-conditioning of low solubility reagents [2, 3], for froth destruction as well as during the flotation [4, 5, 6].

The ultrasound effect at slurry conditioning on sulphide ores flotation results has been investigated by Aldrich and Feng [4]. Among all the parameters (temperature, solids content, conditioning time), the duration of ultrasound treatment has been the most significant. It has been demonstrated that ultrasound treatment allows reducing the depressant consumption.

Jordan [7] has demonstrated that in an impeller less flotation machine ultrasound affects collision and attachment of particles to bubbles, bubble coalescence, and other flotation sub processes in the mixing chamber. The drawback of this design is the direct location of the vibrator in the slurry, which reduces operation reliability.

The literature review [8] show that in the most of studies carried out, whole cell volume has been exposed to the ultrasound. In fact the intensity of the oscillations drops rapidly in the pulp from the wave source. Indeed, only a weak proportion of the slurry was subjected to US treatment even in a laboratory cell as was shown by Vargas-Hernandez et al. [9].

The goal of the present research is to study the role of ultrasound treatment on the bubble-particle attachment in a cell of reactor-separator type. Reactor-separator has distinct zones for air flow dispersion into bubbles in the sparger, bubbles-particles attachment/detachment in the reactor, bubble-mineral aggregates removal into froth in the separator [10]. Thus, it possible to chose only a unique stage of flotation process to be affected by US treatment.

In other types of flotation cells, all these processes occur simultaneously in the same zone, which makes problematic the application of ultrasound treatment at a large scale. The setup used in this work allows application of ultrasound to a selected reactor zone in continuous slurry stream. It is important to note that in this case bubbles with maximum mineral load enter the ultrasound treatment.

2. Experimental

2.1. Experimental Set-up

The experimental set-up included a separation cell, a bubble-sparger and an ultrasonic emitter placed on the reactor. The principles of experimental set-up operation can be described in Figure 1. Pulp prepared after ore grinding was poured into sump (1) for reagent conditioning. At the end of conditioning step, feed pulp is pumped into the reactor (4) through bubble generator (3), creating a high-pressure jet, which disperses the forced air feed in fine bubbles. Due to high mixing velocity and a high gas/liquid interfacial area, rapid collision between bubbles and particles occurred.

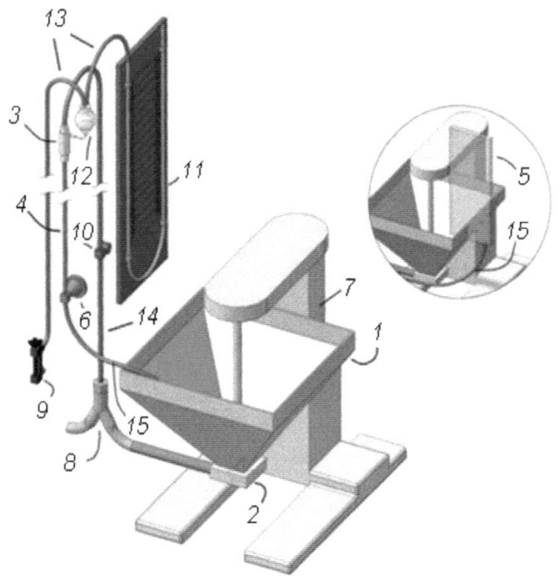

1 – sump
2 - centrifugal pump
3 - bubble generator
4 – reactor
5 - separation cell
6 – ultrasound emitter
7 – pump shaft
8 – valve
9 – air flowmeter
10 – manometer
11 – vacuummeter
12 – water trap
13 – air contour
14 – contour for slurry circulation
15 – reactor input

Figure 1. Set-up of the laboratory scale reactor-separator.

The bubble-particles attachment was provided in the reactor (4) that directs the aerated slurry into separation cell (5). Bubbles-particles attachment in the sparger or/and in the reactor zone (4) occurs in terms of particle surface hydrophobicity. The aerated mixture exits from the reactor and enters the separation cell (5). Bubbles/particles aggregates disengage from the pulp in this low turbulent zone producing two flows: floated products (sulfide concentrate) and non-floated products (tailings).

When the slurry level reaches the steady state, froth is removed and tails are recirculated to sump (1). Air flow was measured and regulated by air flowmeter (9). Pressure level at the centrifugal pump discharge was controlled by the manometer (10).

System capacity was 4.2 L/min of slurry.

An ultrasound emitter (6) is attached in the external side of reactor. Ultrasound frequency was regulated from 0 to 27 kHz.

2.2. Materials and research methods

In this work the effects of US on the flotation of a chalcopyrite/quartz mixture was studied.

Sample of 1 kg of ore has been ground in a ball mill. Solid content in the flotation slurry was 25 %.

Potassium butyl xanthate was used as collector for sulfide mineral flotation. The collector and pin oil type frother were added in the sump. Reagents dosage and particle size distribution after grinding are provided in the Table 1. The pH of slurry during conditioning and flotation steps were maintained between 8 and 8.5.

Flotation kinetics curves (mineral recovery *vs* flotation time) and a model to evaluate the experimental results were used. To obtain the flotation kinetics curves froth product has been removed in fractions for 15.5 minutes.

Table 1. Experiment conditions.

Ore type	Size (μm)	Cu content (%)	Reagent regime			
			Butyl xanthate		Frother	
			Dosage (g/t)	Conditio-ning time (s)	Dosage (g/t)	Conditio-ning time (s)
Chalcopyrite /Quartz = 1/3	86 % -0.074	8.6	150	180	100	120

Three parameters fast/slow floating particles model was used [11]:

$$R = (1 - \Phi)[1 - \exp(-K_f t)] + \Phi[1 - \exp(-K_s t)] \tag{1}$$

R is recovery at time (t), K_f, K_s are the rate constants for fast and slow components (min^{-1}), Φ is the fraction of flotation components with slow rate constant.

This equation (1) incorporates two rate terms instead of one rate constant. The model does not include an ultimate recovery parameter but rather the ultimate fractional recovery is assumed to be 1.0. When K_s parameter approaches 0, the parameter (Φ) will then represent the fraction not recovered and the term (1 - Φ) becomes analogous to ultimate recovery and the model reduces to two-parameter model.

Mineral contents in products have been measured by the method of densities (this was possible due to the difference of chalcopyrite and quartz densities):

$$\rho = \frac{m_2}{m_1 - (m_3 - m_2)} \cdot \rho_c \tag{2}$$

ρ=density of sample; ρ_c = density of water; m_1 =weight of water for a volume *Vconst*; m_2 is the weight of sample; m_3 is the weight of initial suspension (*Vsample+Vwater=Vconst*).

The recovery of chalcopyrite in the floated product was calculated from measured yield and mineral content obtained by formula 2.

3. Results and discussion

Flotation kinetics of chalcopyrite recovery is shown in Figure 2. Chalcopyrite recovery with ultrasound increased by 15-20 %, but change slightly with ultrasound frequency variation.

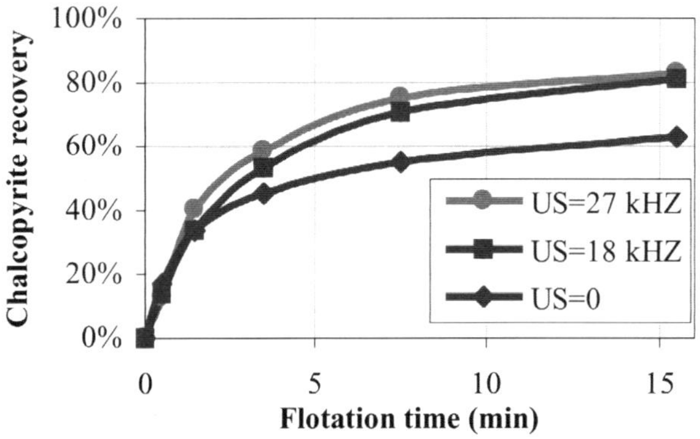

Figure 2. Kinetics of chalcopyrite recovery with ultrasound.

At the same time, yield of concentrate increased from 19.1 % (without ultrasound) to 24.8 % (with ultrasound frequency of 27 kHz). Chalcopyrite content decreased at the end of flotation: without US from 89.3 to 82.5 %, and with US 27 kHz from 93.6 to 83.5 %. Increase of chalcopyrite recovery occurred due to increase of the yield, rather than of the chalcopyrite content in concentrate.

The rate constants of flotation for fast and slow components with ultrasound frequency were calculated using Kelsall's flotation model (figure 3).

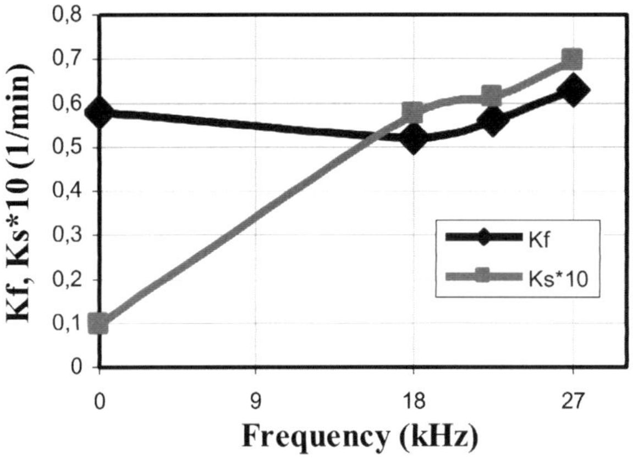

Figure 3. Rate constant of chalcopyrite flotation for fast K_f and slow K_s components *vs*. ultrasound frequency.

The rate constants increase with using of ultrasound treatment. Activation of the floatability of two considered components is noticed. The flotation rate constant of slow fraction $K_s$was multiplied by 6 with US treatment. The flotation rate of fast fraction changed slowly (by 8 %) with ultrasound: at ultrasound frequency variations, considerable changes didn't occur.

The obtained results results show clearly that US treatment can activate the floatability of slow fraction.

Increased kinetics can be explained by three hypotheses:

- the ultrasound allows a selective particle detachment from bubble [12];

- the ultrasound cavitation can modify the surface of mineral particles by micro or nano bubbles formation and facilitate the bubble-particle attachment [13].
- the ultrasound can influence on particle collisions probability.

During the passage of the ultrasonic processing zone (Figure 4), a selective quartz particles detachment from bubble is possible, that allows to clean air bubbles surface and additional chalcopyrite particles can be fixed on the cleaned bubbles surface. Increase of chalcopyrite content in the concentrate can be explained by this effect.

Increase of the concentrate yield (from 19.1 % without US to 24.8 % with US=27 kHz) and the chalcopyrite recovery in the concentrate (Figure 2) can be explained by ultrasonic cavitation. Cavitation bubbles occur mainly on hydrophobic surfaces. Therefore chalcopyrite particles can be activated by the cavitation (Figure 3).

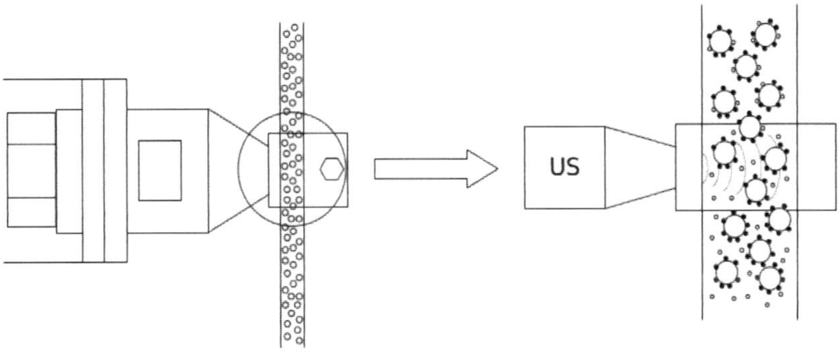

Figure 4. Scheme of flow reactor for ultrasonic activation of particle attachment/detachment sub-processes.
A selective particle detachment from bubble with ultrasound processing occurs. *Flow reactors are more efficient than classical cells because of the lower energy dissipation i.e. the whole flow is subjected to the ultrasonic activation.*

The ultrasound influence on particles collision probability is able to explain the increase of chalcopyrite recovery in the concentrate but this hypothesis needs a direct experimental confirmation.

4. Conclusions

Results show an increase of the flotation efficiency in the presence of ultrasonic processing:

- chalcopyrite recovery increases by 5-20 % with ultrasound treatment.
- an increase in flotation rate constant is obtained when ultrasound frequency varies from 0 to 27 kHz. The significant contribution to this augmentation is related with the “acceleration” of slow component flotation: the rate constant of slow component as defined in Kelsall model, increases 6 times for the ultrasound frequency of 27 kHz.
- it should be noted that the retention time in the ultrasound pan was approximately 0.1 s. Therefore, ultrasound exposure was of the pulse type and its mechanism needs to be studied in more details.
- ultrasound treatment can influence the particle detachment process exclusively. Special zone should be provided for the ultrasound treatment that can be easily designed in impeller less machine as in reactor-separator.

References

[1] Gogate P R, Taayal R K and Pandit A B 2006 *Curr. Sci.* **91 (1)**, 35

[2] Mitome H 2003 Action of ultrasound on particles and cavitation bubbles *Proc. World Congress on Ultrasonics* (Paris), 2342-2346
[3] Ozkan S G and Kuyumcu H Z 2006 *Int. J. Miner. Proces.* **81** 201
[4] Aldrich C and Feng D 1999 *Miner. Eng.* **12** 701
[5] Zaidi S A H 1997 *Fuel Process. Technol.* **53** 31-39
[6] Wenze K, Haixin X and Jun H 2008 coal *Fuel Process. Technol.* **89** 1337
[7] Jordan C E 1991 Ultrasonic flotation system *US Patent 5.059.309.*, Oct 22
[8] Ozkan SG and Kuyumcu H Z 2007 *Ultrason. Sonochem.* **14 (5)** 639
[9] Vargas-Hernandez Y, Gaete-Garreton L and Magne-Ortega L 2002 High Power Ultrasound to Recover Fine Particles in Flotation Process *Revista de Acústica* 33, Special issue
[10] Samyguin V D, Panin V V, Filippov L O and Stenin N Yu 2010 *Metallurgist*, **54**, 394-400
[11] Kelsall D F 1961 Application of probability in the assessment of flotation systems *Bull. Instn. Min. Metall.* **650**, 191-204
[12] Kondratiev S A 2005 Flotation method *Russian Patent 2243824*
[13] Zhou Z A, Zhenghe Xu, Finch J A, Masliyah J H and Chow R S 2008 *Miner. Eng.* **22** 419

Journal of Physics: Conference Series **416** (2013) 012017 doi:10.1088/1742-6596/416/1/012017

Floatability of Fe-bearing silicates in the presence of starch: Adsorption and spectroscopic studies

V V Severov[1,2], I V Filippova[1,2] and L O Filippov[1,2*]

[1] Laboratoire Environnement et Minéralurgie, ENSG-INPL, Nancy-Université, Rue du Doyen Marcel Roubault, B.P.40, F-54501, Vandoeuvre, France

[2] National University of Science and Technology "MISIS", Leninsky Prospect, 4, B-49, 119049, Moscow, Russia

E-mail: lev.filippov@ensg.inpl-nancy.fr

Abstract. Natural polysaccharides such as starch, dextrin, cellulose and their derivatives are promising non-toxic and biodegradable organic flocculants and flotation depressants. This paper presents the investigation of mechanism of adsorption of corn starch on quartz and Fe-bearing amphibole, i.e. pargasite. The direct measurement of starch adsorption on the mineral surfaces shows no difference between quartz and pargasite. However, the starch adsorption on the magnetite is more important. FT-IR spectroscopy studies reports different adsorption mechanism of starch on quartz and pargasite surface. The key changes observed in starch absorption on quartz are the major shifts in C-O stretching frequencies presumed existence of a hydrogen bond between starch and quartz surface. The similar changes were observed in this region of IR-band for pargasite. The appearance and disappearance of the bands in the region 960-920 cm^{-1} corresponds probably to formation of a new chemical bond between starch O-H groups and metal atoms on pargasite surface with formation of a surface complex. This result confirms that adsorption of the starch on the pargasite surface is droved by two mechanism. Hence, existence of strong chemical bond between starch and pargasite surface explains decrease of its floatability compared to quartz in process of iron ore flotation and forces to search new conditioning reagent modes.

1. Introduction

Selective flocculants and depressants are indispensable in any flotation reagent scheme targeting selective separation of different minerals. Natural non-toxic and biodegradable polysaccharides such as starch, dextrin, cellulose and their derivatives are widely applied as selective flocculants and depressants in froth flotation. All of these applications involve adsorption of the polysaccharides onto minerals.

Corn starch is the most used flocculant and depressant in reverse cationic flotation of iron ores due to its selectivity and availability in large amounts. As confirmed by investigations [1,2], starch adsorption is selective for quartz and iron oxides. Adsorption layer of starch on quartz is some times less dense that on hematite. N S Mikhailova [2,3] determined that starch adsorption on quartz and hematite is different not just by magnitude, but also by attachment strength. At pH 8 to 11, washing of inactivated quartz allows to remove 80% of adsorbed starch while starch retains intact on hematite surface while washing it. High attachment strength of starch on hematite proves specific interaction

Journal of Physics: Conference Series **416** (2013) 012017 doi:10.1088/1742-6596/416/1/012017

between starch and iron oxide surface take place. It is also confirmed by fact that amylopectin (starch consists two components – amylose and amylopectin) reacts so strongly with iron (III) in solution [4]. Therefore, the specific affinity of iron minerals for starch can complicate the flotation separation between iron oxides and Fe-bearing silicates as micas and amphiboles, which are presented in the Banded Iron Formations (BIF). In this paper, we report on the adsorption of starch onto magnetite, quartz and Fe-bearing amphibole, i.e. pargasite.

Several mechanisms for the adsorption of starch and related polysaccharides onto minerals have been proposed. In the early studies, hydrogen bonding was proposed as the primary adsorption mechanism of polysaccharides on mineral surfaces [1]. Hydrophobic bonding was suggested by Wie and Fuerstenau [5]. Adsorption by means of electrostatic interaction of starch with quartz and hematite surface is also discussed [1]. Later, chemical interaction was proposed as significant adsorption mechanism [4,6-8]. The aim of this study was to investigate the interaction between corn starch and pargasite, and to make a comparison with the adsorption mechanism onto quartz. This involved diffuse reflectance infrared Fourier transform spectroscopy (DRIFT) and adsorption experiments.

2. Experimental

2.1. Materials

Commercial quartz was purchased from Merck, and a sample of natural magnetite was supplied from Lebedinsky deposit, Kursk region, Russia. Pargasite sample is a natural mineral, however, its origin is not known. Quartz and pargasite samples were prepared by dry ground. The ground pargasite sample was repeatedly passed through a magnetic separator to obtain clean pargasite mineral to be used in the experiments. The ground samples of quartz and pargasite were then dry screened through 53 μm BSS sieve. The -53+10 μm size fractions were used for adsorption tests whereas -10 μm fractions, received by sedimentation, were used for DRIFT measurements. Magnetite sample was already ground and represented a magnetic concentrate of -53+10 μm size fraction. The specific surface areas, chemicals and mineralogical compositions of -53+10 μm size fraction of the samples were determined by BET method with nitrogen as the adsorbate, as well as by ICP-AES and XRD methods. All these methods are summarised in table 1.

Table 1. Characterisation of the samples.

Property	Magnetite	Quartz	Pargasite
Specific surface area (m^2/g)	0.8	1.0	1.1
Chemical analysis (%)			
Fe_2O_3	66.9		13.3
FeO	29.3		
SiO_2	2.8	100	40.5
Al_2O_3	0.1		14.1
MgO	0.2		13.0
TiO_2			1.7
CaO	0.1		11.0
Na_2O			1.9
Mineral composition	Magnetite, hematite, amphiboles, quartz	Quartz	Pargasite, chlorite

Starch used in studies is corn starch, purchased from Sigma. Its amylopectin and amylose contents are 73% and 27%, respectively. Starch is dissolved by alkali gelatinisation with adding NaOH and stored in 0.1 g/L concentration. The starch solution was freshly prepared daily to avoid the degradation. Reagent grade NaOH is used for pH adjustment. Milli-Q water is used in all experiments.

2.2. Methods

2.2.1 Adsorption tests. In these experiments, 0.5 g of mineral particles is placed in 50-mL adsorption bottles with an aqueous solutions of starch of various concentration, at pH 10. After a 12 hours of conditioning on digital horizontal shaker at room temperature, the particles were separated and liquid was centrifuged at 12000 G in order to obtain the supernatants required for the starch analysis. The starch residual concentrations in the obtained supernatants were determined by total organic carbon (TOC) analysis using the TOC-V_{CSH} Shimadzu analyzer. The TOC value is the difference between measured total carbon (TC) and inorganic carbon (IC). The starch adsorbed was determined from the difference between the initial and residual starch concentrations.

2.2.2 DRIFT spectroscopy. The FTIR spectra are obtained with a Bruker Equinox 55 spectrometer with its own diffuse reflectance attachment. Typical spectrum was an average of 200 scans measured at 4 cm^{-1} resolution. The following experiment was carried out, and corresponding infrared spectra were recorded: 0.2 g of mineral powder is placed in contact with an aqueous solution of starch, with a concentration of 0.1 g/L, at pH 10. After a 60-min agitation using a digital horizontal shaker at room temperature, the particles were separated by filtration using filter paper and dried at room temperature on air. KBr pellets were prepared, containing suitable amounts of mineral particles, which are expected to carry adsorbed species on their surfaces.

3. Results and discussion

3.1. Adsorption studies

Adsorption density of starch on magnetite, quartz and pargasite as a function of its initial concentration in the solution at pH 10 is shown in figure 1. It can be observed that adsorption density on magnetite is higher than on quartz and pargasite, both of them adsorb close amounts of starch. Maximum adsorption on magnetite reaches 6.38 mg/m^2 and it is 4.54 mg/m^2 on quartz. At concentrations below 200 mg/l, starch adsorption on magnetite and quartz is linear as indicated by the data. This confirms theory [9] that free starch molecules adsorb at active sites of mineral surface until saturation is reached.

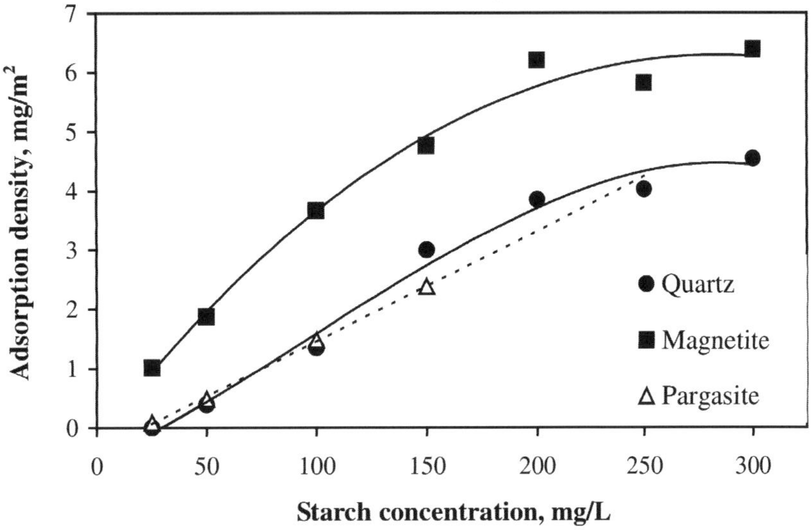

Figure 1. Starch adsorption density of magnetite, quartz and pargasite as a function of starch concentration, pH 10.

3.2. Infrared diffuse reflectance spectral studies

Analysis of IR-spectra before and after the starch treatment indicates that major changes during the treatment are found in the high-frequency area. Intensity increase and shift of the wide band centred at 3385 to 3379 cm^{-1} (-6 cm^{-1}) which corresponds to O-H stretch as well appearance of the band at 2929 cm^{-1} corresponding to C-H stretching [10] confirm starch adsorption on quartz (figure 2). The position decrease of O-H stretching band can suggest a increase in the number of hydrogen bonds. However, it is difficult to prove that hydrogen bonds are formed between starch hydroxyl groups and quartz hydroxyl groups because intermolecular hydrogen bonds (band at 3350 cm^{-1}) are already present in starch molecules. Nevertheless, the presence or absence of hydrogen bonds between starch and mineral surface can be estimated by infrared bands in the region 1080-1000 cm^{-1} associated with the C-O stretch coupled to the C-C stretch and O-H deformation of starch molecules.

Hence, the band shift from 1042 to 1033 cm^{-1} and a certain band deformation presumes existence of a hydrogen bond between starch and quartz surface. No other changes of IR-bands were observed, although a weak band at 800 cm^{-1} can be detected corresponding to O-H bending in silanol groups [11,12].

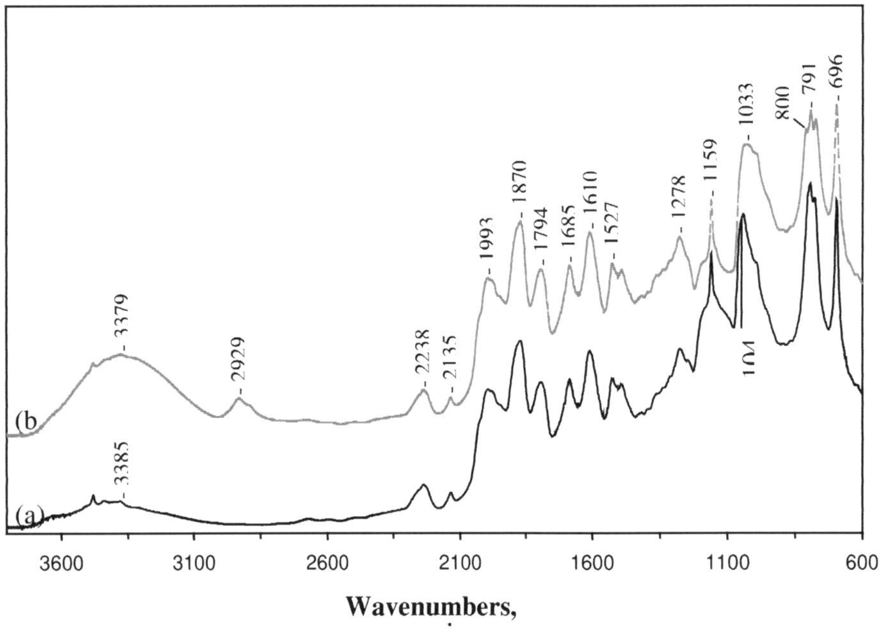

Figure 2. DRIFT spectra of (a) quartz and (b) starch adsorbed on quartz.

Starch adsorption on pargasite is confirmed by the reduction of O-H stretching frequency from 3552 to 3546 cm^{-1} (-6 cm^{-1}) and from 3408 to 3394 cm^{-1} (-14 cm^{-1}), increase of O-H deformation frequency for water molecules from 1637 to 1647 cm^{-1} (+10 cm^{-1}) as well as appearance of adsorption bands for C-H stretching at 2932 cm^{-1}, O-H in-plane deformation at 1419 cm^{-1} coupled with C-H deformation at 1329 cm^{-1} (figure 3). The appearance of the band at 953 cm^{-1} and its disappearance at 924 cm^{-1}, correspond probably to formation of a new chemical bond between starch hydroxyl groups and metal atoms (Fe, Al, Ca) on pargasite surface with formation of a surface complex [4,6,7].

Certain band shift from 1012 to 1005 cm^{-1}, corresponding to C-O stretch coupled with C-C stretch and O-H deformation, may point to the presence of a hydrogen bonds between starch hydroxyl groups and surface hydroxyl groups, such as silanols, unable to participate in surface complexation.

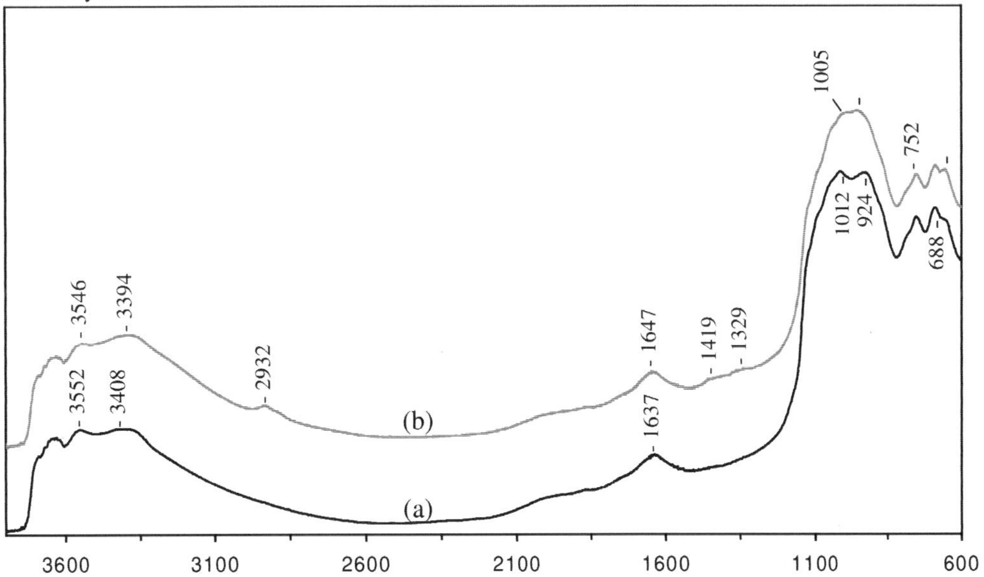

Figure 3. DRIFT spectra of (a) pargasite and (b) starch adsorbed on pargasite.

4. Conclusions

Adsorption and infrared spectroscopy studies have shown that starch is adsorbed both on quartz and pargasite. The main band shift in the region of 1080-1000 cm^{-1} for quartz shows that starch attaches to the surface by means of hydrogen bonds between starch hydroxyl groups and surface silanols of quartz. In case of pargasite, two starch absorption mechanisms are possible: by means of hydrogen bonds with the surface silanols of amphibole as well as by means of formation of chemical complex with metal atoms on the surface.

Hence, existence of strong chemical bond in comparison with hydrogen bond between starch and pargasite surface explains decrease of its floatability compared to quartz in process of iron ore flotation and forces to search new conditioning reagent modes.

References

[1] Balajee S R and Iwasaki J 1969 *Trans. SME/AIME* **244** 401
[2] Mikhailova N S 1972 *Mineral Processing* **6** 20
[3] Mikhailova N S 1973 *Mineral Processing* **1** 21
[4] Weissenborn P K, Warren L J and Dunn J G 1995 *Coll. Surf., A. Physicochem. Eng. Asp.* **99** 11
[5] Wie J M and Fuerstenau D W 1974 *Int. J. Miner. Process.* **1** 17
[6] Khosla N K, Bhagat R P, Gandhi K S and Biswas A K 1984 *Colloids Surf.* **8** 321
[7] Liu Q and Laskowski J S 1989 *Int. J. Miner. Process.* **27** 147
[8] Raju G B, Holmgren A and Forsling W 1997 *J. Colloid Interface Sci.* **193** 215
[9] Khosla N K and Biswas A K 1984 *Colloids Surf.* **9** 219
[10] Brinker J and Sherer W 1990 *Sol–Gel Science: the Physics and Chemistry of Sol–Gel Processing* (New York: Academic Press) pp 515–609
[11] Morrow B A and McFarlan A J 1992 *J. Phys. Chem.* **96** 1395
[12] Lang S J and Marrow B A 1994 *J. Phys. Chem.* **98** 13314

Stress-time context of fault permeability at the Krasnokamensk Area SE Transbaikalia

V.A. Petrov[1], M. Lespinasse[2], V.V. Poluektov[1], M. Cuney[2], R.M. Nasimov[3], J. Hammer[4], S.I. Schukin[5]
[1]Institute of Geology of Ore Deposits, Petrography, Mineralogy and Geochemistry (IGEM) Russian Academy of Sciences, Staromonetny per. 35, Moscow, 119017 Russia
[2]UMR G2R (7566), Nancy-Université, BP 239, 54506 Vandœuvre-les-Nancy, France
[3]Institute of Physics of the Earth, Russian Academy of Sciences, Bol'shaya Gruzinskaya ul. 10, Moscow, 123995 Russia
[4]Bundesanstalt für Geowissenschaften und Rohstoffe (BGR), Geozentrum Hannover, Stilleweg 2, D-30655 Hannover, Germany
[5]Priargunsky Industrial Mining and Chemical Operations (PPGKhO), Krasnokamensk, Zabaikalskiy Region, 674665 Russia

Email: vlad@igem.ru

Abstract. The main aim of the contribution is to combine data on the consecutive stages of deformation, inflow and migration of palaeofluids and accumulation of mineral filling with uranium traces within the faulted-fractured environment at the Krasnokamensk Area, SE Transbaikalia, Russia. Object of examination is a framework of fault zones transecting the Proterozoic-Paleozoic granitic unit to the extent of northwestern part of uranium-bearing Streltsovskaya caldera of Mesozoic age. Considerations of stress- and permeability-time relationships in faulted-fractured zones were taken with account of stress and strain dependencies within fluid saturated rock massifs at crustal seismogenic level. Stress-time consecution of fault zone permeability was developed using set of fieldwork and lab tests including structural-geological survey, fault slip data analysis, mineral-chemical diagnostics, microstructural observations, and radiographic studies. Practical applications of obtained data for solving uranium mining and environmental issues are indicated in conclusion.

1. Introduction

Crystalline rocks, such as granites, are hosted economic uranium deposits and practically viewed as suitable settings for constructing underground facilities with a view to spent nuclear fuel (SNF) long-term storage or high-level wastes (HLW) disposal. The evaluation of mineral resources and safety of the facilities directly depends on the identification of the most probable pathways of fluid flow in connection with spatial-temporal evolution of radionuclide transport conditions. Some aspects of tectonic, geochemical and microstructural context of uranium migration in granites of the Krasnokamensk Area were distinguished previously[1]. There were studied three specific sites located at different places and elevation outside (SNF storage potential Sites No. 1 and No. 5) and inside (the Antey uranium deposit) the Streltsovskaya caldera. It was shown that time scale context has crucial importance for modeling of the main features, events and processes existing at different geometric scales arranged in 3D space of variously deformed and alterated fractured porous granites. This context was studied by the example of fault zone framework at the Proterozoic-Paleozoic granitic unit (potential Site No. 1) exposed within the northwestern part of the caldera.

In the general case, it is suggested[2] that the deformation in fluid-saturated rocks develops in the framework of a seismic cycle consisting of four stages (Fig. 1): (i) preseismic, when stresses accumulate steadily against the background of nonlinear deformation of rocks; (ii) coseismic, when deformation occurs immediately after relaxation of stresses (an earthquake); (iii) postseismic, with nonlinear deformation of rocks during some time after stress relaxation; (iv) interseismic (the stage of seismic quiescence), when the deformation is close to linear.

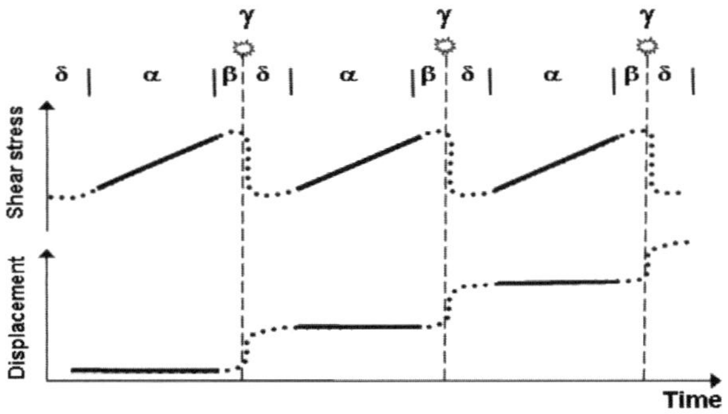

Fig. 1. Hypothetical stress- and displacement-time relationships into the fault zones of seismogenic crustal level[3]. Four stages of seismic cycle: α – interseismic, β – preseismic, γ – coseismic (earthquake), δ – postseismic.

At the interseismic stage, the fault cores undergo less dilatancy and are less permeable in comparison with zones of their dynamic effect, where fluids accumulate and slowly diffuse. At the coseismic stage, a considerable volume of these fluids is squeezed out from the compressed fractures and leaks into the disturbed core of the fault[4], where favorable conditions are created for drainage and circulation of solutions and precipitation of solid matter. Against this background when in a local domain of the geological medium the fluid pressure approaches $P_f=\sigma_e+T_S$ (σ_e – effective stress, T_S – tensile strength) the tensile cracks develop along heterogeneities (plane $\sigma_1\sigma_2$) favorably oriented in the stress field. The spatiotemporal cyclic relationships between pore pressure and favored paths of fluid migration are considered in the fault valve (pumping) model[5,6], which causes the stage-by-stage formation of mineral assemblages in the process of solid matter deposition. The efficiency of the fault valve mechanism of mineral formation is determined by the geometry of pore-fracture space, stressed-strained state of rocks, physicochemical parameters (composition) of solutions and their equilibrium with host rocks.

The first successful attempts to link hydraulically active fracture formation and fluid flow state during tectonic events were made in the University Nancy-1[7]. It was suggested to use fluid inclusions in minerals (mainly quartz) as markers of thermobaric and physicochemical conditions of fluid migration, and fluid inclusion plains (FIPs) as structural indicators of changes in geometry of fluid-conducting fractures and geotectonic setting of permeability. FIPs start to form as tensile microcracks which change orientation in response to rearrangement of the stress field. They are oriented perpendicular to the least compression axis σ_3 while vector of maximum permeability lies in plane $\sigma_1\sigma_2$ forming direction of dominant fluid migration. The new tectonic setting gives rise to a new stage of deformation that is inevitably reflected in the orientation of FIPs of the second generation, etc.

Hence, three complementary aspects were taken into account during consideration of stress- and permeability-time relationships in fault zones of the area under investigation:

(a) Deformation of fluid saturated rock massifs within crustal seismogenic level is essentially non-linear[3] during tectonic events and variation of stressed-strained states;

(b) There is coupling between main phases of fluid inflow into and mineral accumulation inby different domains of fault zones due to fault valve (pumping) effect against the backdrop of seismic deformation cycle leads to local mixture of hydrothermal and meteoric solutions;

(c) Local mineral associations into fault zones do not have univocal correlation due to various *P-T* conditions in distant domains (dilational jogs vs frictional fault planes) along even single fault zone.

2. Methods and Results

The SNF storage potential Site No. 1 occupies ~45 km^2 at the northwest part of the Streltsovskaya caldera where 20 outcrops (reference points) were investigated. The fieldwork procedures were realized as follows: (1) Structural-geological and mineral-petrographic survey of the main fault zone domains, their architecture, morphology, component strikes, dips, mineral filling, etc; (2) Determination of the mean principal stresses orientation, fault kinematics (dextral or sinistral displacements), and faulting regime (normal, strike-slip, and thrust) using fault slip data analysis; (3) Sampling of positioned lumps for studying rock mineral-chemical composition, FIPs characterization and fission-track radiography (FTR) analysis. The lab tests included: (1) Optical microscopy and mineral-chemical studies (XRF and ICP-MS) of rock varieties, nature and intensity of hydrothermal-metasomatic alteration and oxidizing transformation; (2) Microfracture space digitalization for distinguishing stress dependent FIPs and open micro discontinuities; (3) Evaluation of content and specific features of uranium distribution in rock varieties using FTR data. Some structural, mineral-chemical, and radiographic data are mentioned in[1]. Hereafter we would like to focus on stress-time context of fault permeability at the area under investigation.

This area is situated within the Archaean-Proterozoic anticline that was repeatedly intruded by magmatic melts during Late Proterozoic with formation of a granitic gneiss cupola subjected to subsequent granitic injections during Caledonian and Variscan tectonic-magmatic cycles (TMC). Later on, during Early Jurassic to the Early Cretaceous period of regional tectonic-magmatic activation (TMA) the central part of the cupola underwent subsidence due to eruption of a large volume of acid volcanic products from several palaeovolcanoes, devastation of the magmatic chamber and, as a result, formation of the piecemeal-type Streltsovskaya caldera with three main units of volcanic and sedimentary rocks. Concerned granitic Site No. 1 situates at the NW flank of the caldera and its tectonic position is determined by intersection node of the NE-SE, NNE-submeridional, NW-SE and W-E long-lived interblock regional fault zones.

Five elements of internal constitution of the Site No. 1 are identified: (1) original magmatic foliation of granitic rock; (2) NE-SW fault zone of initial ductile deformation (schistosity); (3) NNE-submeridional fault zone of semibrittle deformation; (4) NE-SW, NNE-submeridional, and NW-SE normal and strike-slip faulted-fracture zones; (5) steeply- and gently dipping fractured zones of various strike.

Contour, dimensions and inner structure of granitic massif are defined using strike and dip components of original magmatic foliation: in latitudinal cross-section this is a cupola with steeply-dipping limbs and saddle shaped central part; in meridional cross-section this is a buckle folding structure results from lateral compression. The main peculiarity of the NE-SW fault zone of about 10 m width is ductile rock deformation stated in schistosity, blastic elongation of quartz and feldspar grains into the mylonite quartz-chlorite-carbonate matrix. The width of NNE-submeridional fault zone of semibrittle deformation is about 20 m. The main peculiarities of the zone are several sutures (about 25 cm width) with cataclastic material of quartz-feldspar composition, crush breccias and some necking stage boudinage. Normal and strike-slip faulted-fracture zones of NE-SW, NNE-submeridional, and NW-SE directions consist as a rule of one suture (about 1 m width) with precise striation (indicators of displacement) at the planes and cataclastic material (gouge) as fault filling. The main fault planes are accompanied by veins of milky-white quartz (about 5 cm width) and tensile fractures with quartz-carbonate aggregates.

Reconstruction of geodynamic circumstances throughout Pre-Mesozoic and Mesozoic tectonic-magmatic events shows that activation (reactivation) of interblock regional fault zones occurred on a step-by-step basis during succeed TMCs: Proterozoic (terminated about 600 Ma), Caledonian (~520-430 Ma), Variscan (~360-210 Ma), and Late Mesozoic (~150-100 Ma). Preferred faulting regimes of TMCs were indicated using fault slip data analysis[8] combined with paleogeomorphological analysis[9] and data on relative age of fault infill.

It was founded that magmatic body formation took place during Proterozoic TMC in regime of general tension with complementary component of submeridional compression. The main indicator of this regime is original magmatic foliation of the granitic rocks. During Caledonian TMC the principal axes of stress σ_1 and σ_3 were subhorizontal while the principal axis of stress σ_2 was oriented in subvertical direction (Fig. 2). This is characteristic feature of the strike-slip faulting regime. This regime was gradually changed by normal faulting regime with ENE-WSW direction of general tension axis during Variscan TMC. However at the time border between Paleozoic and Mesozoic eras, probably during T_3-J_2 time[10], all geological structures in SE Transbaikalia underwent inversion of tectonic movements due to global reorganization of the stress-strain field. In this connection the strike-slip faulting regime (transpression, $\sigma_v=\sigma_2\approx\sigma_3$) prevailed. As a result most intensive strike-slip displacements and shearing were developed along NNE-submeridional and WNW- latitudinal fault zones.

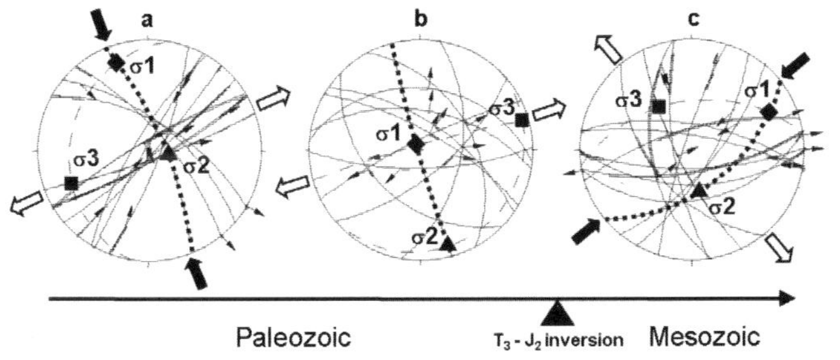

Figure. 2. Dynamics of faulting regime changes at the NW granitic frame of Streltsovskaya caldera. Stereograms (L-hemisphere) present projections of 38 fault planes, direction of displacement (thin arrows) and averaged course of stress principal axes ($\sigma_1 \geq \sigma_2 \geq \sigma_3$) during Caledonian (a) and Variscan (b) TMC, and Late Mesozoic TMA (c). Dominant directions of fluid flow (planes $\sigma_1\sigma_2$) are shown as dotted lines.

In the tectonic context it is very important to define orientation in space the $\sigma_1\sigma_2$ plane, which assigns direction of prevailing fluid flow and migration of the matter along pathways. The dynamics of fluid permeability of the fault zones was reconstructed using spatial distribution and orientation of FIPs (the AnIma video screen method[11]) in connection with data on faulting regimes. Comparative analysis shows that orientation of FIP and fault networks coincides. That denotes the unified mechanism of their formation during different tectogenenic events. Three groups of FIPs prevail: NE-SW, NNE-submeridional, and NW-SE (Fig. 3). Herewith the NE-SW-trending FIPs form considerable cluster within the NE-SW fault zone of initial ductile deformation (schistosity). Preliminary microthermometric measurements show that constituent fluid inclusions are characterized by increased salinity that affirmed data on ancient origin of this zone.

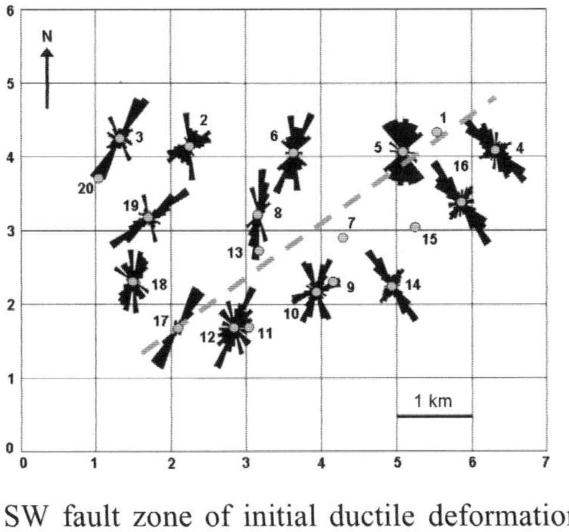

Figure 3. Rose-diagrams of FIPs orientation within the Site No. 1 at the NW granitic frame of Streltsovskaya caldera. 20 outcrops (reference points) and NE-SW fault zone (dotted line) of initial ductile deformation (schistosity) are shown.

Intending changes in faulting regime, dominant directions of fluid flow at fault scale (planes $\sigma_1\sigma_2$) and relationships between rock mechanical basis of discontinuity formation at macro and micro scale, it is possible to conclude that emphasized groups of FIPs were formed throughout different tectonic-magmatic cycles. For instance, NE-SW-trending FIPs were fashioned during Proterozoic TMC (FIPs 1) into the NE-SW fault zone of initial ductile deformation (schistosity) as well as during Late Mesozoic regional TMA (FIPs 4) inside the reactivated brittle segments of the zone. NNE-submeridional and NW-SE-trending FIPs were gradually formed most likely during Caledonian (FIPs 2) and Variscan (FIPs 3) TMCs due to planes $\sigma_1\sigma_2$ were positioned along these directions.

Integrated consecution of the faulting periodicity and hydrothermal events at the NW granitic frame of the caldera is presented at stress-time diagram (Fig. 4). The periodicity of structural, deformation and hydrothermal processes coincides on a first approximation with stages of the seismic cycle when long-continued period of the stress accumulation (I) concurs with interseismic stage while relatively compact sequence of stress relaxation (II) accedes with preseismic, coseismic, and postseismic stages. The level of structural destruction (dotted line in Fig. 4) is governed by magnitude of compression, tension and shearing stress, ultimate yield and shear strength of rocks, fluid pressure and thermal gradient.

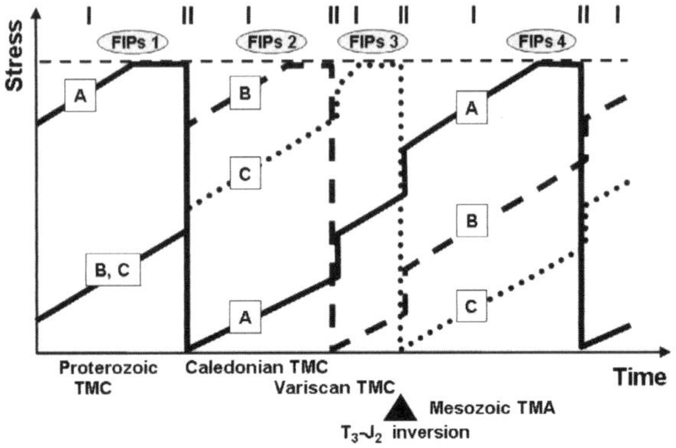

Figure 4. Stress-time dependence of fluid permeability for the fault zones: A – NE-SW, B – NNE-submeridional, C – NW-SE. Periods of stress accumulation (I) and relaxation (II) are accompanied by inflow of multiple-aged fluid portions committed to FIPs generations (from 1 to 4) during various tectonic-magmatic cycles (TMC) and regional tectonic-magmatic activation (TMA).

To reveal linkage between fracture repartition and uranium distribution in spatial-temporal context the fission-track radiography technique was utilizes (Fig. 5). Obtained data clearly indicate mobilization of significant part of uranium into the shearing core of NE-SW fault zone of initial ductile deformation (T17b) and NNE-submeridional zone of oblique-slip movements (T4) while outside the master planes of the faults (T20) the uranium distribution is in stockwork-like manner.

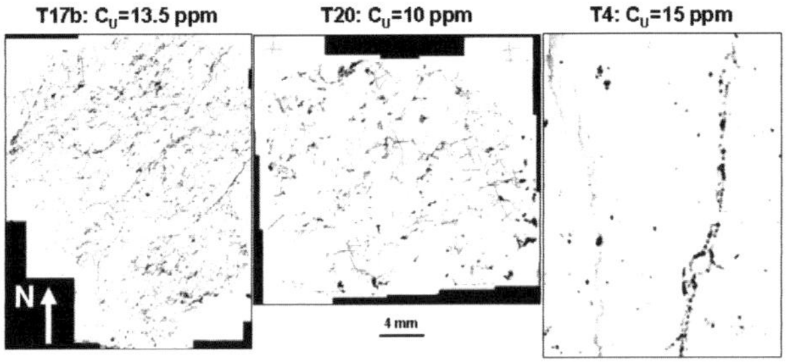

Figure 5. Uranium distribution patterns for variously deformed granitic rocks from reference points (T). Uranium content (C_U) is measured using ICP-MS (CRPG, Nancy).

The main concentrators of uranium are accessory and mafic minerals, and in a less degree, thin veinlets along grain boundaries. Intensive accumulation of uranium (probably in sorbed form) in oxidizing surface conditions is connected with Fe-Mn and Ti oxyhydroxides which fill microfracture networks.

Conclusions

Thus, the example of granitic environment of the Krasnokamensk Area shows long-term circulation of uranium-bearing fluids within the fault zones from Late Proterozoic and Paleozoic to Mesozoic and probably Cenozoic. The periodicity of the uranium-bearing hydrothermal solutions inflowing into the fault zones is connected with alternative repulsion phases of structural and deformation processes during tectonic-magmatic cycles coupled with preseismic, coseismic, postseismic and interseismic stages of the seismic cycle. For considering the stress- and permeability-time relationships for faulted-fractured zones a set of fieldwork and lab tests were used including structural-geological survey, fault slip data analysis, mineral-chemical diagnostics, microstructural (FIPs) observations, and radiographic studies. This approach allowed us to suggest the following results and practical applications:

(1) Four main geodynamic cycles characterizing by inherent stressed-strained state of the rock massifs were identified namely Proterozoic, Caledonian, Variscan, and Late Mesozoic. The generation and activation (reactivation) of interblock regional fault zones occurred on a step-by-step basis during these cycles.

(2) The main phases of the hydrothermal solution inflow to the optimally oriented segments of the fault zones accede with coseismic and postseismic stages of the seismic cycle while durable processes of hydrothermal-metasomatic alteration and mineral deposition concur with interseismic stage of transient and low-amplitude displacements.

(3) Intersection nodes of the NE-SW fault zone of initial ductile deformation and the NNE-submeridional oblique-slip zones form areas of long-term volumetric circulation of uranium-bearing solutions. Here there are prerequisites for revealing Proterozoic-Early Paleozoic vein-disseminated uranium mineralization modified, probably, during Variscan episode of the caldera basement formation and/or during Mesozoic regional TMA.

Practical applications of obtained data on stress- and permeability-time relationships into the fault zones for solving uranium mining and environmental issues at the Krasnokamensk Area are as follows:

(1) Consideration of Pre-Mesozoic, Mesozoic, Cenozoic and current faulting regimes and stressed-strained state of the granitic massifs allows us to develop conceptual and numerical rock mechanical models for localization the rock fall critical areas and organization of the mining operational safety monitoring.

(2) Definition of intersection nodes of the regional long-lived and multiply activated fault zones allows us to identify the areas for long-term volumetric circulating of uranium-bearing solutions and thus to identify localities for further surveying with the aim of reconnaissance the uranium mineralization which is more ancient than Mesozoic (135 Ma) economic ores of the Streltsovskaya caldera.

(3) Detection of hydraulically inactive blocks outside the intersection nodes of the regional fault zones allows us to recognize localities most favorable for constructing underground facility with a view for SNF long-term storage into the granitic environment.

Acknowledgements
We gratefully acknowledge our colleagues form IGEM RAS, Moscow, CRPG, Nancy and BGR, Hannover for the XRF and ICP-MS measurements as well as for comments and review of the contribution. This study was supported by the Presidium of the RAS fundamental research program No. 15, projects 02.740.11.0049 and NSh-4458.2008.5, grant of PPGKhO No. 12238, CNRS project No. 19079 and ARCUS I-II.

References

1. Petrov V A, Lespinasse M, Poluektov V V, Schukin S I, Golubev V N, Cuney M 2008 Tectonic, geochemical and microstructural context of uranium migration in granites of the Krasnokamensk area, Russia. *Proc. NAMES-2007 Seminar*, Metz, France, pp. 115-118.
2. Scholz C H 1990 The Mechanics of Earthquakes and Faulting. *Cambridge Univ. Press*, Cambridge.
3. Sibson R H 1986 *Ann. Rev. Earth Planet. Sci.* **14** 149.
4. Wood Muir R 1994 Earthquakes, strain cycling and the mobilization of fluids. In: Geofluids, Origin, Migration and Evolution of Fluids in Sedimentary Basins. *Geol. Soc. Spec. Publ.*, **78** 85.
5. Cowie P A 1998 *J. Struct. Geol.* **20 (8)** 1075
6. Nguyen P T, Cox S F, Harris L B, Powell C McA 1998 *J. Struct. Geol.* **20 (12)** 1625
7. Lespinasse M, Cathelineau M 1990 *Tectonophysics* **184** 173
8. Angelier J *J. Geophys. Res.* N 89 (B7). 1984. P. 5835-5848.
9. Mercier J L, Carey-Gailhardis E, Sebrier M 1992 *Paleostress determinations from fault kinematics: application to the neotectonics of the Himalayas-Tibet and the Central Andes.* In: R.B. Whitmarsh, M.H.P. Bott, J.D. Faihead, N.J. Kusznir (Eds.). Tectonic Stress in the Lithosphere. *Philos. Trans. R. Soc. London.* **337** pp. 41-52.
10. Delvaux D, Moeys R, Stapel G, Melnikov A, Ermikov V 1995 *Tectonophysics* **252** 61
11. Lespinasse M, Désindes L, Fratczak P, Petrov V **2005** *Chemical Geology Spec. Issue.* **223** 170

Journal of Physics: Conference Series **416** (2013) 012019 doi:10.1088/1742-6596/416/1/012019

Reflectometry simulation as a tool to explore new schemes of characterizing the fusion plasma turbulence

S. Heuraux[1], F. da Silva[2], E. Gusakov[3], A. Yu Popov[3], N. Kosolapova[3], K. V. Syisoeva[3]

[1]Institute Jean Lamour UMR 7198 CNRS-Université Henri Poincaré BP 70239 F-54506 Vandoeuvre cedex, France

[2]Associação EURATOM/IST, Instituto de Plasmas e Fusão Nuclear- Laboratório Associado, 1046-001 Lisboa, Portugal

[3]Ioffé Institute, St.Petersburg 194021, Russia

E-mail: stephane.heuraux@lpmi.uhp-nancy.fr

Abstract. A first step towards the measurement of turbulence characteristics or transient events required for the understanding of turbulent transport is to build an interpretative model able to link the measurements of a given diagnostic to a wanted parameter of the turbulence, and simulation helps us to do that. To obtain density fluctuation parameters in fusion plasmas, microwaves can be used. However, the interpretation of the received signals requires generally sophisticated data processing to extract an exact evaluation of the wanted parameters. Simulations of electromagnetic wave propagation in turbulent plasmas permit to identify the main processes involved in probing wave-fluctuations interaction and the reflectometry signature of the expected events, which gives ideas to model them. It is shown here how simulations have permitted to exhibit the role of resonances of the probing wave induced by turbulence and to explain part of phase jumps seen during reflectometer measurements. The multi-scattering phenomena can be modelled by a photon diffusion equation which can be used to provide information on the turbulence at density fluctuations levels higher than allowed by usual methods. The reflectometry simulations show that at high level of turbulence a competition between the resonances generation mechanism, able to maintain the probing depth, and the Bragg backscattering exists. Its consequences on turbulence characterisation are discussed.

1. Introduction

The knowledge of turbulent transport is one of the key issues for the future of fusion plasma devices. Reflectometry is a versatile microwave diagnostic, suited for ITER, also able to provide qualitative and quantitative measurements of density fluctuations [1]. X-mode fast sweep reflectometry is an useful tool to obtain the entire density profile and to provide information on the plasma fluctuations, assuming the magnetic field structure is known [2]. Up to now, fixed frequency reflectometry has been traditionally devoted to density fluctuation measurements assuming that the main contribution comes from the vicinity of the cut-off layer for a given k-spectrum, which can be approximated using a single wavenumber value in the relationship between phase variation amplitude and density fluctuation level. This effective wavenumber value is given in an ad hoc way [1]. Simulations have permitted to interpret the fluctuation reflectometry signals and thus to obtain the radial density fluctuation profile without the ad-hoc choice [3].

The software developed for reflectometry simulations can be used to recover and to explain the frequency spectrum associated to the Alfven cascades seen with fixed frequency reflectometers working without cut-off layer in the probed tokamak plasma [4]. New possible ways to extract the properties of a turbulent wavefront and its dynamics are explored using the ultra-fast frequency sweep reflectometer [5] and for a reflectometer working as a backscattering diagnostic at fixed frequency. For this scattering diagnostic, the basic mechanisms, with a highlight on the forward scattering, and the technical requirements will be found in [6]. All these studies, helpful to enhance the interpretation of reflectometry signals are impossible until now without computation, especially for studies on the non-linear responses induced by high level of turbulence [7]. The meaning of non-linear responses corresponds to cases out of the scope of Born's approximation for which few analytic expressions for reflectometry exist [8]. Born theory is the starting point for an explanation of the enhanced scattering existing in an artificial resonant cavity and the background needed to understand the mechanisms able to generate resonances in a turbulent plasma. The aim of this paper is to extract from the simulations, based on the Helmholtz' equation or/and wave equation or/and Maxwell's equation solvers [9], shown here the key points for a relevant theoretical description. For example (a) on Fig.1 the Bragg backscattering dominates and forbids the probing wave to reach the cut-off layer, (b) on Fig.2 the probing electric field exhibits resonances that permit the probing wave to go through the plasma despite the thickness of the probing zone and the Bragg backscattering. These described configurations can appear in any plasma except for the presence of resonances which requires a typical size needed to fulfil the resonance conditions. Looking at the peak widths shown on the Fig.2, gives an idea on the requirements to build a local amplification of the probing wave.

Next paragraph is devoted to the resonance modelling and the last one is a discussion on the competition between multi-scattering, seen as a photon diffusion process, and the Bragg backscattering and the possible ways to model the wave propagation in inhomogeneous turbulent plasma for reflectometry, that is to say, to have the tools for strong turbulence characterization.

2. Description of Resonances

A first theoretical study has been established to describe the amplification of the probing wave induced by a coherent density perturbation [8]. This theory requires that the spatial scale, on which the probing wave is amplified, is greater than the density fluctuation wavelength. This intermediate spatial scale, which can be smaller than the radial width of the density fluctuations, is greater than the density fluctuation wavelength, which is on the order of the local value of the probing wavelength due to the Bragg's rule, and smaller than the density gradient length. This constraint forces to study rather big plasmas ($L > 50\lambda_o$). Using this theory it is possible to predict in the framework of Born's approximation that the phase variations induced by very small perturbations are enhanced by the square of amplification factor, compare to the those obtained without amplification of the probing electric field [10]. Recently we have shown theoretically and numerically that this amplification is connected to an enhanced scattering process, looking at the scattered field induced by a tiny density perturbation (see Fig.3). The fact that the amplitude of scattered fields increases linearly with the amplitude of the density perturbation corresponds to a standard response deduced in the framework of Born's approximation. When the saturation is reached (above Born's approximation), the probing wave structure begins to be modified. In the resonant case when the probing electric field is amplified by a factor α, the saturation arises earlier as function of the amplitude of the density perturbation. Compare

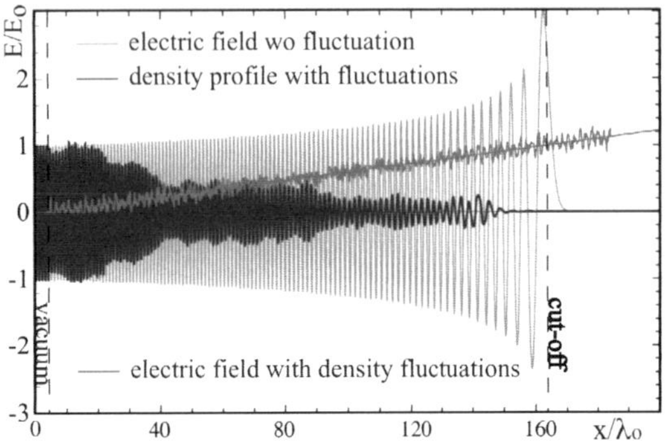

Figure 1. Case with cut-off : normalized probing electric field as a function of the plasma radius with (blue) and without density fluctuations (light blue). The linear density profile normalized by cut-off density n_c (magenta) on which a homogeneous density fluctuation level of 0.05 n_c (RMS values) is superimposed.

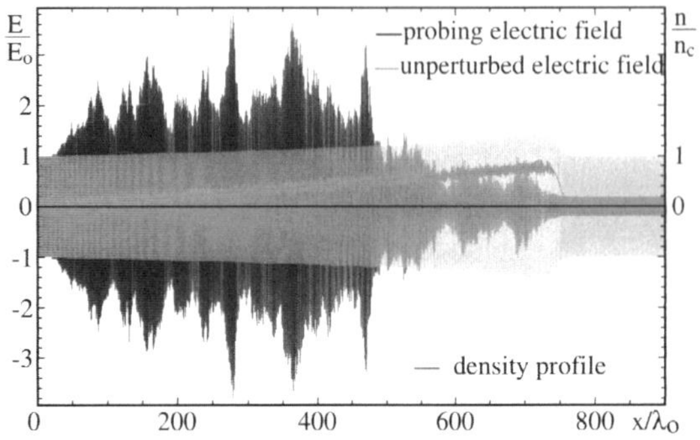

Figure 2. ITER size case without cut-off: normalized probing electric field as a function of the plasma radius with (blue) and without density fluctuation (light blue). The density profile normalized by cut-off density n_c (magneta) on which a homogeneous density fluctuation level of 0.05 n_c (RMS values) is superimposed.

to the non-resonant case, the ratio of amplitude of probed density perturbation resonant over non-resonant situations is also equal the factor α to have the same value of the scattered power. This behaviour of the scattered field evolution following a straight line is observed while Born's approximation stays valid. The theoretical work on this subject is under publication processes. It is interesting to notice that, at higher density perturbation amplitude, the maximum value of the amplification factor is shifted compared to the maximum value of the scattered field due to a non-linear phase shift, which changes the conditions of the resonance of the probing electric field between the density perturbation and the cut-off in which the probed density fluctuations exist. However, at very high density, the probed density fluctuations are able to fulfil the resonance by itself and to generate a second cavity. So the notion of the scattered field is lost in this case but also in the case of the probing wave fully Bragg backscattered, considered here as reflected wave, before reaching the density fluctuations region. The mechanism able to generate local resonances, using a set of coherent density perturbations, described analytically in [8], exhibits an intermediate spatial scale defined localy at x_j by $L_{Bragg j} = |dk/dx|_{x=x_j}^{-1/2}$ needed to have a significant Bragg backscattering to build a cavity in which the probing wave is trapped.

The mechanism giving a local amplification of the probing wave is associated to the multi-scattering effects. However a phase relationship between trapped waves is needed to fulfil the resonance conditions corresponding to constructive interferences between the density perturbation and the cut-off layer. In Fig.2, such resonances are shown in the case of a homogeneous turbulence all over the probing zone. Looking at the electric field structure in the resonance zone one can see that the wave has a coherent structure. A small variation of the phase between the modes used to compute the turbulence destroys the resonance but this can also generate another one elsewhere. In the following paragraph we show that the average of the root mean square (RMS) of the square electric field provides information on wave propagation in turbulent plasmas and permits to know if the probing wave is able to reach the cut-off layer or to diagnose all of the plasma.

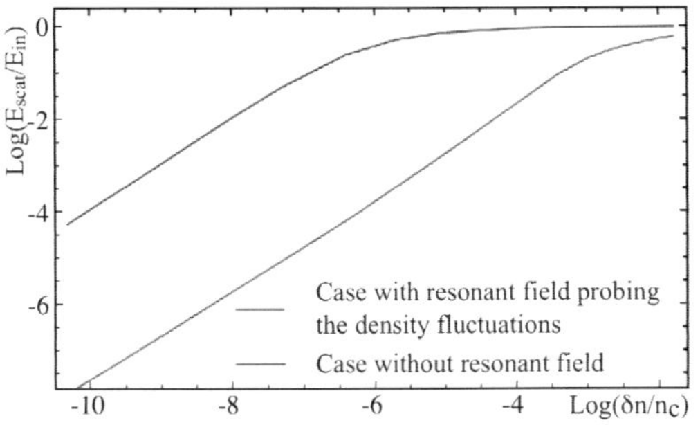

Figure 3. Logarithmic evolution of the field scattered by a Bragg resonant coherent perturbation in a resonant electric versus the Logarithm of the amplitude of the density fluctuation normalized by n_c (blue) and in Airy's like electric field (case without amplification, no Bragg resonant density perturbation) (red). The resonant field is obtained, using a Bragg resonant perturbation with $\delta n/n_c = 5\%$ in a linear density profile with $L_n = 160\lambda_o$

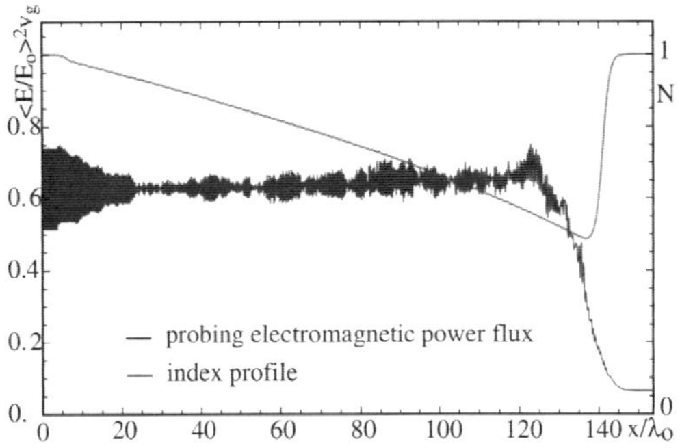

Figure 4. Averaged probing electromagnetic flux over 10^4 samples of density fluctuations profile, versus the plasma position (light blue). The index of the truncated linear density profile with $L_n = 160\lambda_o$ normalized by cut-off density n_c (magenta) over which a homogeneous density fluctuation amplitude of 0.05 n_c (RMS values) is superimposed.

3. Modelling and Discussion

A look at Figures 1-2 shows that various structures (exponential decrease followed by local amplification or global amplification with local amplifications) can be obtained in the case of intermediate values, which corresponds to amplitudes above Born approximation without secondary cut-off as defined in [8], of the homogeneous turbulence covering all the probing zone. At this moment the only ways found to extract information from simulation results is to make an average of the square of the electric field, or more precisely, of the electromagnetic flux of the probing wave (see Fig.4) or to use an iterative method [11]. This average value seems to be constant for intermediate turbulence levels (up to 5-10%, depending on plasma parameters). For higher turbulence levels, the probing wave is unable to reach the cut-off layer (see Fig.5-6). The probing width decreases as a function of the RMS value of density fluctuations. This suggests that a competition exists between two physical processes, the first one is clearly the Bragg backscattering and the second one is associated to multi-scattering effects able to generate resonances. The multi-scattering process can be seen as a photon diffusion process, and it has been shown that the solutions for the electromagnetic (EM) flux correspond to constant EM flux [12]. This modelling finds again the results obtained with full-wave simulations for intermediate values of turbulence levels. This is no longer valid for higher values of turbulence (see Fig.5 and 6). To extract the average properties of the turbulence (amplitude, wavenumber spectrum,...) we need to have a significant averaging able to produce results for which the random phase assumption becomes relevant to remove coherency induced by resonances in EM flux. To establish the number of samples needed to fulfil the random phase assumption, we need to note that the resonances give localised coherent structures with typical sizes 1/10 of the probing width, that the probability to have a resonance is of the order of 1/10, and that to mix the phase requires at minimum 100 events at the same position. These data require at minimum 10^4 samples to compute the average values. The existing differences between Figures 5 (10^3 samples) and 6 (10^4 samples) illustrate this remark. Nevertheless the results shown on Fig.5 represent significantly the behaviour the probing wave which permits to conclude first that the probing zone width decreases with an increase of the turbulence level, that there are two processes in competition for the wave propagation in the strong turbulent plasma, and that the resonances play an important role on the probing wave propagation properties. The status of the research on the turbulence characterisation is the following: on one part the resonant mechanism, based on Bragg scattering, driven enhanced scattering is modelled and on other part a photon diffusion equation is able to describe the behaviour of the probing for intermediate turbulence levels. A complex work has to be done to obtain a theoretical model able to describe the competition between the photon diffusion-like effect (multi-scattering effects including resonance generation) and Bragg backscattering. This competition seems to be associated to two typical spatial lengths, one associated to L_{Bragg_j}, the other, the length required to build resonance in a turbulent plasma for which there is no analytical expression. A 2D theoretical description will be welcomed but can not be presently validated by simulations due to the computation times needed for the requirements on the averaging and the mesh size imposed to fulfil the theoretical assumptions [8].

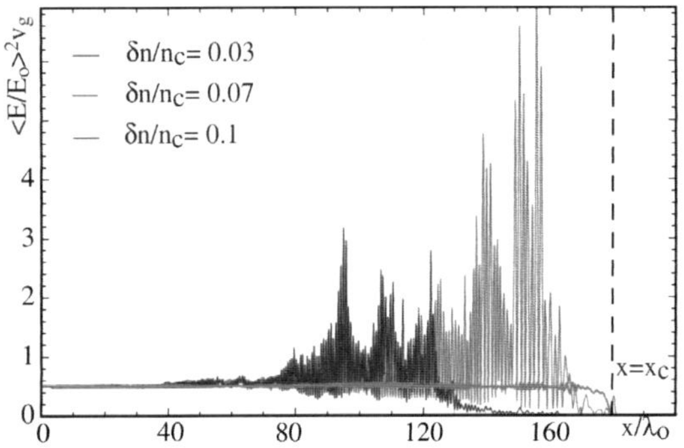

Figure 5. Averaged electromagnetic flux of the probing wave over 10^3 samples in a linear density plasma (density gradient length of $160\lambda_o$ with a superimposed homogenous turbulence with different values of $\delta n/n_c$ (green 3%, red 7% and blue 10%) and a flat k-spectrum with $k_f < 2k_o$, all k contributing to the Bragg's backscattering .

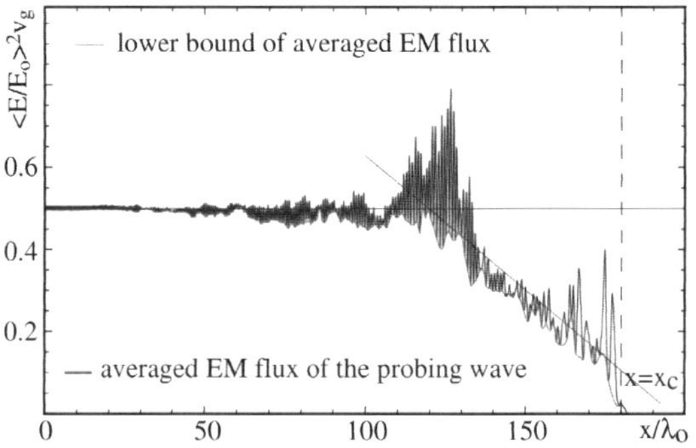

Figure 6. Averaged probing electromagnetic flux over 10^4 samples, versus the plasma position (blue) for a linear density profile on which is superimposed an homogeneous turbulence with an amplitude of 0.075 n_c (RMS values) and a flat Bragg resonant wavenumber spectrum.

Acknowledgments
The authors want to thank the referee for his remarks which have helped us to improve the readability of the paper. This work, supported by the European Community under the contract of Association between EURATOM, CEA, and the French Research Federation for fusion studies, was carried out within the framework of the European Fusion Development Agreement. The views and opinions expressed herein do not necessarily reflect those of the European Commission. Financial support was also received from the ARCUS Lorraine-Russie program, RFBR (grants 09-02-00453, 07-02-92162-CNRS), NWO-RFBR Centre of Excellence on Fusion Physics and Technology (grant 047.018.002), scientific school grant-6214.2010.2, Saint-Petersburg Government and by Dynasty Foundation.

Journal of Physics: Conference Series **416** (2013) 012019 doi:10.1088/1742-6596/416/1/012019

References

[1] R. Sabot, C. Bottereau, J.-M. Chareau, F. Clairet, F. Gabillet, P. Hennequin, S. Heuraux, C. Honoré, G. Leclert, 2004 International Journal of Infrared and Millimeter Waves, **25**, 229-246.
[2] S. Heuraux, F. Clairet, 2009 Instrum. Mes. Metrology **9**, 59-86.
[3] M. Schubert, A. Popov, S. Heuraux, E. Gusakov, T. Gerbaud, 35th EPS Conf. Plasma Phys Cont. Fusion 9-13 June 2008, Hellas Crete, **32D**, P1.096.
[4] F. da Silva, S. Heuraux, S. da Graça, G. Conway, ASDEX-U Team, 2010 Rev. Sci. Instrum. **81**, 10D927 .
[5] F. Clairet, S. Heuraux a, M. Schubert, R. Sabot, C. Bottereau, J.C. Giacalone, D. Molina, L. Ducobu, F. Leroux 2010 Rev. Sci. Instrum. **81**, 10D903.
[6] F. da Silva, S. Heuraux, E. Gusakov, A. Popov, 2010 IEEE Trans. Plasma Sci **38**, 2144.
[7] S. Hacquin, S. Heuraux, M. Colin, G. Leclert, 2001 J. of Computational Physics **174**, 1.
[8] E. Z. Gusakov, S. Heuraux, A. Yu. Popov, 2009 Plasma Phys. Cont. Fusion **51**, 065018.
[9] F. da Silva, S. Heuraux, S. Hacquin, M. Manso, 2005 J. of Computational Physics **203**, 467-492.
[10] S. Heuraux, E. Z. Gusakov, A. Yu Popov, F. da Silva, M. Irzak, 2010 IEEE Trans. Plasma Sci **38**, 2150.
[11] T. Gerbaud, F. Clairet, A. Sirinelli, R.. Sabot, S. Heuraux, L. Vermare, G. Leclert, 2006 Rev. Sci. Instrum. **77**, 10E928.
[12] E.Z. Gusakov, S. Heuraux, A.Yu. Popov, K.V. Syisoeva, 2010 37th EPS Conf. on Plasma Physics,Dublin, Irland. Vol. **34A** (ISBN 2-914771-62-2.) P5-110.

Numerical modelling of tokamak plasma micro turbulence wave number spectrum reconstruction using radial correlation reflectometry data

N V Kosolapova[1,2], E Z Gusakov[1] and S Heuraux[2]

[1]Ioffe Institute, Saint Petersburg, Russian Federation

[2]IJL, UMR-CNRS 7198, UHP, Nancy, France

n.kosolapova@mail.ioffe.ru

Abstract. Radial correlation reflectometry (RCR) is used technique providing information on the tokamak plasma micro turbulence. The data interpretation in this technique using for probing simultaneously different frequencies is based on the hypothesis of high measurements localization. In essence, it is similar to that used in probe correlation method where the signal cross-correlation function (CCF) is identified with the turbulence CCF. Unfortunately according to a strict theoretical analysis, both numerical and analytical, this assumption is incorrect; nevertheless the analytical theory proposed here shows that it is possible to reconstruct the turbulence radial wave number spectra using the RCR experimental data. It is also demonstrated that the turbulence CCF is reconstructed more precisely than spectra. It is illustrated by numerical computations for FT-2 (Russian Federation) and Tore Supra (France) tokamaks. Capabilities of spatial inhomogeneous micro turbulence spectra reconstruction is investigated as the transport barrier characterization for example. Information is also reported on the limitations of this method.

1. Introduction

Reflectometry has attracted the attention of researchers as a technique providing information not only on the plasma density profile but on the tokamak plasma low frequency turbulence as well [1]. Technical simplicity, as well as experimental geometry allowing operation at a single access to plasma, is among its attractive merits. However the interpretation of fluctuation reflectometry results, in general, is a very complicated task even in one dimensional model. To determine wave number spectrum or at least the turbulence correlation length a more complicated radial correlation reflectometry (RCR) was proposed [1]. The RCR uses simultaneously two frequencies for probing; the coherence decay between two reflected signals with growing difference of probing frequencies is studied by such diagnostics. It is supposed that the distance between cut offs at which the correlation of two reflectometry signals is suppressed is equal to the turbulence correlation length. Unfortunately according to both numerical and analytical studies [2-3], this assumption is incorrect. As it was shown already in 1D numerical computations performed in the Born approximation [2] that scattering signal cross correlation function (CCF) decrease is much slower than the decay of turbulence CCF. This slow decay of RCR CCF was attributed to the contribution of small angle scattering of very long scale fluctuations. Later this observation was confirmed also in full-wave 1D numerical modeling for small level of turbulent density fluctuations and at the same time that non-linear effects induce a reduction of the correlation length [4]. However no simple theoretical description of this behavior was provided in 1D geometry, let alone the ways to overcome this difficulty in determining the turbulence correlation length or spectrum.

Recently [5] an analytical integral formula expressing the RCR CCF in terms of turbulence radial wave number spectrum has been obtained and a procedure of its correct inversion for the spectrum determination from the CCF was proposed in a given framework [6]. This procedure feasibility check was started in the framework of 1D numerical modeling in [6] where the possibility of reconstruction of multi Gaussian spectra parameters was shown. Some possibilities of this method were studied in conditions relevant for experiments. The procedure accuracy dependencies on probing frequency range, number of frequency steps, as well as on the poor statistics, and presence of noise were investigated [7]. Also the experimental noise was taken into account when modeling the scattering signal and spectrum reconstruction in case of narrow probing frequency range, large step and low statistics.

In the present paper the results of numerical modeling of micro turbulence wave number reconstruction using RCR data in one dimensional model for the tokamaks Tore Supra (France), FT-2 (Russian Federation) are presented. In these cases, it is shown that turbulence CCF is reconstructed more closely than spectra. To give an idea of the reconstruction method capabilities, a spatial inhomogeneous micro turbulence spectra reconstruction modeling transport barrier is investigated as well and, according to the modeling framework, the use limitations are discussed.

2. The turbulence wave number spectrum reconstruction background

Supposing probing wave propagation strictly in the direction of plasma density gradient we use 1D model to treat the RCR problem describing the O-mode probing by equation

$$\left\{\frac{d^2}{dx^2}+\frac{\omega^2}{c^2}-\frac{4\pi e^2\left[n(x)+\delta n(x)\right]}{m_e c^2}\right\}E_z(x,\omega)=0 \tag{1}$$

$n(x)=n_c\, x/L$ is the background linear density profile; $\delta n(x)=\frac{1}{2\pi}\int \delta n_\kappa e^{-i\kappa x}d\kappa$ stands for turbulent density perturbations assumed statistically homogeneous, κ is a radial wave number; ω is a probing frequency define E_z total field of the probing wave, e electron charge, m_e electron mass and c velocity of light. In the analytical section we suppose the transparent plasma size $L_0(\omega_0)$ and the turbulence correlation length l_c to be large enough to treat (1) in the WKB approximation. Under the assumption that density fluctuation level is small $\delta n(x)/n_c \ll 1$ the reflected wave perturbations are small and the solution could be found using the perturbation theory methods (Born approximation). The scattering signal amplitude in linear (Born) approximation can be obtained with the help of straightforward approach based upon the receptivity theorem in the form introduced in [9] which takes in one-dimensional case the form: $A_s(\omega)=\frac{i\omega\sqrt{S_i}}{16\pi}\int_0^\infty \frac{\delta n(x)}{n_c}E_0^2(x,\omega)dx$ where S_i is an incident wave energy flux density and $E_0(x)$ is calculated in the absence of density fluctuations and gives the distribution of the probing wave electric field in plasma for $S_i=1$. Finally the expression for CCF takes a form

$$CCF(L)=D\int_{-\infty}^{+\infty}\frac{d\kappa}{|\kappa|}n_\kappa^2 e^{i\kappa\Delta L}erf\left(\sqrt{i\kappa L_0}\right)erf^*\left(\sqrt{i\kappa L}\right) \tag{2}$$

where $\Delta L = L_0 - L$ is the cut-off separation, the coefficient $D = \frac{S_i}{4\pi^2}\sqrt{L_0 L_1}\frac{(\omega_0\omega_1)^2}{c^4}$, $erf(s) = \int_0^s e^{-\varsigma^2} d\varsigma$ is a Fresnel integral and the turbulence wave number spectrum $\tilde{n}_\kappa^2$ is related to the density fluctuation correlation function by expression $2\pi\langle\delta n(x')\delta n(x'')\rangle = \delta n^2 \int_{-\infty}^{\infty} \tilde{n}_\kappa^2 \exp[i\kappa(x'-x'')]d\kappa$. The $1/\kappa$ factor entering the integral in (2) is responsible for underlining the contribution of small angle scattering off long scale fluctuations into the RCR signal. This singularity saturated only for $\kappa L_0 < 1$ due to the Fresnel integral behavior leads to a very slow decrease of CCF mentioned above that complicates the RCR data interpretation. The way to exclude this singularity and reconstruct the turbulence wave number spectrum is provided by the similarity of (2) and Fourier transform. Namely, it is possible to show that the relation inverse to (2) expressing the turbulence spectrum in terms of CCF takes a form

$$n_\kappa^2 = \frac{2}{\sqrt{\pi}} e^{i\frac{\pi}{2}(1+sign(\kappa))} \frac{|\kappa|}{erf^*\sqrt{i\kappa L_0}} \int_{-\infty}^{+\infty} \overline{CCF}(\Delta L) e^{i\kappa\Delta L} d\Delta L \qquad (3)$$

where the normalized $\overline{CCF}$ of reflectometry signals is determined as

$$\overline{CCF}(\omega_1) = \frac{\left\langle \left(A_s(\omega_0) - \langle A_s(\omega_0)\rangle\right)\left(A_s(\omega_1) - \langle A_s(\omega_1)\rangle\right)^* \right\rangle}{\sqrt{\left\langle \left(A_s(\omega_0) - \langle A_s(\omega_0)\rangle\right)^2 \right\rangle \left\langle \left(A_s(\omega_1) - \langle A_s(\omega_1)\rangle\right)^2 \right\rangle}} \qquad (4)$$

where averaging is held over random phase samples .

3. Numerical reconstruction of the micro turbulence spectrum and CCF

3.1. Linear plasma density profile

Here we shall analyze the accuracy of the proposed inversion procedure (3) using the CCF computed numerically from (1) in the frame of full-wave modeling or Born approximation. According to experimental results the plasma density profile in FT-2 tokamak could be assumed linear [10].

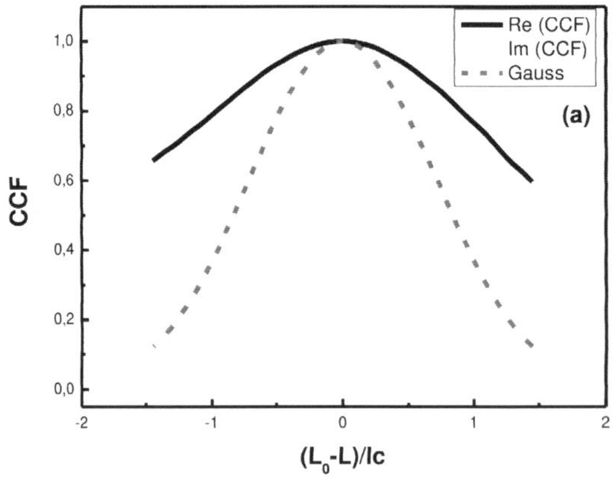

Figure 1. Signal CCF compared to turbulence CCF

The calculation parameters for FT-2 tokamak are as follows: reference frequency $f_0 = 31.7GHz$, reference cut off $L_0 = 0.05m$; probing frequency range for O-mode $24.5GHz < f < 37.5GHz$ which corresponds to probing interval $0.03m<L<0.07m$; Gaussian radial wave number spectrum $\tilde{n}_\kappa^2 = \sqrt{\pi} l_c e^{-l_c^2\kappa^2/4}$ [11], correlation length $l_c = 2cm$. The superposition of $m=10^4$ harmonics $\delta n(x) = \delta n_0 \sum_{j=1}^{j=m} \cos(jqx+\varphi_j)\sqrt{\frac{2q}{\pi}\tilde{n}_{jq}^2}$ possessing wave numbers jq, random phases φ_j, and amplitude distributed in accordance with the turbulence spectrum is used in the analysis. The averaging is performed over ensemble of typically 500 random phase samples. The signal CCF is shown in figure 1 by the solid curve the real part and doted curve the imaginary part, dashed curve shows the input Gaussian turbulence CCF. It can be clearly seen as it was mentioned before that the behavior of the signal CCF and the initial turbulence CCF is different.

The turbulence spectrum reconstructed from the CCF shown in figure 2(a) (solid curve is a real part, doted curve in an imaginary part) in comparison to input Gaussian spectrum (dashed curve) represents the possibility of turbulence spectrum reconstruction with a high accuracy. The turbulence CCF is also reconstructed, utilizing Fourier transform, rather precisely as it is shown in figure 2(b). Importantly that the probing range must be no less than 4–6 probing wavelengths to obtain a good reconstruction on FT-2 tokamak. The discrepancy between the reconstructed turbulence CCF and the input one could be explained as a consequence of small parameter $\Delta L / \lambda_0 = 2$, where λ_0 is a reference frequency wavelength. According to the framework of the proposed method it is applicable for higher densities (4.0 – 8.0 10^{19} m^{-3}) subject to utilizing higher frequencies up to 80GHz and probing range $\Delta L / \lambda_0 \approx 4 \div 6$.

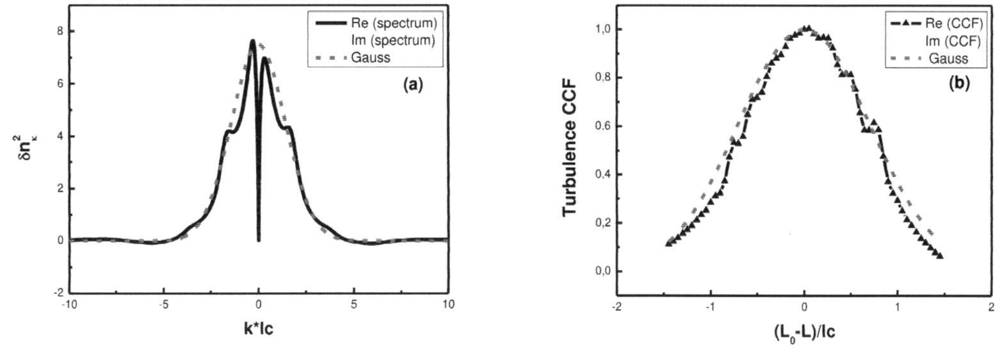

Figure 2. (a) Reconstructed spectrum compared to Gaussian shown in relative units; **(b)** reconstructed turbulence CCF compared to the input Gaussian

3.2. Arbitrary plasma density profile

On the purpose to hold numerical computations modeling the experiment on the tokamak Tore Supra we take the synthetic plasma density profile in the following form $n(x) = n_c \cdot 0.5\left(1+th\left(\alpha\left(x-x_a\right)\right)\right)\cdot\left(1-\left(x-x_b\right)^2/a^2\right)$ where $\alpha = 100m^{-1}$, $x_a = 0.045m$, $x_b = 0.8m$, $a^2 = 2.55$ (figure 3(a)). Such a density profile can be obtained on Tore Supra [12], JET [13] and ITER [14] machines.

The calculations for Tore Supra tokamak have been made in two principal cases. Firstly we simulate the probing in the interval $0.01m<L<0.128m$, reference frequency $f_0 = 73.3GHz$, reference cut off $L_0 = 0.069m$; radial wave number spectrum is shown in figure 3(b) [15]. Turbulence CCF

Journal of Physics: Conference Series **416** (2013) 012020 doi:10.1088/1742-6596/416/1/012020

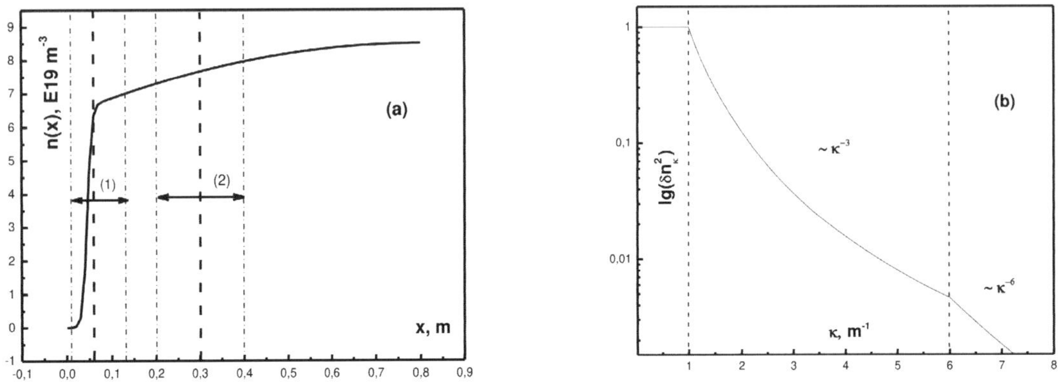

Figure 3. **(a)** Plasma density profile, dashed lines show the cut off positions $L_0 = 0.069m$ (1) and $L_0 = 0.3m$ (2), dash-dotted lines show the corresponding probing range boundaries; **(b)** radial wave number spectrum

compared to the initial one is shown in figure 4(a). Due to strong plasma density profile asymmetry we obtain some discrepancy between these two functions, nevertheless it can be asserted that the turbulence CCF is reconstructed rather precisely. Secondly we apply the simulations to the case when probing is held in the interval *0.2m<L<0.4m,* reference frequency $f_0 = 78.6GHz$, reference cut off $L_0 = 0.3m$ (figure 4(b)). Comparing we draw a conclusion that the proposed procedure is applicable in both cases, however in the asymmetry case the inaccuracy is higher.

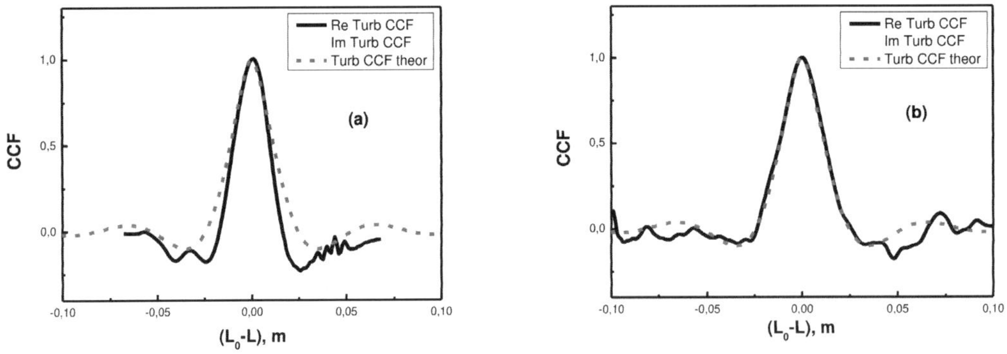

Figure 4. **(a)** Turbulence CCF reconstruction, $L_0 = 0.069m$; **(b)** $L_0 = 0.3m$

4. Conclusions

Summarizing the paper results we would like to stress that a 1D analytical study of the RCR has shown that the RCR CCF is a function of the cut off separation different from the turbulence CCF. This difference is explained by enhancement of reflectometry scattering efficiency at small radial wave numbers. However it is demonstrated that analytical expression for the RCR CCF obtained for different density profiles possesses a typical form of Fourier transform of a function proportional to the turbulence radial wave number spectrum. It is remarkable that the proposed procedure for the turbulence spectrum reconstruction from the RCR experimental data has its restrictions due to the linear Born approximation and does not take into account nonlinear effects. The reconstruction becomes rather complicated in case of flat plasma density profile and/or small ratio $\Delta L / \lambda < 1$. Nevertheless the method is also applicable to detect transport barriers and thin layers as it was described in [8]. Thus due to the turbulence spectrum sudden change assumed in experiments [16] it is possible firstly to detect the signature of a thin layer in the signal CCF as a flat part possessing a barrier width and shift the reference frequency so as to probe the turbulence characteristics inside the transport barrier. So far as the method is rather sensitive to local turbulence characteristics [8] the wave number spectrum could be determined.

Based on simple and universal representation the proposed procedure of turbulence wave number spectrum reconstruction looks very promising. Further 2D computations modeling reflectometry experiments and studies of the possibilities of turbulence characteristics reconstruction from real experimental data are planned.

5. Acknowledgements
The work is done in the frame of ARCUS program and subsidized by RFBR (grants 09-02-00453, 07-02-92162-CNRS), NWO-RFBR Centre of Excellence on Fusion Physics and Technology (grant 047.018.002), ANR_blanc_06_0084, scientific school grant-6214.2010.2, Saint-Petersburg Government and by Dynasty Foundation.

References
[1] N. Bretz, Physical Fluids **B4(8)** 2414 (1992)
[2] H. Hutchinson, Plasma Phys. Control. Fusion **34** 1225 (1992)
[3] E. Gusakov and A. Popov, Plasma Phys. Control. Fusion **44** 2327 (2002)
[4] G. Leclert, et al. Plasma Phys. Control. Fusion **48** 1389 (2006)
[5] E. Gusakov, et al. 34th EPS Conf. on Plasma Phys. Warsaw, 2 – 6 July 2007 ECA Vol.**31F**, P-5.100 (2007)
[6] E. Gusakov, et al. 35th EPS Conf. on Plasma Phys. Hersonissos, 9 – 13 June 2008 ECA Vol.**32D**, P-1.082 (2008)
[7] E. Gusakov, et al. 36th EPS Conf. on Plasma Phys. Sofia, June 29 – July 3 2009 ECA Vol.**33E**, P-4.217 (2009)
[8] N. Kosolapova, et al. proc. Intl. Conf on Plasma Diagn., Pont-a-Mousson, France, 12 – 16 April (2010)
[9] A. Piliya, A. Popov, Plasma Phys. Control. Fusion **44** 467 (2002)
[10] D. Kouprienko, et al. Plasma Physics Reports **36** 371 (2010)
[11] A. Gurchenko, et al. Nucl. Fusion **47** 245 (2007)
[12] F. Clairet, private communication (2010)
[13] C. Bottereau, et al. proc. 9th Intl. Reflectometry Workshop, Lisbon, Portugal, 4–6th May (2009)
[14] L. Zeng, et al. Plasma Phys. Control. Fusion **49** 1277 (2007)
[15] P. Hennequin, et al. Plasma Phys. Control. Fusion **46** B121 (2004)
[16] G. Conway, et al. Plasma Phys. Control. Fusion **43** 1239 (2001)

Integrated transformations of plant biomass to valuable chemicals, biodegradable polymers and nanoporous carbons

B N Kuznetsov,[1,2] N V Chesnokov,[1,2] V E Taraban'ko, [1,2] S A Kuznetsova,[1,2] A V Petrov[1]
[1]Institute of Chemistry and Chemical Technology SB RAS,
Krasnoyarsk, 660049, K. Marx street., 42, Russia
[2]Siberian Federal University, Krasnoyarsk, 660041, Svobodny prospect, 79, Russia

E-mail: bnk@icct.ru, inm@icct.ru

Abstract
Integrated transformations of wood biomass to valuable chemicals and materials are described. They include the main biomass components separation, the conversion of cellulose to glucose, levulinic acid, biodegradable polymers and lignin – to nanoporous carbons. For wood fractionation on pure cellulose and low molecular mass lignin the methods of catalytic oxidation and exploded autohydrolysis are used. The processes of acid-catalysed hydrolysis of cellulose to glucose and levulinic acid were optimized. New methods of biodegradable polymers synthesis from lactone of levulinic acid and nanoporous carbons from lignin were suggested.

1. Introduction

Plant biomass annual stock addition significantly exceeds the yearly demands of a mankind in the fuels and chemical products [1].

Russia has around 23 % of world forest resources which can serve as a relevant source of raw material for production of a big variety of needed chemicals, materials and alternative fuels.

Wood biomass contains 40–50 % of cellulose, 15–30 % of hemicelluloses and up to 30 % of lignin [2]. Cellulose is a linear polymer, constructed from $C_6H_{10}O_5$ units. Hemicelluloses are branched polysaccharides, containing C_5–units with shorter chains as compared to cellulose. Lignin is non-regular polymer of aromatic nature composed of phenylpropane fragments.

Traditional industrial technologies of wood deep processing are inefficient and dangerous for the environment and they give only the limited range of products. Perspective directions in the development of innovative technologies of plant biomass processing to valuable products are connected with a design of integrated processes which ensure the total utilization of all main components of a biomass.

In this paper some results of study the integrated transformations of wood components to valuable chemicals, biodegradable polymers and nanoporous carbon materials are presented.

2. Experimental

Air–dry sawdust (fraction 2–5 mm) of Siberian origin wood: aspen (Populus tremula L.), birch (Betula pendula Roth.), abies (Abies sibirica Ledeb.), larch (Larix sibirica Ledeb.) were used as the initial raw materials. The chemical composition (% wt. on abs. dry wood) of aspen wood: 46.3 cellulose, 21.8 lignin, 24.5 hemicelluloses, 7.8 extractive substances; birch wood: 41.3 cellulose, 19.9 lignin, 30.3 hemicelluloses, 8.4 extractive substances; abies wood: 50.3 cellulose, 27.7 lignin, 15.4 hemicelluloses, 6.8 extractive substances; larch wood: 34.5 cellulose, 26.1 lignin, 27.2 hemicelluloses, 13.0 extractive substances.

Method of catalytic oxidation by hydrogen peroxide was used for the fractionation of wood biomass on cellulose and low molecular mass lignin. The delignification medium was composed of acetic acid, hydrogen peroxide, sulphuric acid catalyst and distilled water.

Oidative delignification of wood sawdust with 2 % mas. of H_2SO_4 catalyst was carried out in metal shaking reactor of 200 cm^3 volume at variation of temperature, process time, liquid/wood ratios, concentrations of CH_3COOH, H_2O_2 and catalyst. Such parameter as a residual lignin content in cellulosic product of wood delignification was used to evaluate the delignification activity of catalyst. The yield of cellulosic product was estimated by weight method.

The exploded autohydrolysis of wood by overheated steam was carried out in experimental installation with stainless reactor of 800 cm^3 volume at temperatures 187–240 °C, pressure 1.2–3.4 MPa [3].

Wood sawdust were heated at 105 °C and then loaded in the heated reactor. The overheated steam was injected into the sealed reactor. The duration of steam treatment was 1–5 min. and then the pressure was dropped through a spherical stopcock. This treatment resulted in transformation of initial wood cheaps to fibrous material.

The contents of cellulose and lignin in cellulosis products were determined by standard chemical analysis. The structural characteristics of cellulose were studied by FTIR (Vector–22, Bruker) and γ–ray (Dron–4) methods.

Nanoporous carbon materials were synthesized by alkaline thermal activation of lignin and cellulose. The mixture of air–dried raw material (fraction 0.25–1.0 mm) was placed into a horizontal reactor made from stainless steel and was subjected to pyrolysis in argon atmosphere. The procedure of a thermal treatment included the nonisothermic heating of sample up to 800 °C with the rate 5 °C/min and the isothermic soaking at final temperature during 1 hour. Thermally treated samples were washed by water, then treated by 0.5 M HCl, washed again by water at 50 °C and dried at 100-105 °C up to the constant weight.

The texture characteristics of porous carbons were measured by BET method (N_2 adsorption at 77 K) using ASAP 2420 V2.02J and Soptometer–M.

Hydrolysis of cellulose to glucose was carried out with 60–80 % sulfuric acid at 25 °C, liquid to solid ratio 3–5 and time 1–2 hours, with following inversion of hydrolyzate at 100 °C, liquid to solid ratio 30–40 during 1–2 hours. Then the solution of glucose was neutralized by $Ca(OH)_2$ and purified by active carbon. The content of C_6– and C_5–sugars in hydrolyzates was measured by photocalorimetric method with o–toluidine blue at 355, 460, 630 nm.

The conversion of cellulose to levulinic acid (LA) was accomplished at autoclave conditions at temperature 200–250 °C. The cellulose was impregnated by sulfuric acid catalyst (5 % mas.). Synthesis of LA from glucose, fructose and sucrose was carried out at 98 °C and atmospheric pressure in the presence of acid catalysts HCl, H_2SO_4, H_3PO_4.

Angelicalactone was obtained from levulinic acid in vacuum still at 67–72 °C, 3.3 Pa and was purified by fractional melt crystallization.

Polyesterification of angelicalactone was carried out in the presence of sodium butylate and sodium hydroxide at temperatures 25–122 °C. Molecular weights of the obtained polymers were estimated by viscosimetry method. The chain–length distribution was determined by phase–reversal chromatography with the use Nova-Pak CIS column, octadecyl–coupled silica gel sorbent; vaporizing light scattering detection device, model 500 (Alltech Corporation, USA). ^{1}H NMR spectroscopy was also used for study the polymers structure.

Biodegradation ability of the obtained polymers was estimated by liveweight gain of cultures of microorganisms Sacharomices cerevisae, Streptomyces chrysomallus and Streptomyces lividans and by detection of polymers mass decreasing.

3. Results and discussion

The study of integrated transformation of wood sawdust to cellulose, glucose, levulinic acid, biodegradable polymers and nanoporous carbons was accomplished according to the scheme, presented in figure 1.

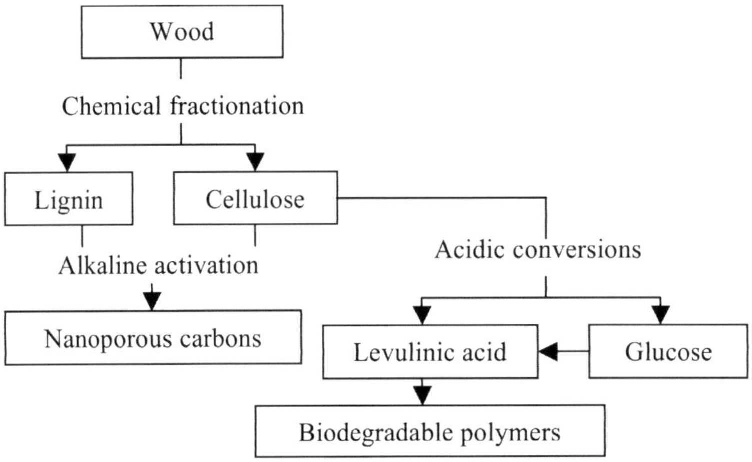

Figure 1. Scheme of integrated transformations of wood to valuable chemicals and materials

The first stage of studied integrated transformations is the wood separation on cellulose and low molecular mass lignin. The following methods of wood separation were compared: catalytic oxidation by hydrogen peroxide in acetic acid–water medium and exploded autohydrolysis with overheated water steam.

The complete separation of wood biomass on cellulose and soluble lignin was realized by catalytic oxidation with hydrogen peroxide in acetic acid–water medium in the presence sulfuric acid catalyst.

It was found that the growth of oxidation process temperature from 110 to 140 °C increases the cellulose content and degreases the lignin concentration in cellulosic product. At the same time, the yield of cellulosic product decreases because the oxidative destructions of lignin and polysaccharides are accelerated with temperature. Similar effects were observed when the initial hydrogen peroxide concentration is increased from 2.0 to 10.2%. The optimal conditions of aspen–wood and birch–wood oxidation were found, which supply the rather high yield of cellulose (48–50 % relative to a. d. wood) with very low content of residual lignin (0.3–0.4 % mas.): temperature 120–130 °C, H_2O_2 concentration 4–6 % mas., H_2SO_4 concentration 2 % mas. In order to obtain high–quality cellulose from abies or larch wood, which contain more lignin than aspen or birch wood, it is necessary to use the more active catalysts, such as TiO_2 and H_2MoO_4 [4].

The process of high temperature hydrolysis of wood cellulose by diluted mineral acids is used in industry and it demands the low expenses of acid catalyst. But the low–temperature hydrolysis of cellulose by concentrated acids has its own advantages: higher yield of glucose and possibility to carry out the hydrolysis process at atmospheric pressure.

The data on hydrolysis of cellulose from aspen wood with concentrated sulfuric acid are presented in table 1.

Table 1. Influence the conditions of aspen cellulose hydrolysis and inversion stages on the yield of glucose.

Concentration of H_2SO_4, %	Conditions of hydrolysis and inversion stages						Yield of glucose, % **
	Hydrolysis			Inversion*			
	Liquid to solid ratio	Temperature, °C	Time, h	Liquid to solid ratio	Temperature, °C	Time, h	
80	5	20	1	30	100	2	97,3
70	3	20	2	40	90	2	97,9
60	5	20	2	40	90	1	98,5
80	3	20	2	40	90	1	80,6
70	5	20	1	30	100	1	93,7
60	3	20	1	30	100	2	85,2

* hydrolysis of oligosaccharides obtained at the first stage of cellulose hydrolysis, ** from mass of a.d. cellulose

Obtained solutions have no C_5–sugars, which inhibit the fermentation of glucose to alcohols, lactic acid, polyhydroxyalcanoates etc.

The effective methods of lignin depolymerization and hemicelluloses removal from wood is the short–time treatment of wood with overheated water steam with following fast drop of the pressure (exploded autohydrolysis) [5].

The significant reduction of hemicelluloses concentration was observed after aspen wood treatment at 220 °C at conditions of exploded autohydrolysis (Figure 2).

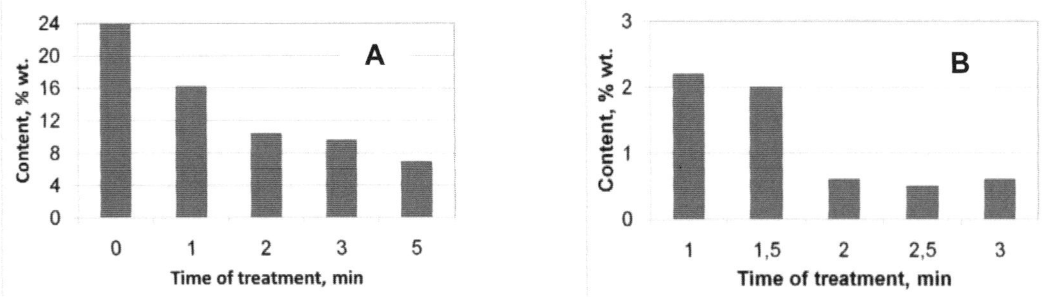

Figure 2. Content of hemicelluloses in aspen wood, autohydrolyzed at 187 °C (A) and 220 °c (B)

The same results were obtained after treatment of birch wood, abies wood and larch wood. The optimal conditions of wood autohydrolysis were selected (220–230 °C, 3 min) which allow to obtain solution of glucose with concentration of C_5–sugars less 1 %. C_5–sugars are completely absent in solutions of glucose obtained via stages of autohydrolysed wood catalytic oxidation to cellulose and of subsequent cellulose hydrolysis by concentrated H_2SO_4.

The other valuable product of acid-catalyzed conversions of carbohydrates is levulinic acid (LA). The high–temperature and low–temperature methods are used for LA synthesis. In the presence of sulfuric acid catalyst the highest yield of LA from cellulose reaches 35–40 % mas. at 240 °C.

The relative reactivity of different carbohydrates at 98 °C decreases in the order: fructose > sucrose > inulin > glucose > cellulose. Glucose is of 20–40 times less reactive, than fructose, and cellulose is of 30 times less reactive, than glucose.

In glucose conversion to LA the hydrochloric acid is twelve times more active, than sulfuric acid, but the selectivities of the LA formation are practically the same in both cases. In the presence of phosphoric acid catalyst the maximum yield of levulinic acid does not exceed 5 mol. %, and the rate of process is low.

Lacton of LA (α-angelicalactone) is an attractive substance for syntheses the new types of biodegradable polymeric materials.

There are two possible ways of angelicalactone polymerization: – disclosing of an olefinic linkage with formation of a polyfuranone and disclosing of the lactone cycle with formation of a polyester.

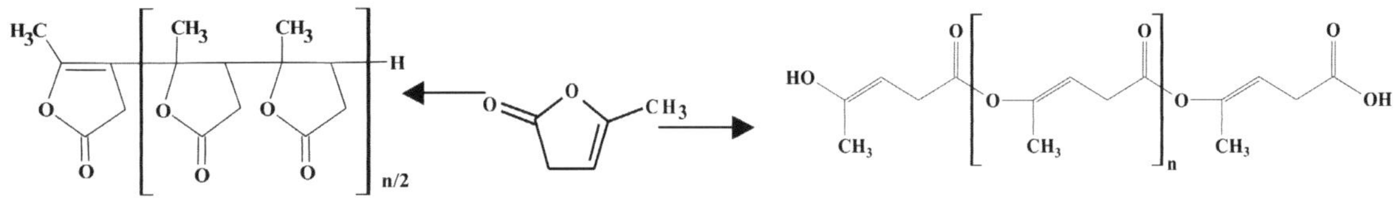

Polyfuranone α-Angelicalactone Polyester

Polyesterification of angelicalactone is the most interesting way for producing of new biologically compatible polymers [6].

Polyesters of alpha-angelicalactone were obtained with the use of alkali based catalysts – sodium butylate and NaOH. The obtained polymers are light yellow resins or solids, water-insoluble and soluble in polar organic solvents. Molecular weight of the obtained polymers is up to 2000 amu.

According to ^1H NMR spectroscopy data the polymer of alpha-angelicalactone is a polyester having the "head to tail" structure. The part of polyester intermonomeric bonds in the obtained products of polymerization reaches 68–80 %.

The obtained polymers are exposed to full biodegradation by microorganisms Sacharomices cerevisae during 5–15 day, and by Streptomyces lividans and Streptomyces chrysomallus - during 20–30 day.

For producing the nanoporous carbons from various type of carbon–containing raw materials the different methods of alkaline thermal activation are used [7]. It was found that the melted sodium hydroxide and potassium hydroxide promote the significant development of nanoporous structure of carbon materials from lignin and cellulose.

Such characteristics of obtained nanoporous carbons as a specific surface area, the volume and size of pores are depend on the nature of initial raw material and alkali, ratio raw material/alkali and temperature of thermal treatment.

The specific surface area of nanoporous carbons, obtained by alkaline thermal activation of cellulose and lignin in melted KOH goes through a maximum with the increase of ratio KOH/raw material (Fig. 3). Maximal surface area of carbons from cellulose (1170 m^2/g) corresponds to the content of KOH in a mixture of 67 % mas. In the case of lignin the maximal surface area (2035 m^2/g) of obtained carbons corresponds to KOH content 75 % mas.

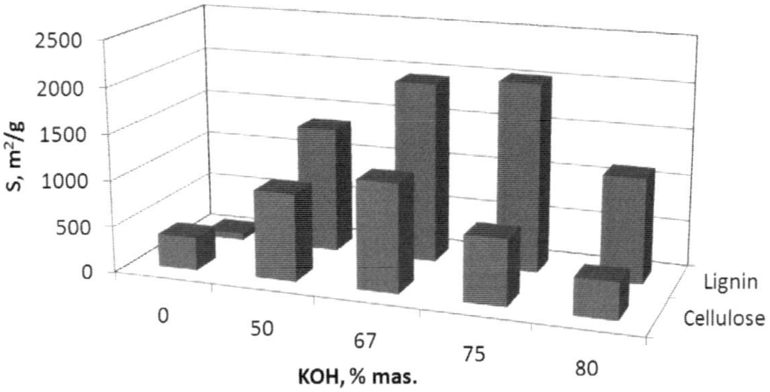

Figure 3. Influence of KOH content in the mixture on specific surface area of porous carbons, obtained by alkaline thermal activation of cellulose and lignin.

Nanoporous carbon materials obtained by alkaline thermal activation of cellulose and lignin have high sorption activity to hydrogen (sorption capacity 3 % mas. at 77 K and 5 MPa H_2), to volatile organic compounds (hexane, $CHCl_3$, benzol, butanol etc.). Also they are able to separate the mixtures He–CH_4, H_2–CH_4 (separation factor 3.6–3.8).

4. Conclusion

As a result of the accomplished study, the integrated process of wood biomass conversion to cellulose and low molecular mass lignin with their subsequent transformations to glucose, levulinic acid, biodegradable polymers and nanoporous carbons was developed. For the each stage of developed integrated process the optimal reaction conditions were selected which allow to produce the target product with the high yield and with the required characteristics.

5. References

[1] Perlack R D, Wright L L, Turhollow A, Graham R L, Stokes B, Erbach D C 2005 *Report No. DOE/GO-102995-2135 Oak Ridge TN* http://www.osti.gov/bridge
[2] Fengel D, Wegener G 1984 *Wood: Chemistry, Ultrastructure, Reaction* (Berlin: Walter de Gruyter)
[3] Kuznetsov B N, Efremov A A, Levdanskii V A, Kuznetsova S A, Polezhaeva N I, Shilkina T A, Krotova I V 1996 *Bioresource Technology* **58** 181
[4] Kuznetsov B N, Taraban'ko V E, Kuznetsov S A 2008 *Kinetics and Catalysis* **49** 517
[5] Li J, Henziriksson G, Gellerstedt G 2007 *Bioresource Technology* **98** 3016
[6] Taraban'ko V E, Kaygorodov K L, Chernyak M Yu 2008 *Journal of Siberian Federal University* **2** 118
[7] Lillo-Ródenas M A, Marco-Lozar J P, Cazorla-Amorós D, Linares-Solano A 2007 *J. Anal. Appl. Pyrolysis* **80** 166

New families of carbon gels based on natural resources

Andrzej Szczurek[1], Gisele Amaral-Labat[1], Vanessa Fierro[1], Antonio Pizzi[2], Alain Celzard[1*]

[1]Institut Jean Lamour – UMR UHP - CNRS 7198, ENSTIB, 27 rue Philippe Séguin, F-88051 Epinal cedex 9

[2]LERMAB, EA 4370, ENSTIB, 27 rue Philippe Séguin, BP 1041, F-88051 Epinal cedex 9

*Institut Universitaire de France. E-mail: Alain.Celzard@enstib.uhp-nancy.fr

Abstract. Carbon gels are versatile materials which can be used for many applications. They are extremely expensive, because generally prepared from resorcinol – formaldehyde (RF) resins first gelled and next dried with supercritical carbon dioxide. In the present work, resorcinol has been substituted partly or completely by tannins, a family of molecules extracted from mimosa tree barks. Tannins are natural, non-toxic products, typically thirty times cheaper than resorcinol. Their chemical resemblance with the latter makes them be often called natural resorcinol. Using tannins not only substantially decreases the cost but also allows preparing materials in a much wider range of pHs than that usually employed for RF gels. Consequently the main pore size and the fraction of given families of pores, controlling the carbon gels' properties, are tuned in an easier way, and a much wider range of pore structures is obtained. Finally, two alternative ways of drying are suggested for further decreasing the cost: freeze-drying and supercritical drying in acetone. Both are shown to lead, in some conditions described below, to materials having similar characteristics to those of expensive RF carbon aerogels previously dried in supercritical CO_2.

1. Introduction

Carbon gels are porous materials which have been suggested as adsorbents, porous electrodes for electric double-layer supercapacitors, filling material for HPLC columns, catalyst supports and heat insulators at high temperature. These solids indeed present both a suitable mesoporosity (i.e. pores whose width ranges from 2 to 50 nm) for all these applications and a high chemical resistance. Usually, such materials are prepared by pyrolysis in inert atmosphere of porous organic gels based on resorcinol – formaldehyde (RF) resins. RF formulations, prepared in water, are very expensive but are also very easy to handle, leading to nice, monolithic, big and reproducible elastic hydrogels. The latter need to be dried in a suitable way, for both removing the liquid saturating the porosity and preventing the collapse of the pores. For that purpose, convective, supercritical or freeze drying can be used, leading to xerogels, aerogels and cryogels, respectively.

Xerogels present a rather high bulk density, whereas aerogels and cryogels have a much higher porosity. Producing aerogels is a costly procedure, since drying of organic hydrogels by supercritical carbon dioxide is involved. Freeze-drying is much cheaper, and is thus a very interesting alternative drying route. Pore structures similar to those of aerogels have indeed been reported if water in the hydrogels has been exchanged by t-butanol beforehand, a solvent having a higher vapour pressure and presenting much smaller volume changes than water when frozen. In the present work, low-cost carbon gels have been produced according to two simultaneous ways: (i) substitution of resorcinol by flavonoid tannins, and (ii) use of cheap drying modes: supercritical drying with acetone or freeze-drying. We show that, in a number of cases, carbon gels having characteristics very similar to those of their expensive RF counterparts dried in supercritical CO_2 can be obtained.

2. Materials and methods

2.1. Preparation of tannin-based gels

Information about tannins is extensively given in another paper from the present issue [1]. Briefly, tannins are commercial raw materials extracted from mimosa tree barks and widely used in leather, chemical and food industries. These are phenolic molecules able to lead to the same kind of reactions as resorcinol does. Especially, their cross-linking with formaldehyde in water produces high-quality resins, already used and commercialised for wood joining. In the present case, the formulations known for the preparation of adhesives, for which hard and highly cross-linked materials are desired, have been modified for obtaining highly porous gels. Two types of formulations have been investigated: tannin – resorcinol – formaldehyde (TRF) and tannin – formaldehyde (TF).

2.1.1. *Organic tannin-based hydrogels.* TRF hydrogels were prepared from a TRF resin whose composition was optimised in a former work [2], and to which formaldehyde was added. In a typical RF formulation, the stoichiometric R/F molar ratio is 0.5. In the present case, 2/3 in weight of the resorcinol was replaced by tannins. The pH of the initial TRF solution was 8.15. Small amounts of 99.8% acetic acid were added for decreasing and adjusting the pH at different values ranging from 2 to 8. Such a broad range of pH is very unusual for RF gels but, as shown below, allows obtaining rather different pore structures.

TF hydrogels were prepared by mixing tannins, first dissolved in a 20% water solution of methanol, and 37% water solution of formaldehyde, in such a way that the weight ratio T/F was 1.35. The pH of the initial TF solution was 4.3 and was increased or decreased by addition of 33% aqueous solution of NaOH or 99.8% acetic acid, respectively. Again, various solutions having a wide range of pH, ranging from 3.3 to 8.3, were obtained.

All these solutions of known pH were poured into sealed glass tubes and left for gelling and ageing in an oven at 50 and 85°C for TRF and TF formulations, respectively. Gelation of TRF solutions occurred within a few hours, whereas that of TF required several days. Because the gel point was quickly reached for TRF, its determination was easily carried out through a careful examination of the tubes. The gelation was assumed to be complete when the solution no longer flowed when tilted at an angle of 45°. A strong influence of pH was observed, see Figure 1. Because the reactions in TF solutions were much slower despite the higher temperature, gelation times of TF formulations could not be measured. Reactivity of T is lower than that of R, because electrophilic substitution is always easier with two -OH groups (case of resorcinol) than with one -OR group (case of flavonoid tannin).

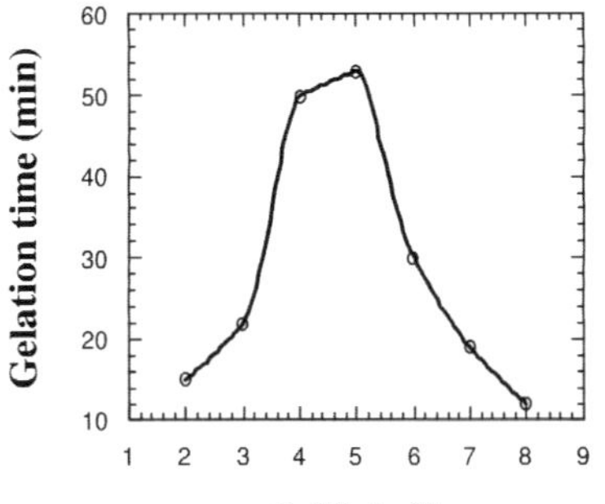

Figure 1. Gelation time of TRF solutions at 50°C as a function of pH.

2.1.2. *Tannin-based carbon gels.* After a few days of curing, during which the stiffness of the hydrogels was improved, the latter were removed from their tubes, cut into cylindrical pieces, and soaked into dry solvent. For TRF gels, the solvent was acetone, which was replaced 24 h later by t-butanol, next changed by fresh one three consecutive times every three days. Doing this, the solvent saturating the porosity was assumed to be pure t-butanol. For TF gels, the solvent was only acetone, replaced every day by fresh one, three times.

Then, exchanged TRF and TF wet gels were freeze-dried and supercritically dried, respectively. Temperature and pressure conditions used for each process are given in Figure 2. Advantages of using acetone for supercritical drying are that acetone is typically 1000 times cheaper than liquid CO_2 on one hand, and that only a few cm^3 are required instead of a few hundreds of cm^3 in the case of CO_2, further reducing the costs. The only drawback is the use of much higher temperature and pressure.

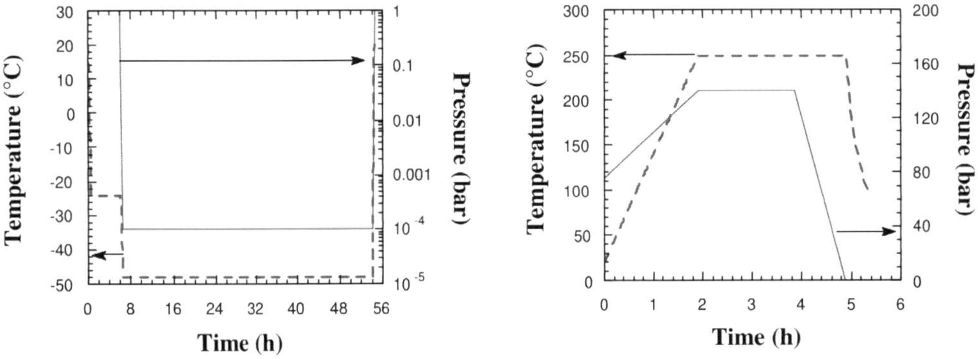

Figure 2. Temperature and pressure conditions used for: (left) freeze-drying TRF organic gels; (right) drying TF organic gels in supercritical acetone.

After drying, samples were installed inside a quartz boat placed at the center of a fused silica tube heated by a horizontal furnace and flushed with nitrogen. Heating rate and set point (maintained for 2h) were 5°C min^{-1} and 900°C, respectively. After cooling under nitrogen flow, TRF carbon cryogels and TF carbon aerogels were recovered.

2.2. Investigation of carbon gels derived from tannins

Carbon gels were characterised in terms of total, micro (i.e. pores narrower than 2 nm) and mesoporosity, surface area and pore-size distribution. These features were obtained through the use of different techniques such as helium pycnometry and nitrogen adsorption at 77K.

2.2.1. *Pycnometry.* Helium pycnometry allowed determining the skeletal density of the carbon constituting the gels. For avoiding any bias related to possible closed porosity, the samples were finely crushed in a mortar before any measurement. Monolithic carbon gels having perfectly known dimensions and regular shapes were weighed in order to measure their apparent, or bulk, density. The total porosity was deduced from the two values of density.

2.2.2. *Adsorption.* BET surface area, micro and mesopore volumes were calculated from the nitrogen adsorption isotherms obtained at 77K according to the techniques extensively detailed elsewhere [3]. Pore-size distribution within the range of micro and mesopores was calculated by the DFT method [4].

3. Pore texture of TRF and TF carbon gels

3.1. Porosity and surface area

Whatever the tannin-based precursor, the skeletal density of the resultant carbon was always close to 2 g cm^{-3}, as expected for a disordered carbon prepared at 900°C. In contrast, the bulk densities were found to range from 0.10 to 0.74 g cm^{-3}, leading to total porosities within the range 56 – 95 %, see

Figure 3. The highest porosity values are typical of RF-based carbon aerogels prepared by drying of their organic precursors in supercritical CO_2, i.e., close to 95%. The same applies to the BET surface area, shown in Figure 4, whose highest values are, again, very similar to those of classical carbon aerogels, typically ranging from 500 to 700 $m^2\ g^{-1}$.

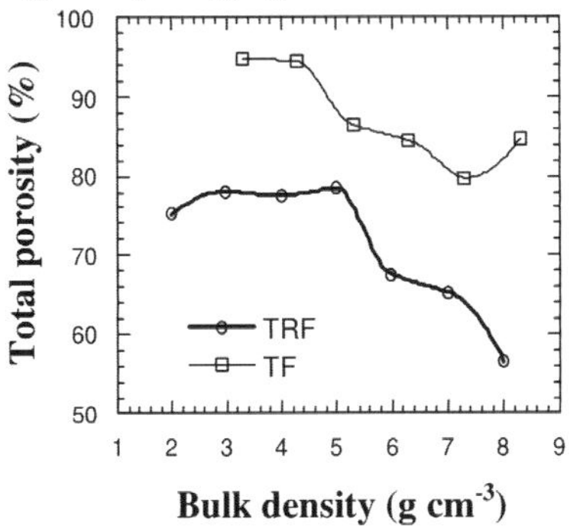

Figure 3. Total porosity of TRF carbon cryogels and TF carbon aerogels.

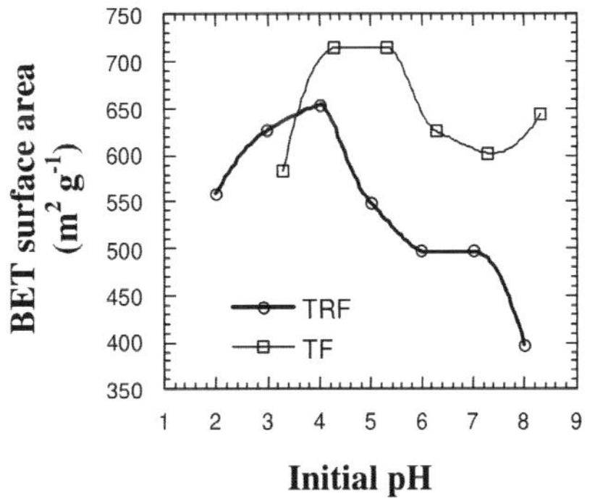

Figure 4. BET surface area of TRF carbon cryogels and TF carbon aerogels.

It is interesting to note that the maximum of both porosity and surface area clearly corresponds to the minimum of reactivity of the precursor monomers, i.e. to the pH at which the gelation time was found to be the highest: close to 5 (see again Figure 1). It can thus be assumed that fast gelation leads to strongly cross-linked polymer network and finally to dense materials. However, the strongest effect is observed at high pHs, for which gels having the lowest porosity are obtained. The higher surface area and porosity of TF gels prepared at pH 8.3 are one exception due to the NaOH used in higher amount for increasing the pH. NaOH remains in the carbon and participates to its activation.

3.2. Micro and mesoporosity

For clarity, only a few nitrogen adsorption isotherms obtained with TRF- and TF-derived carbon gels are presented in Figure 5. All are of type IV according to the IUPAC classification, and mean that the materials are essentially mesoporous. Desorption branches are parallel to, and almost merged with, adsorption curves. Such narrow hysteresis loops are H1-type, and are typical of cylindrical mesopores in which capillary condensation occurs.

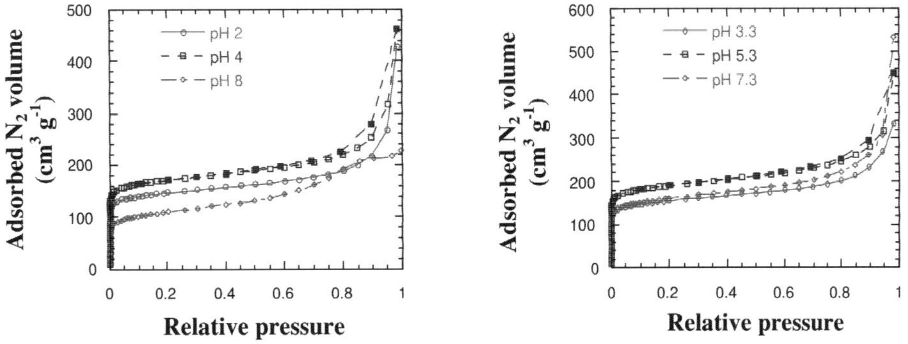

Figure 5. Nitrogen adsorption isotherms at 77K for: (left) TRF carbon cryogels; (right) TF carbon aerogels. For clarity, only one hysteresis loop is shown in each plot (at pH 4 and 5.3): adsorption (empty symbols) – desorption (full symbols).

From these curves, in addition to the BET surface areas presented above, micropore (V_μ) and mesopore (V_m) volumes as well as average micropore widths (L_0) could be determined. V_μ was derived from the Dubinin – Radushkevich method (see [3] and refs. therein), whereas V_m was assumed to be the difference between the adsorbed volume of liquid nitrogen at a relative pressure of 0.99 minus V_μ. L_0 was calculated from the equation of Stoeckli et al. [5]. All these data are gathered in Table 1, as a function of the synthesis pH.

Table 1. Micropore (V_μ) and mesopore (V_m) volumes, and average micropore widths (L_0), of TRF carbon cryogels and TF carbon aerogels.

pH	TRF carbon cryogels			TF carbon aerogels		
	V_μ (cm^3 g^{-1})	V_m (cm^3 g^{-1})	L_0 (Å)	V_μ (cm^3 g^{-1})	V_m (cm^3 g^{-1})	L_0 (Å)
2	0.215	0.45	6.7			
3	0.243	0.23	6.6			
3.3				0.224	0.29	6.4
4	0.252	0.46	6.7			
4.3				0.276	0.36	7.2
5	0.213	0.22	6.1			
5.3				0.277	0.42	5.9
6	0.192	0.39	7.8			
6.3				0.241	0.51	6.8
7	0.19	0.40	7.5			
7.3				0.229	0.59	7.0
8	0.155	0.20	9.0			
8.3				0.246	0.89	8.9

The changes of the pore texture parameters listed in Table 1 are in agreement with the trends shown in Figures 3 and 4. For TRF carbon cryogels, pore volumes at high pH are much lower than those of materials prepared in acidic medium. At intermediate pH, a maximum is observed. The average pore width tends to increase with pH. These findings are consistent with obtaining materials of low porosity at high pH, and agree with previous studies of the effect of pH on the porosity of RF carbon xerogels carried out, however, in a much narrower range of pH [6,7].

The behavior of TF carbon gels is not exactly the same when compared to that of TRF materials. The micropore volume also goes through a maximum near pH 4 – 5, and the average micropore width also tends to increase with pH. But the mesopore volumes always and considerably increase with pH. The fraction of mesoporosity thus changes from 56 to 78% at pH 3.3 to 8.3, respectively. Such big variations of pore structures have never been observed before with RF carbon gels, whatever the way they have been prepared.

Pore-size distributions derived from Figure 5 by application of DFT are presented in Figure 6. Again, for clarity, not all synthesis pHs are shown. In the case of TRF carbon cryogels, increasing the synthesis pH has two simultaneous effects: broadening of the distribution of pore widths and shifting it towards narrower pores. These effects are modest below the pH of lower reactivity, close to 4, for which the pore texture parameters were rather similar to each other, especially the surface areas, as seen in Figure 4. But the effects of pH are strongly marked above pH 5. Thus, trough the use of various synthesis pH, it is possible to tune easily and accurately the porosity. RF formulations, for which the pH cannot be changed so much, don't allow this at such extent.

In the case of TF aerogels, the pore-size distribution is neither shifted nor broadened with pH, but the relative proportions of pore families change considerably. In first approximation, two main peaks may be considered, one corresponding to micropores and centered on 0.5 nm, the other one corresponding to wide mesopores and centered on 35 nm. When the pH increases, the centers of the peaks are unchanged, but their heights increases for mesopores, and go through maximum for micropores. These features are consistent with the values of Table 1, but Figure 6 brought the additional information that these porosity changes occurred at constant pore width.

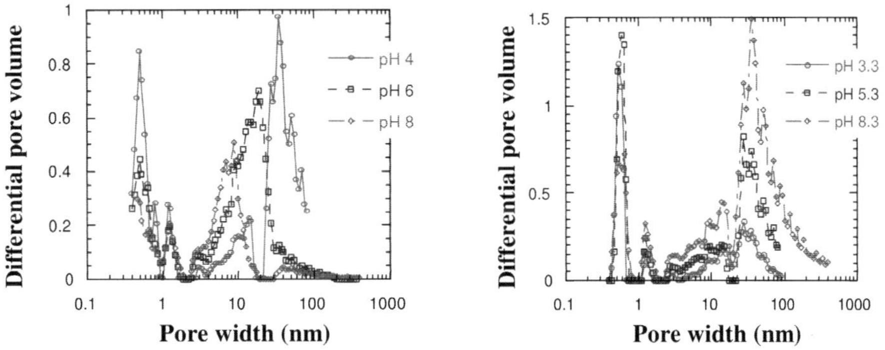

Figure 6. Micro-mesopore-size distributions derived from Figure 5 by application of DFT for: (left) TRF carbon cryogels; (right) TF carbon aerogels.

4. Conclusion

Using flavonoid tannins (T) for partial or complete substitution of resorcinol (R) allowed obtaining very different carbon gels and observing very different behaviors as well. A range of synthesis pHs much wider than that typically employed for RF gels could indeed be used, leading to the preparation of carbon gels having either pore textures close to those of classical RF aerogels, or having much lower porosity and different mesopore fractions.

If 2/3 in weight of R was replaced by T, gelation of tannin – resorcinol – formaldehyde formulations in aqueous solution was possible from pH 2 to 8. Both broadening and shifting of the pore-size distributions towards narrower pores were observed at increasingly high pH. Thus, tailoring the pore texture is easier in the case of TRF carbon gels than in the case of classical RF gels.

Complete replacement of R by T also allowed preparing new materials in a wide range of pH from 3 to 8. In this case, the pore-size distributions were not shifted with pH but the relative proportions of micro and mesopores were strongly changed instead. At constant pore diameters, mesoporosity increased considerably with pH, whereas microporosity presented a maximum near pH 5.

Tannins not only lead to carbon gels whose pore texture is more easily tailored than that of RF gels. Tannins are also non-toxic, environment-friendly, natural products which are typically thirty times cheaper than resorcinol. Combining their use with that of cheaper drying routes, such as supercritical drying in acetone and freeze-drying, low-cost carbon aerogels and cryogels, respectively, could be prepared. Such new materials are serious substitutes to expensive RF carbon aerogels for the same applications, provided that the synthesis pH was suitably chosen.

Acknowledgments

The authors gratefully acknowledge the financial support of the CPER 2007–2013 “Structuration du Pôle de Compétitivité Fibres Grand'Est” (Competitiveness Fibre Cluster), through local (Conseil Général des Vosges), regional (Région Lorraine), national (DRRT and FNADT) and European (FEDER) funds.

References

[1] Celzard A, Fierro V, Pizzi A and Zhao W 2011 *J. Phys. Conf. Ser.* present issue
[2] Pizzi A and Roux G 1978 *J. Appl. Polym. Sci.* **22** 1945
[3] Fierro V, Zhao W, Izquierdo MT, Aylon E and Celzard A 2010 *Int. J. Hydrogen Energy* **35** 9038
[4] Tarazona P 1995 *Surf. Sci. B* **331-333** 989
[5] Stoeckli F, Slasli A, Hugi-Cleary D and Guillot A 2002 *Micropor. Mesopor. Mater.* **51** 197
[6] Job N, Pirard R, Marien J and Pirard JP 2004 *Carbon* **42** 619
[7] Lin C and Ritter JA 1997 *Carbon* **35** 1271

Journal of Physics: Conference Series **416** (2013) 012023 doi:10.1088/1742-6596/416/1/012023

Multifunctional porous solids derived from tannins

Alain Celzard[1*], Vanessa Fierro[1], Antonio Pizzi[2], Weigang Zhao[1]

[1]Institut Jean Lamour – UMR UHP - CNRS 7198, ENSTIB, 27 rue Philippe Séguin, F-88051 Epinal cedex 9

[2]LERMAB, EA 4370, ENSTIB, 27 rue Philippe Séguin, BP 1041, F-88051 Epinal cedex 9

*Institut Universitaire de France. E-mail: Alain.Celzard@enstib.uhp-nancy.fr

Abstract. Tannins are extremely valuable, non toxic, wood extractives combining reactivity towards aldehydes, low cost, natural origin and easy handling. When polymerized in the presence of suitable chemicals including blowing agent, ultra lightweight rigid tannin-based foams are obtained. If pyrolyzed under inert gas, reticulated carbon foams having the same pore structure and the same density are obtained. The most remarkable features of tannin-based foams are the following: mechanical resistance similar to, or higher than, that of commercial phenolic foams, tuneable pore size and permeability, infusibility, very low thermal conductivity, cheapness, ecological character, high resistance to flame and to chemicals. Carbon foams have even better properties and are also electrically conducting. Consequently, various applications are suggested for organic foams: cores of sandwich composite panels, sound and shock absorbers and thermal insulators, whereas carbon foams can be used as porous electrodes, filters for molten metals and corrosive chemicals, catalyst supports and adsorbents.

1. Introduction

Condensed tannins belong to the family of flavonoids. Characterized by aromatic rings bearing hydroxyl groups, they present reactivity rather similar to that of resorcinol or phenol, two synthetic molecules with which resins are actually prepared and commercialized. The flavonoid tannins used in the present study were extracted from mimosa barks, and are "green", cheap but commercial raw materials that can replace resorcinol and phenol in a number of applications.

The aim of the present work was studying ultra lightweight rigid foams derived from tannins. Such materials could be used as such or after pyrolysis at 900°C under inert atmosphere. The former are organic materials which could replace the commercial synthetic phenolic foams, whereas the latter are reticulated vitreous carbon (RVC) foams with potential applications in the field of energy and environment. Preparation, main characteristics and physical properties are described in this paper.

2. Preparation of organic and RVC tannin-based rigid foams

Tannins were supplied by the company SilvaChimica (Italy) in the form of a light brown powder. When mixed with water, formaldehyde and an acidic catalyst, cross-linking occurs within a few minutes during which heat is released, leading to a hard black resin. Addition of furfuryl alcohol, another chemical of natural origin, both enhances the amount of heat produced and improves the mechanical resistance of the resultant resin. Such phenomenon is used to produce the foaming of the resin, through boiling of a liquid previously incorporated to the formulation: the blowing agent.

The amount of boiling agent, here diethyl ether, controls the final density of the foam, ranging from 0.04 to 0.16 g cm^{-3}. The formulation, see Table 1, is optimized in such a way that hardening and foaming occur simultaneously, so neither cracks nor pore collapse are observed. Foam preparation requires less than 2 minutes, is very easy, and any shape is possible, depending on that of the mould in which the liquid precursor has been poured. Finally, these are materials natural at the 95% level that are obtained, from which RVC foams can be prepared as follows.

Heating organic tannin-based foams in nitrogen at 4°C min^{-1} up to 900°C leads to uncracked, flawless, highly porous carbonaceous materials. Given the composition of the initial resin, non-graphitizable, glasslike, carbon is produced. During heating, typically half of the mass as well as half of the volume of the material are lost. Because shrinkage is exactly compensated by weight loss, the bulk density of RVC foams is strictly equal to that of their tannin-based precursors.

Table 1. Typical formulation of organic tannin-based foams.

Compound	Amount	Role
Tannin	30	Base of the resin
Furfuryl alcohol	6 - 15	Strengthener
Diethyl ether	1.5 - 5	Blowing agent
Water	6	Solvent
Formaldehyde (37% water solution)	7.4	Cross-linking agent
Para-toluene-4-sulphonic acid (65% water solution)	11	Catalyst

3. Main features and physical properties of tannin-based rigid foams

3.1. Characteristics of tannin-based foams

Bulk and skeletal densities, pore structure and cost have been determined as follows.

3.1.1. Density and porosity. Bulk density, being defined as the weight of one volume unit of material, was measured by careful weighing of parallelepiped samples of perfectly known dimensions. Bulk density is, as expected from simple calculations [1] and as shown in Figure 1, inversely proportional to the amount of blowing agent. Skeletal density is defined as the density of the solid from which the foam is made, and was measured by helium pycnometry. That of organic foam was found to be 1.59, whereas that of derived carbon materials was 1.98 g cm^{-3} [2,3]. From the two kinds of density values, total porosity could be calculated and is presented in Figure 2.

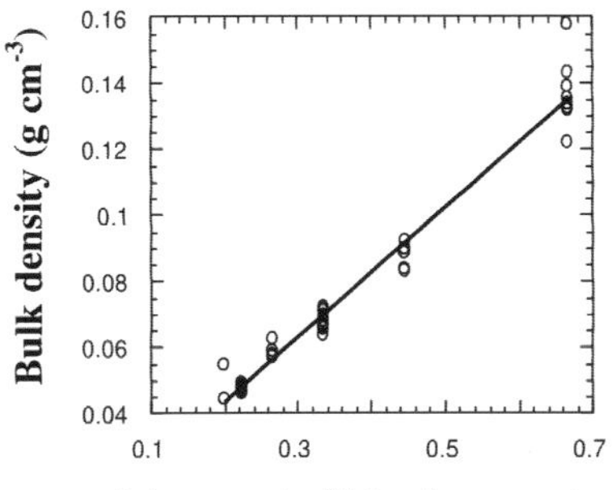

Figure 1. Bulk density of organic foams versus the inverse of the amount of blowing agent.

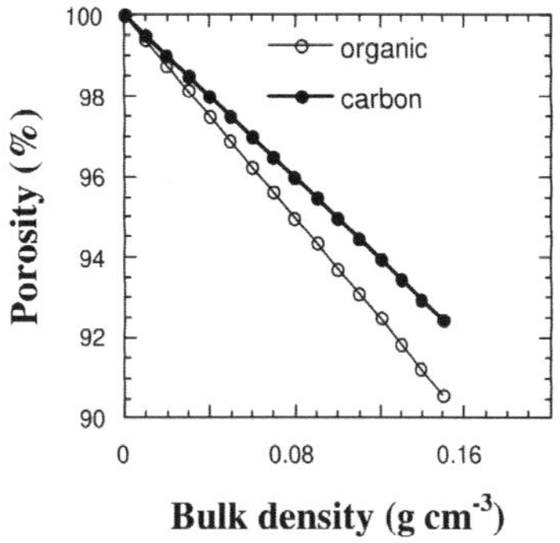

Figure 2. Porosity of organic and carbon foams as a function of their bulk density.

Journal of Physics: Conference Series **416** (2013) 012023 doi:10.1088/1742-6596/416/1/012023

3.1.2. *Pore size.* Pores of carbon foams were thoroughly observed by Scanning Electron Microscopy (SEM). Secondary electrons best revealed cell diameters and pore walls, whereas backscattered electrons best allowed visualizing struts and windows connecting the cells. Figure 3 shows that the linear cell density (expressed in number of pores per inch: ppi) depends on the density, and Figure 4 presents the correlation between diameter and bulk density. Simple theoretical considerations justify that cell size is inversely proportional to bulk density [1].

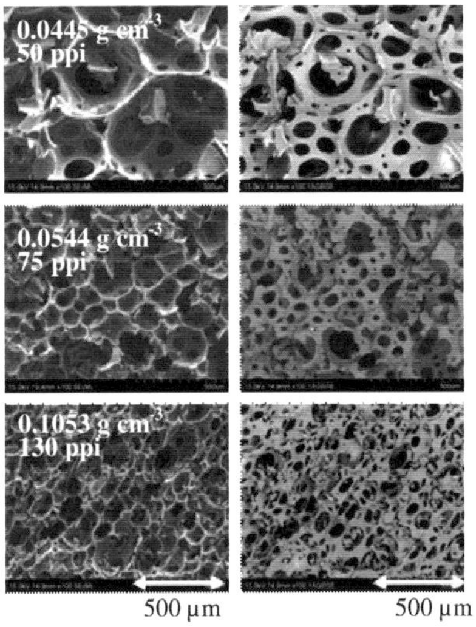

Figure 3. SEM pictures of carbon foams of various bulk densities. Left: secondary electrons; Right: backscattered electrons.

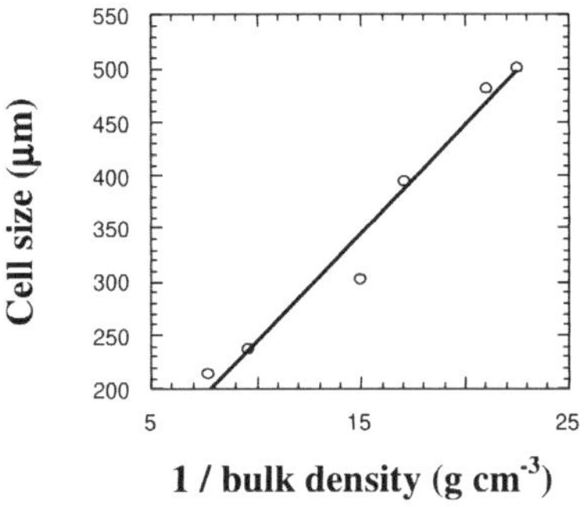

Figure 4. Pore size of carbon foams versus the inverse of their bulk density.

3.1.3. *Anisotropy.* Because the foams were grown vertically, anisotropy occurred. The cells are slightly elongated along the vertical direction, here referred to as the z-axis. The cells remain roughly circular in the orthogonal plane, here referred to as the xy-plane, see Figure 5. Anisotropy increases with bulk density and explains the slight differences observed between properties measured along z- or xy-directions (see next subsection).

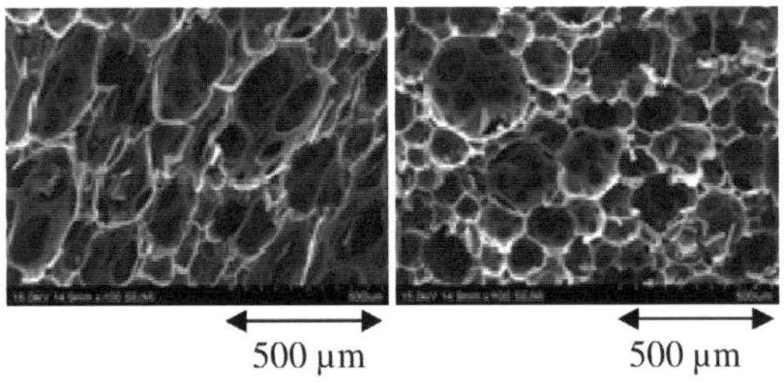

Figure 5. SEM views of the same carbon foam (density 0.067 g cm^{-3}) along two orthogonal directions. Left: z-axis; right: xy-plane.

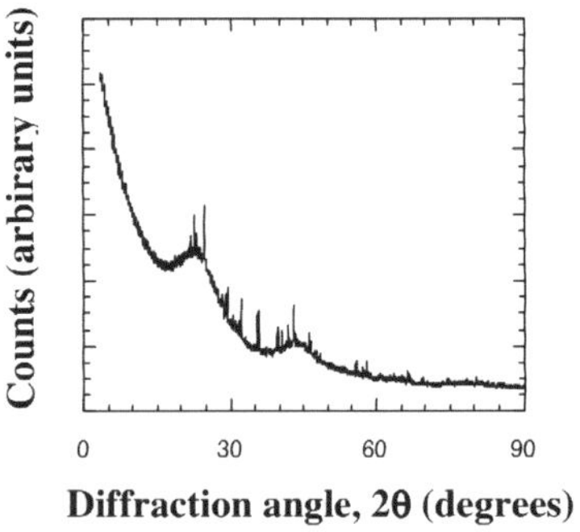

Figure 6. XRD pattern of carbon foam.

3.1.4. *Microscopic structure.* Organic foams were made of tannins and furfuryl alcohol cross-linked with formaldehyde, leading to an insoluble and infusible polymer network based on aromatic rings and oxygen-containing heterocycles. Such phenol-like precursors are known to lead to glasslike carbon after pyrolysis. Figure 6 indeed shows that RVC foams are based on disordered carbon, as evidenced by the X-ray diffraction pattern. Small peaks correspond to impurities naturally present in tannins.

3.1.5. Cost. Tannins are typically six and thirty times cheaper than phenol and resorcinol, respectively. The other chemical used here: furfuryl alcohol, diethyl ether and formaldehyde, have a moderate cost. Thus, foams cheaper than 3€ per kg have been prepared. Given the cost of nitrogen gas and electric power, carbon foams having a cost lower than 10 € per kg are obtained, making them the cheapest RVC foams of the market.

3.2. Physical properties of tannin-based foams

Resistance to fire, permeability, surface area, mechanical properties, electrical and thermal properties have been investigated.

3.2.1. *Resistance to fire.* As expected from phenol-like resins, tannin-based organic foams present outstanding resistance to fire, irrespective to their bulk density. It not only requires heat densities as high as 50 kW m^{-2} for igniting the samples, but the peak heat released is never higher than 12 kW m^{-2}, i.e. two order of magnitudes lower than that of epoxy for example, see Figure 7. Besides, the gas yield of the materials undergoing combustion is extremely low. These findings mean that, if the imposed heat is removed, the materials self extinguish. Resistance to fire of carbon foams, though not measured in similar conditions, is even higher, given that materials submitted to a 3000°C oxyacetylenic flame in air also self extinguish, see Figure 8.

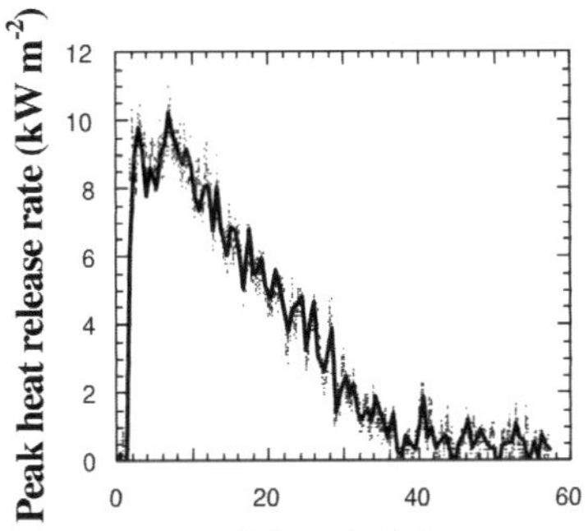

Figure 7. Peak heat release rate of organic foams submitted to a fire of power 50 kW m^{-2}.

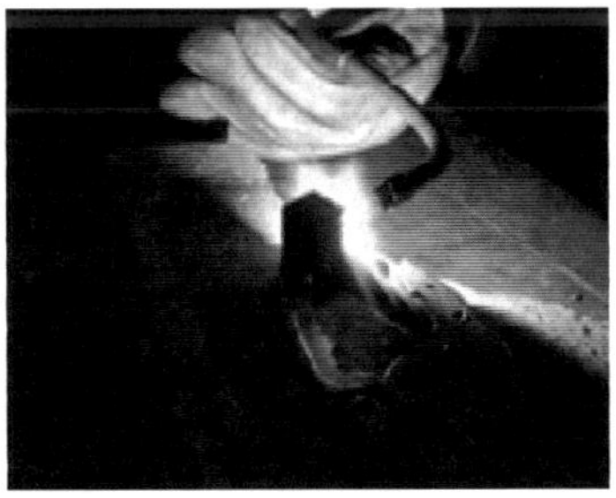

Figure 8. Carbon foam submitted to a 3000°C oxyacetylenic flame in air. Neither flame nor smokes are emitted.

3.2.2. *Permeability and surface area.* Permeability of organic and carbon foams has been determined by flowing water at a know flow rate across samples of known thickness and cross-section under a given pressure. The results are presented in Figure 9, showing that permeability follows a power la w of density with an exponent -2. Such dependence has been justified theoretically [4]. From the results given in Figure 9, the surface area could be calculated, see Figure 10. The same kind of curve, presenting a maximum at a bulk density close to 0.07 g cm^{-3}, was fully recovered when the surface area was estimated from the pore structure, considering spherical pores in close packing [4]. The agreement is good between the two sets of data, and also with the values measured by both krypton adsorption [3] and by X-ray microtomography [5], close to 1 m^2 g^{-1} for a bulk density of 0.07 g cm^{-3}.

3.2.3. *Electrical and thermal conductivities.* Only carbon foams present measurable electrical conductivity, presented in Figure 11. Generalised effective medium theory [6] predicts that the electrical conductivity increases with bulk density through a power law whose exponent depends on the morphology of the cells. Values of 1.5 and higher are expected for spherical and elongated cells, respectively. As seen in Figure 11, finding exponents slightly above 1.5 suggests slightly elongated cells, as expected. Whether the foams are organic or carbonaceous, the thermal conductivity is very low, although that of carbon is a little higher than that of the precursor resin. However, even after pyrolysis, the materials present extremely interesting insulating properties, given that the thermal conductivity of RVC foams of density 0.067 g cm^{-3} is 0.035 and 0.044 W m^{-1} K^{-1} along xy- and z-directions, respectively. These values are less than two times the conductivity of air at room temperature (0.024 W m^{-1} K^{-1}), and are of the order of magnitude of expanded polystyrene.

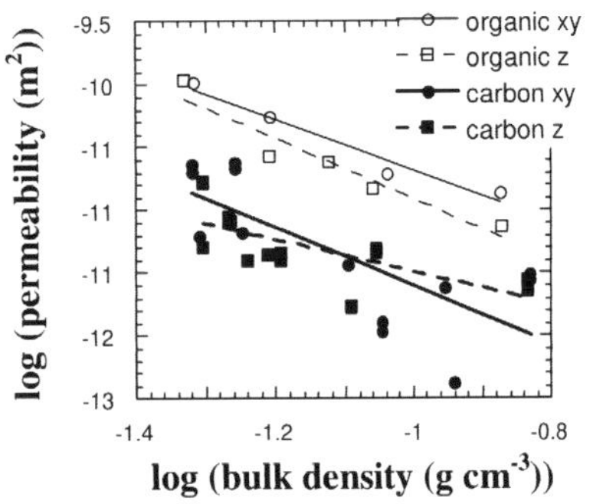

Figure 9. Permeability of organic and carbon foams, plotted in double-log scale so the power law dependence with an exponent close to -2 is evidenced.

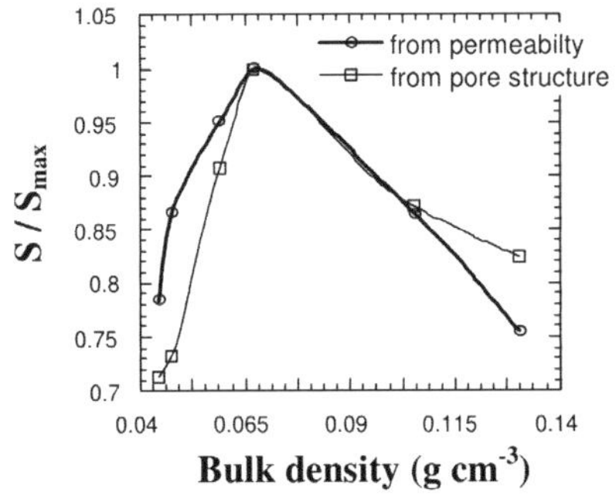

Figure 10. Normalised surface area of carbon foams, calculated by two methods.

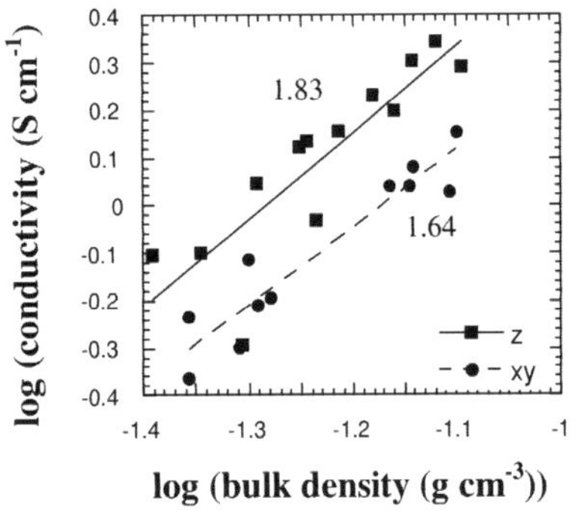

Figure 11. Electrical conductivity of carbon foams as a function of their bulk density, plotted in double-logarithmic scale. The slope of the straight lines fitted to the experimental points is given on the plot.

3.2.4. Mechanical properties. Compression of tannin-based foams leads to the typical behaviour of rigid cellular materials submitted to compression, see Figure 12: a linear elastic region, whose slope is the elastic modulus, followed by a long, serrated, plateau, corresponding to the collapse of successive cell layers, and finally a strong curvature upwards corresponding to the densification of the material once all the pores have been crushed. Elastic modulus and compressive strength both increase with bulk density and, due to the aforementioned anisotropy, present higher values along the z-axis. From these curves, the mechanical energy that can be absorbed by foams' irreversible crushing may be determined as the product of the compressive strength times the densification strain. As an example, the RVC foam of bulk density 0.13 g cm^{-3} can absorb 720 kJ m^{-3}. In other words, one m^3 of this foam may absorb the kinetic energy of a one ton car running at 140 km h^{-1}. Figure 13 shows that compressive strength again follows a power law of bulk density, with an exponent slightly higher than 1.5. According to Ashby et al. [7], exponents of 1.5 and 2 correspond to open cell and closed cell foams, respectively. Finding an intermediate value suggests that, as expected from SEM pictures, most of the cells but not all are open and fully connected with each other.

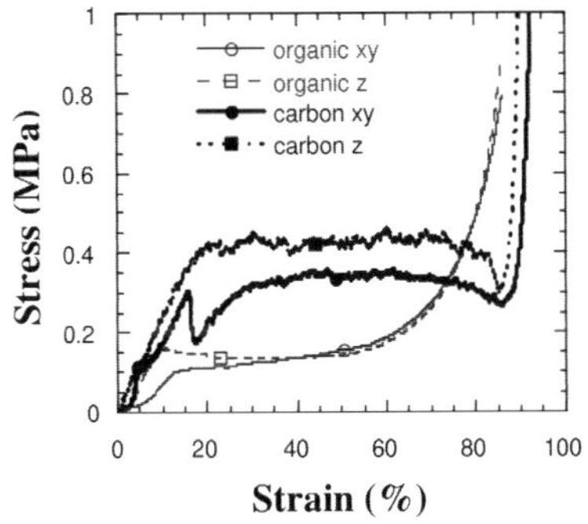

Figure 12. Strain – stress characteristics of organic and carbon foams having the same bulk density of 0.11 g cm^{-3}.

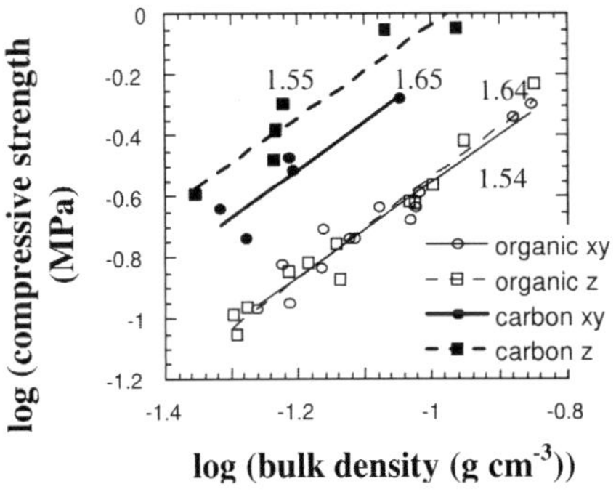

Figure 13. Power law dependence of compressive strength versus bulk density. The slopes are given on the plot.

4. Conclusion

Condensed tannins extracted from mimosa barks are extremely valuable chemicals. Being both low-cost and reliable raw materials, a rare feature for a natural product, tannins allow the preparation of "green" rigid foams having a number of interesting properties and characteristics. Organic foams are extremely efficient thermal insulators and present outstanding fire resistance. These features make them highly suitable for core materials in sandwich panels for marine and aircraft applications. Protection against shocks and sound absorption, presently in test, are also promising applications. It is important to recall that these materials are much cheaper than their phenolic counterparts, derived from the petrochemical industry, and presenting very similar mechanical properties.

After pyrolysis, glasslike (RVC) foams were obtained, having the same, tough narrower, pore structure. Their permeability and very high thermal and chemical resistance make them suitable as filters for hot corrosive fluids or molten metals. Their electrical conductivity allows using them as porous electrodes, and their wide, connected, pore volume, also suggests using these materials as catalyst supports. For that purpose, the inner surface of the cells may be covered by meso or microporous activated carbon through a preliminary step of impregnation with furfuryl alcohol, followed by pyrolysis and activation with steam. Doing so, a supported activated carbon is obtained [8], presenting a narrow surface porosity that is fully and quickly accessible. High adsorption and catalysis kinetics are thus expected. Finally, applications as scaffolds and heat insulation at high temperature are of interest and will be investigated soon.

Acknowledgments

The authors gratefully acknowledge the financial support of the CPER 2007–2013 "Structuration du Pôle de Compétitivité Fibres Grand'Est" (Competitiveness Fibre Cluster), through local (Conseil Général des Vosges), regional (Région Lorraine), national (DRRT and FNADT) and European (FEDER) funds.

References

[1] Zhao W, Pizzi A, Fierro V, Du G and Celzard A 2010 *Mat. Chem. Phys.* **122** 175
[2] Tondi G, Zhao W, Pizzi A, Du G, Fierro V and Celzard A 2009 *Biores. Technol.* **100** 5162
[3] Tondi G, Fierro V, Pizzi A, Du G and Celzard A 2009 *Carbon* **47** 1480
[4] Zhao W, Fierro V, Pizzi A, Du G and Celzard A 2010 *Mat. Chem. Phys.* **123** 210
[5] Tondi G, Blacher S, Léonard A, Pizzi A, Fierro V, Leban JM and Celzard A 2009 *Microsc. Microanal.* **15** 384
[6] Celzard A, Marêché JF, Payot F and Furdin G 2002 *Carbon* **40** 2801
[7] Sanders WS and Gibson LJ 2003 *Mater. Sci. Eng. A* **347** 70
[8] Zhao W, Fierro V, Pizzi A and Celzard A 2010 *J. Mater. Sci.* **45** 5778

Journal of Physics: Conference Series **416** (2013) 012024 doi:10.1088/1742-6596/416/1/012024

High-performances carbonaceous adsorbents for hydrogen storage

Weigang Zhao[1], Vanessa Fierro[1], E. Aylon[2], M. T. Izquierdo[2], Alain Celzard[1]

[1] Institut Jean Lamour – UMR UHP - CNRS 7198, ENSTIB, 27 rue Philippe Séguin, F-88051 Epinal cedex 9

[2]Instituto de Carboquímica (CSIC), Miguel Luesma Castán, 4. 50018 Zaragoza, Spain

E-mail: Vanessa.Fierro@lcsm.uhp-nancy.fr

Abstract. Activated carbons (ACs) with controlled microporosity have been prepared and their H_2 storage performances have been tested in a gravimetric device. Such adsorbents are natural Chinese anthracites chemically activated with alkaline hydroxides, NaOH or KOH. Outstanding total storage capacities of hydrogen, as high as 6.6wt.% equivalent to excess capacity of 6.2 wt.%, have been obtained at 4MPa for some of these adsorbents. These values of hydrogen adsorption are among the best, if not the highest, ever published so far in the open literature. They are well above those of some commercial materials, e.g. Maxsorb-3, considered as a reference of high-performance adsorbent for hydrogen adsorption. Such exceptional storage capacities may be ascribed to a higher volume of micropores (< 2nm).

1. Introduction

Hydrogen is a renewable energy vector frequently proposed as an alternative to the current fossil fuel-based energy system to solve problems pertaining to energy distribution and environmental issues. However, the main drawback of hydrogen arises from its supercritical nature under standard conditions (Tc = 33.19K, Pc = 1.296MPa), giving it a very low density: only 90g m^{-3} at 1bar and 273K. The US Department of Energy set targets for hydrogen storage systems in light-duty vehicles in 2009. These targets correspond to a gravimetric capacity (i.e., the ratio between net useful energy and maximum system mass) of 4.5% or 1.5kWh kg^{-1} (formerly: 6% or 2kWh kg^{-1}) in 2010, and of 5.5% or 3kWh kg^{-1} (formerly: 9% or 3kWh kg^{-1}) in 2015 [1]. These new targets are still difficult to reach at room temperature. However, stationary applications are not limited by their weight and therefore technical solutions could be available in a shorter time.

The success of any future hydrogen economy depends on the development of inexpensive materials with sufficiently high hydrogen-storage capacity. Four alternatives for hydrogen storage can be anticipated: liquefaction, compression, physisorption, and storage in the form of metallic hydrides. Liquefaction and compression strategies have high operation costs. Moreover, compression needs high pressure operations, which causes high cost for both process and storage tank, and gives rise to safety issues related to high pressures Metal hydrides have incomplete reversibility and low charge/discharge kinetics. Porous and high-surface area activated carbons (ACs) are among the most important candidates to store efficiently hydrogen due to their low price, high surface area and tunable nanoporosity. Hydrogen storage by physisorption on activated carbons can be an interesting option in stationary applications, wherein weight and volume are not limiting factors. Expected benefits are lower pressures thus safer operations, fast kinetics and complete reversibility.

Anthracites combine low cost, low ash content and an ideal microtexture for getting a tunable nanoporosity. In this study, we have prepared ACs from Chinese anthracite by varying systematically the activation temperature and the amount of activating agent (NaOH or KOH). We compared their performances to that of Maxsorb-3.

2. Experimental

We have prepared ACs by chemical activation, and the commercial AC Maxsorb-3 (formerly AX21) from Kansai Coke and Chemicals Co., Ltd. (Japan) was used for the sake of comparison. All the materials were characterized by N_2 and H_2 adsorption at 77K.

2.1. AC's synthesis

The precursor of the ACs investigated here was Chinese anthracite from Taisi mine. It was first ground and sieved in order to collect the grains having an average size within the range 100 - 200µm. The anthracite powder was then physically mixed with alkaline hydroxide (XOH): either sodium (X = Na) or potassium (X = K) hydroxide lentils (typical size 5 mm), according to various weight ratios XOH / anthracite, *R*, ranging from 1.5 to 2.5. Additional ACs where synthesised by KOH activation with an *R* up to 5. The resultant mixture was introduced into a nickel crucible, and heat-treated in a horizontal furnace under a stream of nitrogen at constant heating rate of 5K min^{-1} up to the final activation temperature, *T*. *T* was fixed within the range 973 – 1073K and maintained for 1h. The crucible was then allowed to cool down to room temperature under nitrogen flow. More details are given elsewhere [2, 3].

2.2. AC's characterisation

2.2.1. *N_2 adsorption.* N_2 adsorption isotherms were obtained at 77K using an automatic adsorption apparatus, ASAP 2020 from Micromeritics. The samples were outgassed for 48h under vacuum at 523K prior to N_2 adsorption. Surface areas, S_{BET}, were determined by the BET calculation method [4] applied to the adsorption branch of the isotherms. S_{BET} and pore volumes were estimated with a typical, maximum, uncertainty of 3%.The pore-size distributions (PSD) were calculated by application of the DFT model [5] supplied by Micromeritics software, considering slit-shaped pores.

2.2.2. *H_2 adsorption.* Determination of H_2 storage properties at 77 K for the ACs was carried with a gravimetric device from VTI Corporation (Miami, USA) at ICB (Zaragoza, Spain). The system consists of a fully computer-controlled microbalance, which automatically measures the weight of the carbon sample as a function of time, with the same hydrogen pressure and sample temperature under control. Hydrogen isotherms were obtained by setting pressure steps within the range 0 - 8MPa at 77K. For each run, approximately 500mg of activated carbon sample were placed into a bucket made from stainless steel mesh. Hydrogen pressure was gradually increased to prevent disruption of the microbalance, until the desired value was reached. A pressure transducer (accuracy 0.05% of the full pressure range up to 10MPa) was used to monitor hydrogen pressure in the system. More details have been given elsewhere [2, 3]. Repeatability was always found to be very satisfactory, leading to relative errors lower than 3% on each measurement.

3. Results and discussion

3.1. Effect of R and T

Figure 1 and Figure 2 show the effect of *R* and *T*, respectively, on H_2 storage capacity. Absolute (total) H_2 storage capacities are shown. The increase of *R* and *T* produced a nearly linear increase of H_2 storage for both activating agents. The slope of the fitted straight line of the H_2 storage capacity with *T* was higher for KOH than for NaOH.

Journal of Physics: Conference Series **416** (2013) 012024 doi:10.1088/1742-6596/416/1/012024

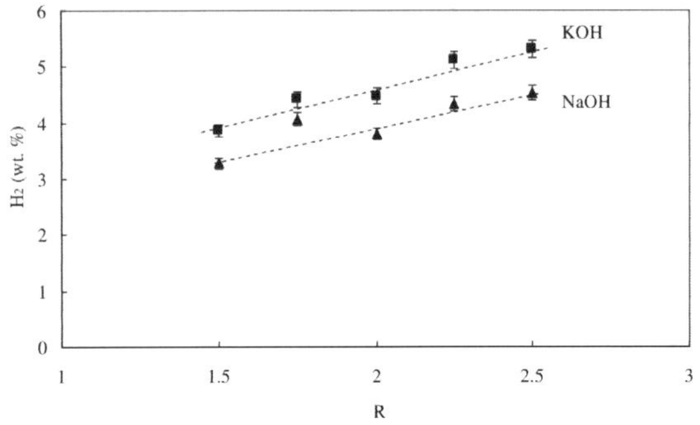

Figure 1. Evolution of H_2 storage capacity with the activation ratio, *R* (▲ NaOH, ■ KOH).

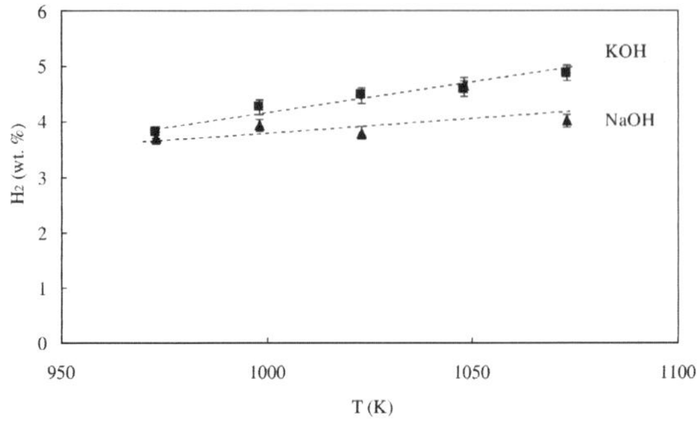

Figure 2. Evolution of H_2 storage capacity with the activation temperature, *T* (▲ NaOH, ■ KOH).

H_2 storage capacity increased with both *R* and *T*, and was always higher on ACs produced by KOH activation. The use of KOH allowed obtaining S_{BET} values ranging from 1688 to 2772m^2 g^{-1}, whereas NaOH led to lower surface areas, within the range 1500 – 2065m^2 g^{-1}.

Figure 3 shows the evolution of the pore-size distribution of three ACs prepared with KOH when *T* increased from 1023 to 1073K, at *R*=2, and when *R* increased from 2 to 2.5, at *T*=1023K. Increasing *R* and *T* increased the micropore volume but also the fraction of wider micropores (0.7-2 nm) and that of mesopores (> 2 nm).

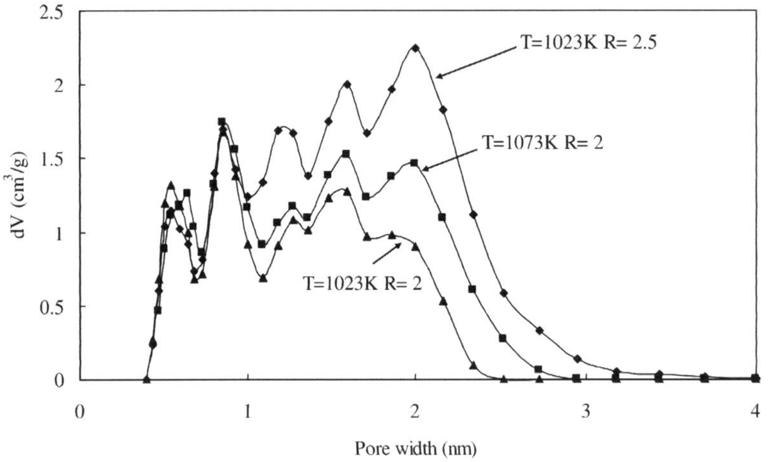

Figure 3. PSDs of three ACs prepared by KOH activation (▲ *T*=1023K and *R*=2, ■ *T*=1073K and *R*=2 and ◆ *T*=1023K and *R*=2.5).

In other words, the development of surface area by KOH activation is always accompanied of pore widening. Therefore, PSDs shifted to wider pores can lead to samples exhibiting higher H_2 adsorption capacities, provided that a high volume of micropores (< 2 nm) still exists.

3.2. Further H_2 storage improvements on KOH-ACs

Based on the previous experiments, we decided to continue our study using only KOH as activating agent and increasing *R* up to 5. We got ACs with S_{BET} as high as 3434 m^2 g^{-1}. These S_{BET} are considered as apparent surface areas, given that it is not physically possible to have carbon materials with a surface higher than 2630 m^2 g^{-1}.

Figure 4 shows the total hydrogen storage capacity of all the ACs prepared by KOH activation. Total hydrogen storage capacity as high as 6.6 wt.%, equivalent to 6.2 wt.% excess capacity, were obtained. The US DOE target for 2010 was 5.5 wt.% for hydrogen storage systems (AC + tank). However, the performances reached in this study are among the best reported by the latest reviews [6, 7] and recent literature in general, e.g. [8-10]. Therefore, our materials are among the best reported in the literature so far.

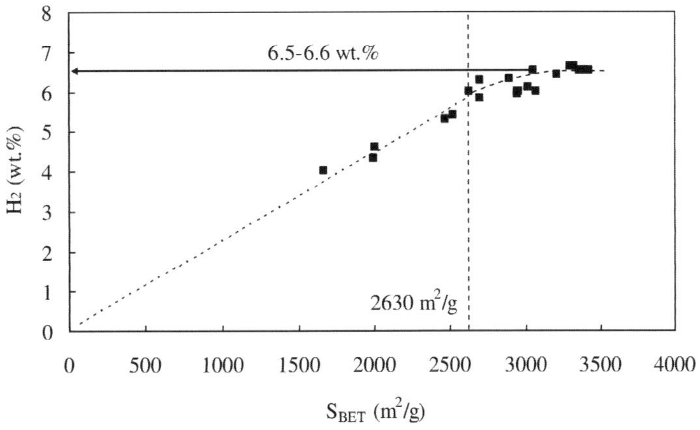

Figure 4. Total H_2 storage capacity (wt. %) as a function of S_{BET}.

We have already shown [2] that a linear relationship is observed between maximum hydrogen uptake at 77K and BET surface area only if the latter is not higher than typically 2630 m^2 g^{-1}. Above such a value, a plateau is reached due to a saturation of H_2 storage capacity at 2630 m^2 g^{-1}, and a slight decrease was found for the highest surface areas probably due to the broadening of the pores.

3.3. Comparison with Maxsorb-3

Maxsorb-3 has been claimed to adsorb about 6 wt. % of hydrogen at 77K and 5MPa, 5 wt. % being already reached at 2MPa [11]. Other results on the same material are about 3 wt. % at 80K and 10MPa [12] and, at room temperature: 1 wt. % at 5MPa [13], and 1.6 wt. % at 70MPa [12].

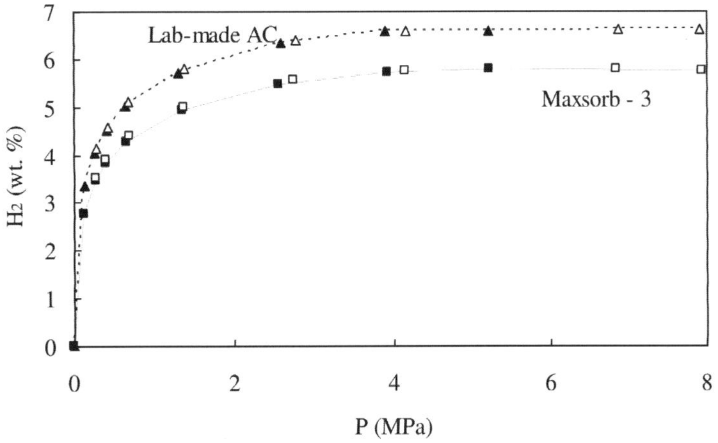

Figure 5. Total H_2 storage (wt. %) isotherms at 77K for Mawsorb-3 and for our best lab-made AC. Adsorption (full symbols) - desorption (open symbols).

Figure 5 shows the H_2 adsorption isotherm for the best of our ACs and Maxsorb-3. Our lab-made AC reached 6.6 wt. % whereas Maxsorb-3 reached "only" 5.8 wt. %. H_2 adsorption was totally reversible and desorption was fast. Our lab-made AC has a higher S_{BET} than Maxsorb-3: 3310 and 3203 $m^2\ g^{-1}$, respectively. However, this small difference in the S_{BET} can not justify such a difference of H_2 storage capacity. The reason of the higher performances of our lab-made AC is based on its PSD shifted towards narrower pores. Our lab-made AC indeed presents a higher faction of micropores (<2nm) calculated by the DFT method.

4. Conclusions

We have prepared highly microporous ACs by activation with NaOH or KOH, which showed outstanding performances for hydrogen storage at 77K. The ACs with the highest H_2 storage performances were obtained by KOH activation, using an activating agent to precursor weight ratio of 4 or higher, and activation temperatures higher than 1023K. Some of these ACs reached a total hydrogen storage capacity of 6.6 wt.%. These performances are well above those of the well-known commercial AC Maxsorb-3. Such exceptional hydrogen storage capacities may be ascribed to a higher volume of micropores (< 2nm) of our ACs when compared to Maxsorb-3.

Aknowledgements

We thank the financial support of the CPER 2007-2013 "Structuration du Pôle de Compétitivité Fibres Grand'Est" (Competitiveness Fibre Cluster), through local (Conseil Général des Vosges), regional (Région Lorraine), national (DRRT and FNADT) and European (FEDER) funds. Pr. Carrott from Universidade de Evora (Portugal) is gratefully acknowledged for providing us with the commercial activated carbon Maxsorb-3.

References

[1] http://www1.eere.energy.gov/hydrogenandfuelcells/storage/pdfs/targets_onboard_hydro_storage_explanation.pdf (consulted 23/03/2011)

[2] Fierro V, Szczurek A, Zlotea C, Marêché JF, Izquierdo MT, Albiniak A, Latroche M, Furdin G and Celzard A 2010 *Carbon* **48** 1902

[3] Fierro V, Zhao W, Izquierdo MT, Aylon E and Celzard A 2010 *Int. J. Hydrogen Energ.* **35** 9038

[4] Brunauer S, Emmet PH and Teller E 1938 *J. Am. Chem. Soc.* **60** 309

[5] Tarazona P 1995 Surf. Sci. **331-333** 989

[6] Yürüm Y, Taralp A and Veziroglu TN 2009 *Int. J. Hydrogen Energ.* **34** 3784

[7] Zubizarreta L, Arenillas A and Pis JJ 2009 *Int. J. Hydrogen Energ.* **34** 4575

[8] Wu HC, Li YY and Sakoda A 2010 *Int. J. Hydrogen Energ.* **35** 4123

[9] Huang CC, Chen HM and Chen CH 2010 *Int. J. Hydrogen Energ.* **35** 2777

[10] Suarez-Garcia F, Vilaplana-Ortego E, Kunowsky M, Kimura M, Oya A and Linares-Solano A 2009 *Int. J. Hydrogen Energ.* **34** 9141

[11] Chahine R and Bose TK. 1994 *Int. J. Hydrogen Energ.* **19** 161.

[12] de la Casa-Lillo MA, Lamari-Darkrim F, Cazorla-Amoros D and Linares-Solano A 2002 *J. Phys. Chem. B* **106** 10930

[13] Ye Y, Ahn CC, Witham B, Fultz B, Liu J, Rinzler AG, Colbert D, Smith KA, Smalley RE 1999 *Appl. Phys. Lett.* **74** 2307

Journal of Physics: Conference Series **416** (2013) 012025 doi:10.1088/1742-6596/416/1/012025

Economical interest of a X-rays vision system in a planing mill production chain

Almecija Benjamin*, Choffel Denise, Daquitaine Renaud***, Bombardier Vincent**, Charpentier Patrick****
*Siat-Braun/CRAN
**CRAN
*** Siat-Braun
E-mail : almecija.b@orange.fr

Abstract. This presentation is the first phase of a more complete study about the interest of the use of an X-ray scanner in the context of a planing mill. This scanner, placed upstream to edging phases of concerned wood products (unedged timber, square edged timber), allows to make visible not perceptible defect of the material by the human operator eye. These new knowledge in correlation with rough materials has to allow an improvement of the valuation of the end products, but also has to indicate the improvement tracks of the production installation. Our first tries aim at validating this hypothesis.

1. Introduction

In wood industry, yield increase and improvement of the qualitative classification represent important financial gains. Various ways to improve the industrial process are followed: automation of jobs, production operators training, implementation of decision-making tools and means of control. The study presented here links the automation of working posts and implementation of decision-making tools. The application case is about SIAT-BRAUN planing mill and aims to improve the qualitative sort out of a production chain.

2. Context of the study

A planing mill makes wooden products with high added value such as brackets or mouldings. The raw material is unedged timber or square edged timber in 15 % of relative humidity. This raw material is also said "Chose wood", i.e., it contains few defects (knots, marrow, compression wood, heart wood, cracks, and tints). It is necessary to valuate at the best this expensive raw material. To do it, two main ways are followed: (1), from a quantitative point of view, reduce the losses to increase the yield and, (2), on a qualitative point of view, estimate the characteristics of every board to maximize the associated financial gain.

To make profitable an industrial installation aiming at these objectives, big volumes must be treated there. Time granted for every product and the decision-making are thus weak.

In the entrance of the planing mill, we find a dryer and an edging chain. This last one consists of:

- A supply post in unedged timber or in square edged timber by variable dimensions,
- A valuation workshop of the board on the length by a purge and a selection of the standard lengths of finished products stemming from the board and the cuttings into sections which ensue from it,
- A valuation workshop of the board on the width by a selection of zones of homogeneous quality of variable width.

Journal of Physics: Conference Series **416** (2013) 012025 doi:10.1088/1742-6596/416/1/012025

The edging patterns are obtained by means of an optimization taking into account quality zones. They allow to obtain wood pieces of small rectangular sections (pre-cut). The pre-cut are then sorted out and manufactured on moulding machine to obtain finished products.

One of the key points of the transformation process is in the definition of homogeneous quality zones on the width of every board.

The study, presented in this paper, concerns this valuation workshop in particular. Zone selection is made by a human operator assisted by technical means. Quality decisions are based on the analysis of the external profile of every board according to five quantifiable criteria:

- Presence or not of a defect,
- Defect type,
- Size of the defect,
- Location on the board,
- Type of finished products which can be made in the thickness of the board.

Evaluation of these criteria for every board is subject, because of the human operator, to decision errors [Buehlmann U., 2002]. These errors are linked to the hardness and the repetitiveness of the task without taking into account the short time to analyse and decide. The objective of the study is to estimate the qualitative and quantitative impact that an intern vision system can have in selection of the homogeneous quality zones.

So, it is necessary to define the various homogeneous qualities used to zoning as well as the current process. Then, it is necessary to justify the choice of an intern vision system able to improve process, and finally, estimate the contribution of this vision system in the application case.

2.1. Defects taken into account for the qualitative sort out

The wood defects are taken into account in the planing mill standards which defines qualities of the sold finished products. Every defect must be analysed and located to predict quality class which will associate to the end product. The qualitative classification (or sorting) is based on six different wood defects. These are, in importance order: (1) knots, (2) curly grain wood, (3) compression wood, (4) warmed wood and tints, (5) cracks and (6) resin pockets.

2.1.1. Knots

Knots are origins of branches (of stronger density than the wood of the trunk) Stemming from buds, knots can be located everywhere in the tree by the tree. Knots can be dead or alive at the moment of the tree cross-cutting. In the dead knots, it is necessary to distinguish those which are linked to healthy wood (white knots) and those which are unhealthy or susceptible to fall from surrounding wood (knots corks of the conifer, the black knots). Knots are locations of strong withdrawal during board drying. This leads to crack or to total unlink of knot. These cracks contribute to unsightly aspect of knots on the finished products. Concerning the influence zone of these knots, peripheral zones of knots where the thread of wood is diverted, they can be the cause of a qualitative displacement but it is about a subjective criterion appropriate for the quality operator.

2.1.2. Curly grain wood

Curly grain wood is mainly located in the footing of the trunk. The aspect of the grain is contorted, of a sinuous grain in irregular and muddled curves. Its presence is avoided in the production of brackets or high-quality mouldings because of its aspect. Furthermore, this wood is breakable and can create difficulties during manufacturing.

2.1.3. Compression wood

This wood results from a natural reaction of the vegetable to compression strength. The physical properties of this wooden type are such as it results deformations due to release of the internal tensions, irreversible collapses in the drying, bad surface aspect, or cracks . Furthermore, this wood is tinted in some species such as Fir or Spruce. All these factors make that a safety margin must be taken during the selection of the homogeneous quality zones.

2.1.4. Warmed wood and tints

Warmed or tinted wood often have fungal origins. Attacks, if they are stopped on time, do not affect the mechanical resistances of the wood but tint the wood. Superficial tints can also be due to a prolonged exposure in the direct sun rays (UV).

2.1.5. Cracks

Cracks are due to a rough change of temperature (frost cracks, check or star shake), to a separation of two growth rings (ring shake), to the sun (sunburn or crack of drought) orto an unsticking of fibers (honeycombing). Frost cracks, checks and star cracks develop in the radial direction. The cracks of drought propagate of the bark up to the sapwood and are characterized by flows of resin leading to a tanning of the wood. Cracks, more at least long and deep, can, besides the unsightly aspect, be a reason to breakage of products after or in the course of manufacturing.

2.1.6. Resin pockets

Resin pockets are present only in evergreen wood. Resin pockets are internal cracks filled with resin. These pockets raise problems more aesthetic than structural even if the resin presents complications during the manufacturing (fouling of tools). These pockets can be inside the end product but don't have to be visible on end products of high quality.

2.2. Qualitative zoning of the production chain

In production, zoning is made by a human operator who pilots a measurement station. This station is linked to optimization software which establish the edging patterns and to a multi-blades edging machine. The operator selects the zones of homogeneous quality through a laser and through a control panel. These homogeneous quality zones are defined, at the same time, by the direction of the grain and by a variable width, i.e., for an unedged timber, 2 different zonings (several parallel zones) are necessary as the Figure 1 shows it.

In the experimental case of the study, there are 5 different qualities:

- The quality "OA", the highest, which contains no defect and outcome exclusively from a straight wood grain on all the length. A little deviation of the grain (subjective criterion) is however accepted,

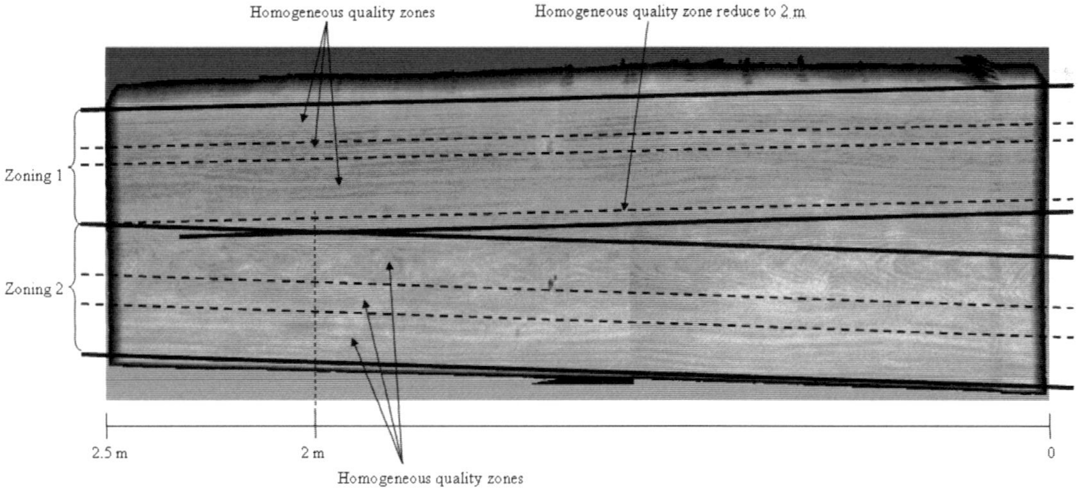

Figure 1: Zonings and zones définitions

- The quality "OB", the second, which can contain black knots of diameter lower than 5 millimetres, knots white with diameter lower than 7 millimetres, with tints, with collapses or crack lower than 0,3 millimetres, compression wood with straight grain and some curly grain wood, resin pockets lower than 2 centimetres, bark pocket lower than 3 millimetres. All these criteria are the maximal tolerances of the quality but they cannot be all combined on the same zone because this one would present too much defects. In that case, the zone is downgraded towards the "Relegated" quality (see after). On the other hand, values of these tolerances decrease according to the dimensions of finished products, or families of products, to which the thickness of the board linked. There is a subjective criterion concerning the general aspect of the zone and the definition of this quality,

- The lowest, "Relegated" quality, which accepts the 6 types of defects regardless to their size,
- The quality "OA to cut" which links the criteria of the quality "OA" with a purge in one side (maximum 50 centimetres) allowing to eliminate a penalizing defect,
- The quality "OB to be cut" which links the criteria of the quality "OB" with a purge in one side (maximum 50 centimetres) allowing to eliminate a penalizing defect.

The qualities "OA to cut" and "OB to cut" are respectively included in the qualities "OA" and "OB" during the analysis of the production.

3. State of art

3.1. Contribution of the means of visions

Plentiful literatures tries to show that the use of industrial vision means during the first manufacturing of the wood generate financial gains [Thawornwong S., 2003; Rinnhofer A., 2003; Schmoldt D.L., 2000; Lundgren N., 2007]. These studies are applicable to the manufacturing of boards in pre-cut.

There are two types of industrial vision: (1) the external vision and (2) the internal vision. By misuse of language, the vision initially linked to the sensor "eye" is henceforth said internal when it is a question of obtaining a representation of the internal structure (measure of density, for example) products via a technique of non-destructive control working by transmission in the material.

3.2. External vision

The main external vision means used in industrial environment are laser and digital camera (Manufacturer and designer referents: Microtec-Springer, Luxscan-Weinig, COE Newnes / McGehee). These means of detection allow obtaining images of external shape. Afterward, algorithms of detection and, sometimes, interpretations of the defects are applied. The results concerning the detection of defects depend on algorithms which are associated to it but also dimensions of the scanned products. Funck [Funck J.W., 2003] realized a comparative study where results of correct detection of defects extend from 80 % to 99 % according to the algorithms. There are also new techniques as the fuzzy reasoning method [Bombardier V., 2007; 2009] which proposes to translate the subjectivities of the human language in a numerical language by “fuzzy sensor”. This new kind of classification algorithms have better results than the others and, in addition, it need less training samples, and finally, calculation times match to industrial process times. This type of vision requires necessarily the use of algorithms of prediction (interpretation) for internal analysis of the material. In the case of log, it is possible to interpret the forms of the shape [Mäkelä A., 2003] to suppose (to predict) underlying defects. The results remain however little satisfactory because scanned logs always presented shape defects which perturbed the analysis (missing bark, mistletoe bump). In the case of boards, the interpretation of defect behaviour, visible on the shape, inside the material is much simpler because the dimensional and geographical variation of the defect is limited by the weak thickness of the product and the shape defects are reduced because of the already realized manufacturing. However, external vision never allows to display and to locate non-visible defects on shape such as pockets of resin.

3.3. Internal vision

Internal vision of the material is possible by technological means using the properties of gamma rays [Hampel U., 2007] still in the study, of the nuclear magnetic resonance (NMR) for molecular analyses [Maunu, 2002], ultrasounds [Linen C-J, 2008], microwaves for the detection of defects [Yu G., 2009] or X-rays for the localization and the labelling of defects [Sarigul E., 2003a and 2003b; Longuetaud F., 2005]. These methods of vision have results on a macroscopic or on a microscopic level (analysis of the molecules which constitute some material). Each of these techniques has their advantages and their drawbacks. At the moment, NMR technique is dedicated for the molecular vision while techniques based on X-rays, ultrasounds, microwaves, although useful in molecular vision, are rather dedicated for the industrial world for the macroscopic vision.

According to comparative studies, the X-ray imaging is the most successful and the most promising method on an industrial point of view [Skatter S., 1998; Wei Q., 2009]. Furthermore, towards the context of the study, this method is compatible with the environment of wood manufacturing (dust, vibration, climat).

3.4. The vision by X-ray scanner

The imaging by X-rays allows localization of several wood defects. Combined with detection algorithms, this method allows obtaining a mapping (labelling) of the material which can, in theory, be used in every manufacturing stage of the material.

More specific studies for every type of defect were led. These studies concern mainly improvement or creation of detection algorithms. These algorithms use statistical models, intelligent systems of "on-line" type ("dynamics") or artificial neurones network (ANN), pre-treatments or post-treatments of images, mathematical operators or still, detection results of the defects.

From the X-ray imaging, it is possible to detect and to qualify knots, as concludes it Oh J-K [Oh J-K., 2009] which specifies that a precise determination is not yet possible. The cracks detection, limits between sapwood and hardwood, compression wood, wane, embedded bark or resin pockets is also possible. It is also possible to consider the density of particular zones to analyse causes or consequences [Leban J.M., 2004]. This type of imaging still allows to estimate the mechanical resistance of wood from the characteristics of knots [Öhman M., 1999] or from models, as Brännström M. made by correlating the density, the width of rings, the module of elasticity and the resistance in flexion [Brännström M., 2007].

Reliable and strong algorithms of detection require however long time of calculations and are often useful for one single type of defect at the same time. So, there is a time accumulation for all the detection on every image. On the other hand, the acquisition, in an industrial environment, is generally made with a linear source / detector system (range of X-rays) perpendicular to the scanned piece and not use the reconstruction of the volumes, the tomography, which is much more precise (resolution raised in 3 dimensions).

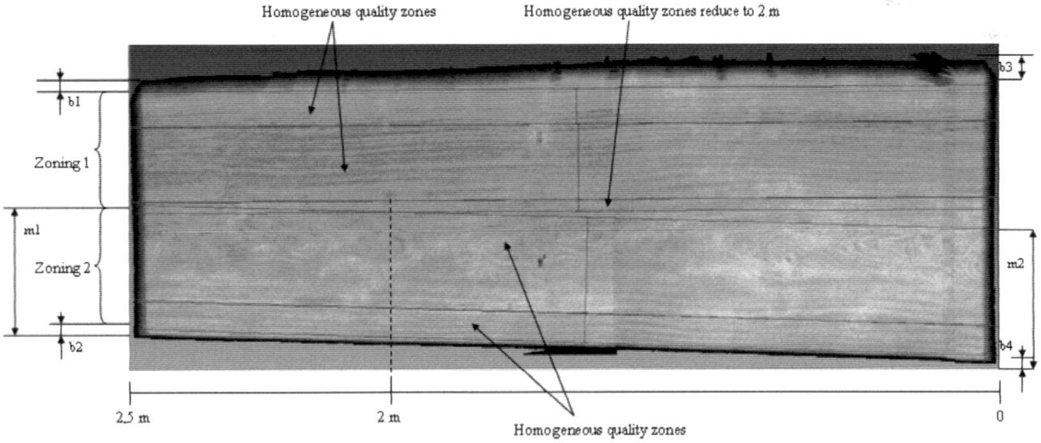

Figure 2: Zoning localisation

Times of acquisition to realize a tomography are indeed long. The study of Magnusson Seger M. in 2003 encourages exploitation of the cone of X radiation in the industrial circles by developing, in laboratory, fast algorithms and able to reconstruct volume with high production speed [Magnusson Seger M, 2003]. It becomes now possible to use reconstructions in 3 dimensions to detect more easily and more exactly the defects.

However, the imaging by X-rays does not allow visualizing tints (aesthetic criteria) because these present a too tiny difference of density.

In this study, X-rays imaging is used for its capacity to detect and to localize internal and external defects of the wood.

4. Experimental method

4.1. Objectives
The study presented here aims to three objectives: estimate the impact of a X-ray scanner in identification and sizing (1) of visible defects by the operator but also (2) of non-visible defects, and finally, consequences due to its introduction in the edging chain (3). To do it, an experiment was led in a real scale: the following paragraphs describe this one.

4.2. Sampling
We selected randomly and among stocks, one parcel of square edged timber in dry mixed Fir-Spruce whose nominal thickness is 24 millimetres and 4 meter long, then one parcel of unedged timber in mixed Fir-Spruce of 32 millimetres thickness and 5 meters long. These two thicknesses represent 50 % of the production of the edging chain.

Boards are cut up in a fixed length of 2.5 meters because it is the length of the most sold products, the most representative. Among these boards, a group of 29 boards is selected for the analysis by X-rays. These boards are chosen for the defects which they contain. The purpose is to have the biggest proportion of defects in a minimum of samples. In this way, among 29 boards, 14 have a 24 millimetres thickness and 15 have a 32 millimetres thickness. Then, the boards of width superior to 50 centimetres are edged (in two) because their passage is impossible in the X-ray scanner (truncated images). This operation carries the number of products scanned in 32.

Boards not selected for the analysis by X-rays (more than 70) are visually analysed to evaluate presence or not of defects and so, establish general statistics based on the proportions of every defect according to the thickness of boards. In this way, the results obtained with experiment samples are generalized at 50 % of the planing mill.

4.3. Measure of the samples by X-ray scanner
X-ray scans are acquired by means of a medical scanner named Brightspeed 4 from General Electrics. Scans are made in two steps because the displacement of the scanner is lower than length of boards. Furthermore, to limit the acquisition time, several boards are scanned at the same time; they are piled and separated by low density foam.

4.4. Treatment of the measures obtained by X-ray scanner
The measures obtained by the scanner were manually processed to:
- Separate the boards of the same scan,
- Put the two half-scans together to obtain an image of the board of all length.

The difficulty of the operation results from variations of orientation and from location due to the manual reversal of piles to scan the second half-length. Every volume (board) is then handled with manual threshold. Thresholds allow to eliminate or to highlight grey level ranges by means of colours or luminosity. The method of thresholding bases itself on the spectres of the levels of grey of boards.

4.5. Qualitative zoning from the X-rays imaging
The widths of homogeneous quality zones are measured with a tool issued from the software as the Figure 2 shows it. Furthermore, beginning of the zonings, whose are located regard to the board edge in both extremities, are the distances b1, b2, b3 and b4. We also note the distances m1 and m2 whose localise the end of the first zoning made on the board.

4.6. Qualitative zoning in production
Location of zones is known only by the computer and is not accessible by the user. However, order and widths of the various quality zones selected in the same zoning are accessible by computer checking production statistics.

Quality zoning, manual and in the course of production, is done in the same way as the zoning made with the imaging by X-rays: distances b1, b2, b3, b4, m1 and m2 are found. It is possible to know quality zones selected on the real board by the operator of production.

5. Results and analyses

5.1. Comparison of both zonings

A dimensional characterization and a geographical localization of all visible defects are made out of production. This statement is confronted with the zoning made in production and with the one made by the X-rays imaging. This comparison highlights:

- Not visible defects from the outside but visible with the X-rays imaging,
- Visible defects from the outside but not seen by the operator,
- Visible defects from the outside but not visible with the X-rays imaging,
- Visible defects from the outside, seen by the operator and visible with the X-rays imaging.

The comparison between quality zones selected in production and with the X-rays imaging is made according to two criteria: (1) the selected quality and (2) the dimensions and the location of zones.

The analysis of zonings, the one in production and the other one based on X-rays imaging allows to obtain the Table 1 and the Table 2.

	Surface (m^2)	Percentage /Zoned surface visually in production	Percentage /Zoned surface by X-rays imaging
Zoned surface visually in production			
OA Quality	2.07	16%	
OB Quality	8.51	66%	
Relegated Quality	2.39	18%	
Total surface	12.96	100%	
Zoned surface by X-rays imaging			
OA Quality	2.37		18%
OB Quality	7.92		59%
Relegated Quality	3.04		23%
Total surface	13.32		100%
Difference between zoned surfaces by X-rays imaging and visually in production			
OA Quality	0.30	2%	2%
OB Quality	-0.59	-5%	-4%
OA/OB addition	-0,29	-2%	-2%
Relegated Quality	0.65	5%	5%
Total surface	0.36	3%	3%

Table 1: Global results of the comparison of both zonings

We notice 3 % deviation between the total surface zoned in production and that one zoned by means of the X-rays imaging (See Table 1). This deviation is due to the various locations of zonings (slopes to follow the straight grain) made either visually and or with the X-rays imaging. Indeed, some degrees of slope furthermore or at least echo on the zoned total surface. Furthermore, this table shows a distribution of the similar qualities with both used means of vision.

		Visual zoning (m^2)		
		OA	OB	Relegated
X-rays imaging zoning (m^2)	OA	0.9850	1.2000	0.1900
		8%	9%	2%
	OB	1.1350	5.3025	1.4300
		8%	41%	11%
	Relegated	0.1125	1.8375	0.8150
		1%	14%	6%

Table 2: Detailed results of the comparison of both zoning

The Table 2 informs causes of the displacements made during the passage from one to another vision means: visual and X-rays. We notice that 45 % of the zoned total surface undergoes a change of quality class according to the used vision mean; the vision means do not thus influence qualitative choice of operator on the rest of the surface.

	Defects: Knots < 5 mm (crosiing or less visible)	Tinits	Compression wood	Resin pockets	Curly grain wood	Diverted grain
Defects by studied board by RX sampling	0.59	0.38	0.53	0.16	0.09	0.09
Defects by studied board by total sampling	0.37	0.36	0.43	0.06	0.11	0.09
Relation between total sampling and RX sampling	0.62	0.96	0.80	0.36	1.19	0.99

Table 3: Relation between total sampling and X-rays sampling

The operator changes 32 % quality on the 45 % of the total surface displaced in spite of visible elements with both means. The causes of these changes of class are mainly: diverted grain (9.7 %), brown spots of strong density (8 %), compression wood (2.8 %), curly grain (1.5 %) and a groupof isolated defects or the material not identified under the other categories (8.4 %). We also notice a deviation from total 1.6 % of the surface due to the accumulation of several variations, some millimetres, from the location of the borders of zones. We explain all these changes of classes by different properties of the material according to the used vision system. Indeed, some defects are more contrasted with the X-rays imaging than visually and conversely.

With X-rays imaging, operator access to internal characteristics of the material; so he downgrades 23 % of the total surface towards qualities lower than that he zoned visually in production. Among these 23 %, displacements are of:
- 8 % are a displacement of quality OA (visual) towards quality OB (by X-rays imaging),
- 14 % of quality OB towards Relegated quality and,
- 1 % of quality OA towards Relegated quality.

Besides displacements due to factors explained previously, 4.7 % of the total surface is downgraded by presence of resin pockets (2.4 % of total zoned surface) and by presence of crossing or little visible knots (2.3 %) detectable only by X-rays imaging.

Conversely, by basing itself on the visual aspect, operator downgrades 22 % of the total surface zoned from X-rays imaging. The main cause of the visual displacement is wood tint (8.9 % of the total zoned surface) because X-rays imaging does not allow to visualize characteristics, as tints, which do not affect material density.

Experimentally, the X-ray scanner allows to improve qualitative classification on approximately 4.7 % of the total surface but it also implies loss of information about the tint which is the cause of the displacement of 8.9 % of surface zoned.

Statistical analysis on presence of defects is showed in the Table 3. It allows generalize the experimental results in 50 % of the production of the planing mill. The displacements due to tints are thus worn in 8.57 % of half production, and those due to internal defects in 2.29 %.

5.2. Additional result: threshold characteristic

As show in example Figure 3, every spectre consists of a strong presence of low levels of grey (black: 0) which represents the background of X-rays images.

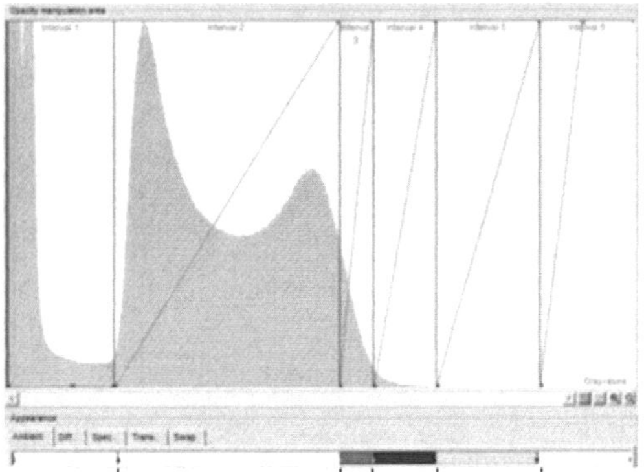

Figure 3: Board grey level spectre

The spectre shows two central peaks representing all the grey levels of the studied board. After analysis, we noticed that the levels of grey correspond to 4 defects of the wood decomposed into 4 ranges: wood without visible defect by scanning, compression wood, knots and resin pockets. Thresholds are defined by the intersection between the curve and the maximal value of peaks multiplied by a coefficient function of the defect ($C_{compression\ wood}$, C_{knot}, $C_{resin\ pocket}$). These 3 coefficients are defined in an empirical way by comparison between human perception and result obtained by threshold.

Range of "wood without defect" begins when the spectre is minimal between the levels of grey of background of the image and the first central peak; the abscissa of this point is named $x_{wood\ without\ defect}$. Other ranges begin in the points of abscissa $x_{compression\ wood}$, x_{knot}, $x_{resin\ pocket}$ according to the following mathematical definitions:

If f is function of the grey levels spectre and M the maximal value of the central peak,

$M = \max(f(x)) = f(x_{max})$, with $x > x_{wood\ without\ defect}$

$f(x_{compression\ wood}) = C_{compression\ wood} \cdot M$, with $x > x_{max}$

$f(x_{knot}) = C_{knot} \cdot M$, with $x > x_{max}$

$f(x_{resin\ pocket}) = C_{resin\ pocket} M$, with $x > x_{max}$

Thresholding analysis showed that spectres of grey levels have a similar profile and that threshold coefficients are equivalent from a sample to another. The analysis allows determining an empirical value for every coefficient:

$C_{compression\ wood} = 0{,}363$, $C_{knot} = 0{,}036$ and $C_{resin\ pocket} = 0{,}009$

However, abscissas of the thresholds of every defect are different and become confused from a sample to another as the Figure 4 shows. Thresholds representing the knots are included in those representing the compression wood. We obtain better result, as shows in Figure 5, by deducting the appropriate threshold of the wood without defect in three thresholds representatives of the defects. The ranges of thresholds are reduced but there is always a strong inclusion between ranges.

These inclusions are essentially due to the fact that the grey levels spectres associated to every part have dimensions which extend from 647 to 3745 levels of grey. On the other hand, a light overlapping is accepted between thresholds because defects have closed densities (grey levels) or even confused in certain cases.

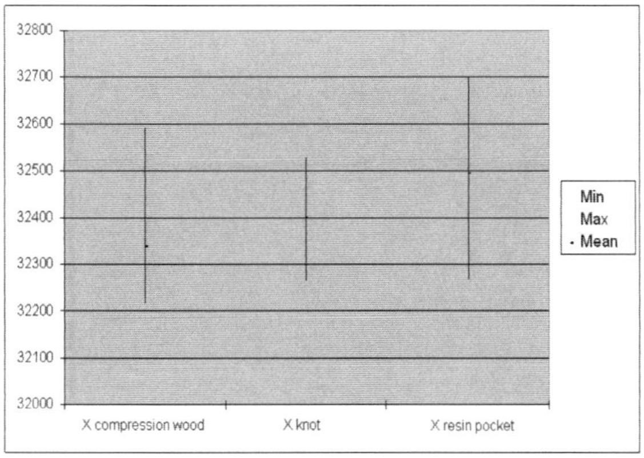

Figure 4: Inclusions of defect's grey level

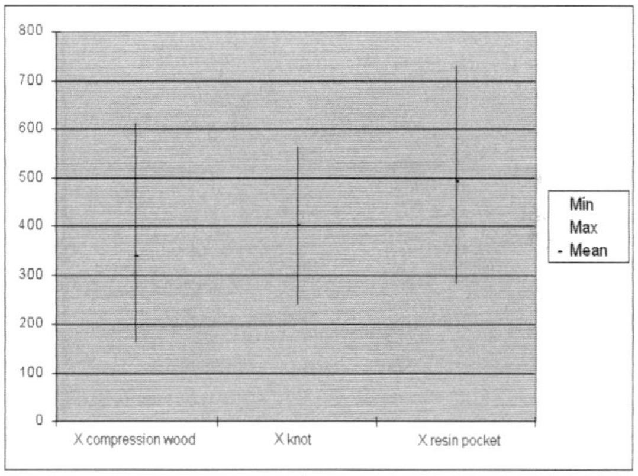

Figure 5: Inclusions of defect's grey level after correction

Thresholding can be improved by calibrating all the spectres on the same range of grey levels and by improving definition method of the characteristic coefficients in thresholds.

6. Conclusion and discussion

Comparative analysis of the qualitative zonings of 32 wooden boards made from two various means of vision allowed to highlight that 45 % of the zoned surface undergoes a change of class according to used vision mean, either human eye or X-ray scanner. The experiment demonstrated that the use of a single X-ray scanner in the production chain is penalizing. Indeed, X-ray scanner does not allow detection of wood tints which are the cause of 8.9 % of the visual displacements. Therefore, if the scanner is combined to a colorimetric vision system, such as digital cameras or human eye, then it allows to improve qualitative classification on 4.7 % of the surface zoned experimentally. The analysis also showed that visible defects by both vision means do not lead obviously to the same allocation of quality (32 % of the surface zoned experimentally) because of the contrasts on the material are not the same visually and by X-rays imaging. Interpolation of experimental results shows that contribution of the X-ray scanner concerns 2.29 % of the surface of 50 % of the production of Siat-Braun planing mill. So, it is necessary to continue a statistical study concerning all the production, more than 9 thicknesses, to know the real contribution of X-ray scanner.

Characteristic thresholds of every defect tried to be unsuccessfully updated: it is necessary to add a step of spectre calibration to succeed. It should be possible, after this stage of calibration, to establish fixed average thresholds which allow generalizing thresholding according to specie and to board thickness. The first calculations, not presented here, realized on the thresholds of a single thickness of board show a decrease of the inclusion of the threshold ranges between them (especially for the 24 mm thickness). However, to conclude on the generalization of thresholding, it is necessary to diversify widths and types of boards of the sampling because that experiment deals with the same type of product of small width.

7. Acknowledgement

The authors gratefully acknowledge the financial support of the CPER 2007-2013 "Structuration du Pôle de Compétitivité Fibres Grand'Est" (Competitiveness Fibre Cluster), through local (Conseil Général des Vosges), regional (Région Lorraine), national (DRRT and FNADT) and European (FEDER) funds. The authors want to thank the INRA (National Institute in Agronomical Research) and, particularly, Jean-Michel LEBAN and Charline FREYBURGER for the use of their X rays scanner.

References

1. Bombardier V. Mazaud C, Lhoste P, Vogrig R 2007 *Computers in Industry* **58** 355
2. Bombardier V, Schmitt E, Charpentier P 2009 *Measurement* **42** 189,
3. Brännström M, Oja J, Grönlund A 2007 *Scandinavian Journal of* ***Forest Research*** **22(1) 60**
4. Buehlmann U, Thomas R E 2002 *Robotics and Computer Integrated Manufacturing* **18** 197
5. Funck J W, Zhong Y, Butler D A, Brunner C C, Forrer JB 2003 *Computers and Electronics in Agriculture* **41** 157.
6. Hampel U, Bieberle A, Hoppe D, Kronenberg J, Schleicher E, Sühnel T, Zimmermann F, Zippe C 2007 *Review of scientific instruments* **78**, 103704.
7. Leban J M, Pizzi A, Wieland S, Zanetti M, Properzi M, Pichelin F 2004 *J. of Adhesion Sci. and Technology*, **18** (6) 673
8. Lin C-J, Kao Y-C, Lin T-T. Tsai M-J. Wang S-Y, Lin L-D, Wang Y-N. Chan M-H, 2008 *Int. biodeterioration and biodegradation* **62**, pp.434-441, 2008.
9. Longuetaud F, Mothe F, Leban J M, 2005 Détection et analyse non destructive de caractéristiques internes de billons d'Epicea commun par tomographie à rayons X », *Ph.D.Thesis*, INRA-ENGREF, France.
10. Lundgren N, Brännström M, Hagman O, Oja J. 2007 *Wood and Fiber Science* **39** 167.
11. Magnusson Seger M, Danielsson E 2003 *Computers and Electronics in Agriculture* **41** 45
12. Mäkelä A, Mäkinen H 2003 *Forest Ecology and Management* **184** 337
13. Maunu S L, 2002 *Progress in Nuclear Magnetic Resonance Spectroscopy* **40** 151.
14. Meder R, Franich R A, Callaghan P T 1999 *Solid state Nuclear Magnetic Resonance* **15** 69
15. Oh J-K, Shim K, Kim K-M, Lee J-J 2009 *Journal of Wood Science* **55** 264
16. Öhman M 1999 Planck grade indicators in radiograph images of Scots pine logs, *Holz als Roh- und Werkstoff* **57**, pp 359-363.
17. Rinnhofer A, Petutschnigg A, Andreu J P, 2003 *Computers and Electronics in Agriculture* **41**, 7.
18. Sarigul E, Abbott A. Lynn, Schmoldt Daniel L, 2003a "Progress in analysis of computed tomography (CT) images of hardwood logs for defect detection", ScanTech, November 3-4, Washington, USA.
19. Sarigul E. et al., "Rule-driven defect detection in CT images of hardwood logs", *2003b Computers and Electronics in Agriculture*, pp.1-19,
20. Schmoldt D L, Scheinman E , Rinnhofer A, Occena L.G, 2000 "Internal log scanning: Research to reality", *Hardwood Symposium Proceedings, Asheville, NC,* May 11-13, 2000.
21. Skatter S, Hoibo O A, Gjerdrum P 1998 "Simulated yield in a sawmill using different measurement technologies", *Holz als Roh- und Werkstoff* **56**, pp.267-274.
22. Thawornwong S, Occena L G, Schmoldt D L 2003 *Computers and Electronics in Agriculture* **41** 23.
23. Wei Q, Chui Y H, Leblon B. Zhang S Y, 2009 *J. of Wood Science* **55** 175-180.
24. Yu G, Kamarthi SV 2010 *Engineering Application of Artificial Intelligence* **23** 196.

Biosorption of model pollutants in liquid phase on raw and modified rice husks

L Toniazzo[1], V Fierro[1], F Braghiroli[1], G Amaral[1] and A Celzard[1,2]

[1]Institut Jean Lamour - UMR CNRS 7198, CNRS - Nancy-Université - UPV-Metz, Département Chimie et Physique des Solides et des Surfaces.
ENSTIB, 27 rue Philippe Séguin, BP 1041, 88051 Épinal cedex 9, France

[2]Institut Universitaire de France

Email: Vanessa.Fierro@lcsm.uhp-nancy.fr

Abstract. We studied the application of rice husk (RH) as a biosorbent and we demonstrated that it can be employed for the treatment of dyeing wastewater streams. RH was obtained from Nile Delta (Egypt) and it was used as received, or after a chemical treatment using HNO_3 or NaOH, or after conversion into activated carbon (RH-AC) using H_3PO_4 as activating agent. A commercial activated carbon GAC 830 provided by NORIT was also tested for comparison purposes. These materials were evaluated by adsorption of methylene blue (MB) with an initial concentration of 20 ppm in an aqueous solution at 30°C. The results showed that alkali-treated and RH-AC were the best sorbents. They got a nearly complete MB removal from water and they had better performance than GAC 830. Therefore, the use of RH for pollutant removal makes this method an environment-friendly option and an economically feasible alternative to treat industrial effluents.

1. Introduction

Rice husk (RH) is a by-product of the milling process of rice crop which is widely cultivated in the world. RH contains suitable characteristics for some pollutant adsorption and it could be discharged as agricultural waste [1]. The major components of RH which may be responsible for sorption are carbon and silica: the former for its high surface area and pore volumes, and the latter for its interaction with organic polar compounds [2, 3].

Different methods can be used for the removal of organic compounds and dyes from wastewater streams. The conventional treatment processes, such as coagulation, sedimentation, and filtration are less effective in removing organic molecules [4, 5]. Biological treatment (aerobic and anaerobic) can reduce organic matter and the type of treatment applied to sludge is usually chosen depending on its nature and environmental factors. However, several technological changes in the system are needed in order to decrease the concentration of some persistent organic pollutants. Therefore, it seems that the adsorption method is the most effective and economically feasible alternative for the removal of organic pollutants [6, 7].

Adsorption on indigenous materials is a relatively simple and low-cost method for the treatment of industrial and domestic effluents. Few studies examined the utilisation of rice by-products for organics adsorption, such as: rice bran being a potent sorbent for the removal of organochlorine compounds and benzene from industrial wastewaters; use of river sand for sorption of Cd (II) from aqueous solutions and also use of rice husk ash for preconcentration of gold [2] and therefore it seems that rice husk is a promising agricultural waste material to be implemented in the wastewater treatment systems.

2. Experimental

2.1. Preparation of treated rice husk and characterisation of natural RH

RH was used without any modification as biosorbent. RH was also treated with HNO_3 (RH-HNO_3) and NaOH (RH-NaOH). We prepared RH-HNO_3 as follows: 250g of RH were soaked in 1L of HNO_3 2M, left for 2 hours at room temperature and then washed with distilled water using a Soxhlet extractor. Finally, it was dried in an oven at 60°C for 24 hours. RH-NaOH was prepared as follows: 300g of RH was soaked with 1L of NaOH 1.5% and left for 2 hours at room temperature. After that, the mixture was transferred to an autoclave at 121°C during 30min and then washed using Soxhlet extractor. Finally, the material was dried in an oven at 60°C for 24 hours.

The proximate characterisation of the natural RH included the percentage of cellulose, hemicellulose, lignin, ash and moisture content. For the determination of lignin, the Klason method [8] was used. The ash content was determined by weighing up natural RH before and after being burnt in a muffle oven at 600°C following the standard procedure (ASTM D 1102-84). All the samples were weighed before and after being placed in an oven for 4 hours at 105°C for the determination of moisture.

2.2. Preparation of activated carbon from rice husk and characterisation of ACs

A series of experiments were carried out to optimize the orthophosphoric acid (H_3PO_4) activation conditions and the results showed that the optimal conditions, in relation to textural properties (surface area and microporosity) were: temperature of 425°C, activation time of 30 min; and impregnation weight ratio (H_3PO_4/RH) of 1.2. The mixture was kept one hour for impregnation and introduced into a vertical oven for activation. The mixture was placed in the thimble to be washed with water for five days using a Soxhlet extractor.

The specific surface areas (S_{BET}) of the commercial activated carbon (GAC 830) and the activated carbon prepared from rice husk (RH) and porosity were determined by N_2 adsorption at 77 K with an automatic instrument Micromeritics - ASAP 2020. The samples were analysed according to the BET method for calculating the specific surface area and the Dubinin-Radushkevich (DR) method [9] was use to calculate the micropore volume, V_{micro}. The total pore volume, $V_{0.99}$ was calculated from N_2 adsorption at a relative pressure of 0.99. The mesopore volume, V_{meso}, was calculated by the difference between $V_{0.99}$ and V_{micro}. The percentage of ash content and humidity was determined following the ASTM (D 2867-04) standard.

2.3. Adsorption experiments

2.3.1. Kinetic study. Adsorption kinetics were studied by mixing 500 mg of AC (NORIT GAC 830, RH-AC) or 2.5 g of biomass (RH-AC, RH-HNO_3, RH-NaOH or raw RH) with 500 mL of solutions containing 20 ppm of methylene blue (MB). The mixture was then put in a thermal bath with water temperature set at 30°C [10,11] and stirred till reaching the adsorption equilibrium.

2.3.2. *Adsorption isotherms.* We determined the MB adsorption isotherms onto the 5 aforementioned materials. We used glass tubes containing 10 mL of 20 ppm of MB solution and we added between 1 and 35 mg of sorbent. The tubes were closed and put in the thermal bath under stirring at 30°C for approximately 8 hours. The MB residual concentration was measured by UV spectrophotometry and in order to describe the adsorption data, three well-known mathematical equations, namely: Freundlich, Langmuir and Tempkin were tested. The main difference between those three models is basically the way the heat of adsorption decreases with the surface coverage. Langmuir assumes no decrease at all, Freundlich assumes a logarithmic decrease, while Tempkin assumes a linear decrease [12, 13].

3. Results and Discussion

3.1. Characterisation of natural RH

Proximate analysis results of the raw RH were: 35% of cellulose; 21% of hemi-cellulose; 20% of lignin; 19% of ash and 5% of humidity, in good agreement with other authors [1, 15].

3.2. Characterisation of ACs

Table 1 contains the following values for the two activated carbons: S_{BET}, $V_{0.99}$, V_{micro}, V_{meso}, % of moisture and ash content. The surface area (976 m^2/g) and the corresponding micropore volume (0.35 cm^3/g) are higher for the commercial activated carbon comparing to the RH: 717 m^2/g and 0.25 cm^3/g, respectively. Whereas the RH-AC presents higher mesopore volume (0.41 cm^3/g) compared to GAC 830 (0.21 cm^3/g) as it can be seen in table 1.

Table 1. Textural parameters, moisture and ash content of the ACs.

	S_{BET} (m^2/g)	$V_{0.99}$ (cm^3/g)	V_{micro} (cm^3/g)	V_{meso} (cm^3/g)	Moisture (%)	Ash (%)
GAC 830	976	0.56	0.35	0.21	1.6	8.0
RH-AC	717	0.66	0.25	0.41	1.0	31.4

3.3. MB adsorption kinetics

Figure 1 shows that MB concentration first rapidly decreased and then remained almost constant. RH-NaOH reached in less than 1 hour MB adsorption equilibrium and MB removal was complete. RH and RH-HNO_3 reached the equilibrium after 2 hours and MB was not completely removed from aqueous solution.

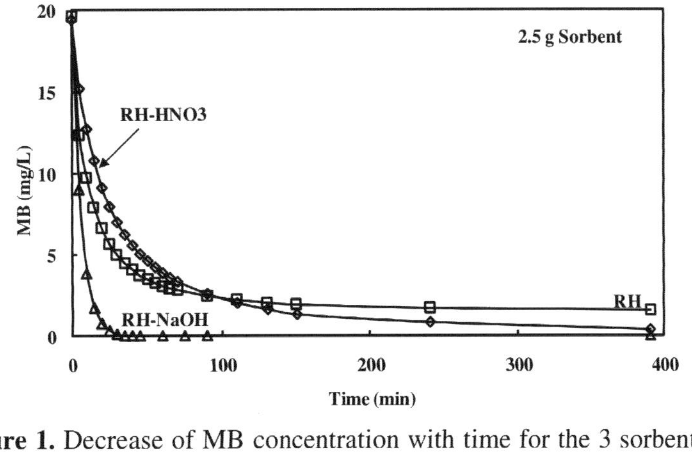

Figure 1. Decrease of MB concentration with time for the 3 sorbents studied: △ RH-NaOH, ◇ RH-HNO_3 and □ RH. (C_0=20 mg/L, V= 500 mL, 2.5g of sorbent, T=30°C).

3.4. Kinetics for MB removal using ACs

Figure 2 shows the performance of the two ACs, GAC 830 and AC-RH, for MB removal. Adsorption studies clearly indicated that the RH-AC possesses excellent adsorption characteristics and has the potential to be used for the removal of dyeing basic compounds. Both ACs are able to remove 100% of MB from aqueous solution but RH-AC reached the equilibrium much faster than GAC 830, approximately in 2 h compared to GAC 830: around 6 h.

The main explanation for finding great adsorption capacity of MB onto activated carbon materials with lower surface areas and microporosity may be the size of molecule which is around 1.5 nm, so this compound is adsorbed only in large micropores and mesopores (>2 nm). Furthermore, the higher uptake capacity may be explained by the surface chemistry of the RH-AC (pH equal to 2.53). MB has a basic character and when it is put in contact with the strong acidic surface of RH-AC, electrostatic interactions occur between electrons from the carbon surface and the dye molecule favouring the adsorption mechanism [4, 12].

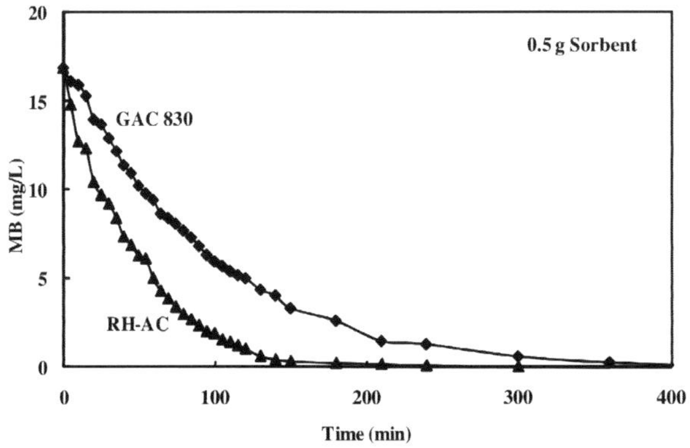

Figure 2. Decrease of MB concentration with time for the 2 ACs studied: ▲ RH-AC and ◆ GAC 830. (C_0=20 mg/L, V= 500 mL, 0.5g of sorbent, T=30°C).

3.5. Adsorption isotherms

Figure 3 shows MB adsorption isotherms onto the 5 materials tested. RH-AC showed the highest and RH the lowest MB adsorption for the same concentration at equilibrium. RH-NaOH demonstrated also to be a very good adsorbent and it had better performances than the commercial AC, GAC 830.

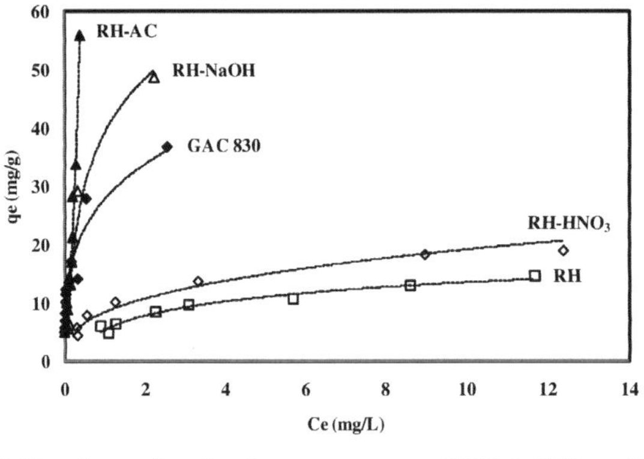

Figure 3. MB adsorption isotherms onto: □ RH-NaOH, ◇ $RH-HNO_3$, □ RH, ▲ RH activated and ◆ GAC 830. (C_0=20 mg/L, T=30°C).

Table 2 shows the maximum uptake per unit mass of sorbent, q_m, determined by the application of Langmuir model. These data confirm that RH-AC had the best performances for MB adsorption using only 0.5 g of ACs compared to 2.5 g of treated and raw RHs.

Table 2. Maximum uptake per unit of adsorbent, q_m, derived from application of Langmuir's model

Sorbent	q_m (mg/g)
RH	16.9
RH-HNO_3	20.8
RH-NaOH	71.4
GAC 830	83.3
RH-AC	217.4

4. Conclusions

RH can be employed efficiently for the removal of basic dyes after NaOH treatment: RH-NaOH, or after H_3PO_4 activation: RH-AC. For biosorbents, the order in MB adsorption efficiency was RH-NaOH (71.4 mg/g) > RH-HNO_3 (20.8 mg/g) > RH (16.9 mg/g). RH-AC showed better performances as adsorbent for MB (217.4 mg/g) than the commercial GAC 830 (83.3 mg/g); this is probably due to the higher mesoporosity and more acidic character of RH-AC. Taking into account the activation costs, RH-HNO_3 is probably the most economical sorbent for MB removal.

Aknowledgements

We thank the financial support of the CPER 2007-2013 "Structuration du Pôle de Compétitivité Fibres Grand'Est" (Competitiveness Fibre Cluster), through local (Conseil Général des Vosges), regional (Région Lorraine), national (DRRT and FNADT) and European (FEDER) funds.

References

[1] Krishnani KK, Meng X, Christodoulatos C and Boddu VM 2008 *J. Hazard. Mater.* **153** 1222
[2] Akhtar M, Bhanger MI, Iqbal S and Hasany SM 2006 *J. Hazard. Mater.* **128** 44
[3] Mahvi AH, Maleki A and Eslami A 2004 *Am. J. Appl. Sci.* **1** 321
[4] Attia AA, Girgis BS and Khedr SA 2003 *J. Chem. Technol. Biot.* **78** 611
[5] Semmens MJ and Ayers K 1985 *J. Am. Water Works Ass.* **77** 79
[6] Gupta VK, Srivastava SK and Tyagi R 2000 *Water Res.* **34** 1543
[7] Snoeyink VL & Chen ASC 1985 *Sci. Total Environ.* **47** 155
[8] The Institute of Paper Chemistry, 1951 Institute Method No. 428. Institute of Paper Chemistry, Appletton, Wisconsin.
[9] Sing, KSW 1989 *Carbon* **27** 5
[10] McKay G, Porter JF and Prasad GR 1999 *Water Air Soil Poll.* **114** 423
[11] Ponnusami V, Krithika V, Madhuram R and Srivastava SN 2007 *J. Hazard. Mater.* **142** 397
[12] Torné-Fernández V, Mateo-Sanz JM, Montané D and Fierro V 2009 *J. Chem. Eng. Data* **54** 2216
[13] Fierro V, Torné-Fernández V, Montané D and Celzard A 2008 *Micropor. Mesopor. Mat.* **111** 276
[14] Hamdaoui O and Naffrechoux E 2007 *J. Hazard. Mater.* **147** 381
[15] Liou T and Wu S 2009 *J. Hazard. Mater.* **171** 693

Thermal convection in a closed cavity in zero-gravity space conditions with stationary magnetic forces

T Lyubimova[1], A Mailfert[2]

[1] Institute of Continuous Media Mechanics, Ural Branch of RAS, Perm, Russia
[2] Université de Lorraine, Laboratoire Environnement et Minéralurgie UMR7569 CNRS, Rue du Doyen Marcel Roubault, F-54501 Vandoeuvre-lès-Nancy, France

Email : lubimova@psu.ru

Abstract. The paper deals with the investigation of thermo-magnetic convection in a paramagnetic liquid subjected to a non-uniform magnetic field in weightlessness conditions. Indeed, in zero-g space conditions such as realized in International Space Station (ISS), or in artificial satellite, or in free-flight space vessels, the classical thermo-gravitational convection in fluid disappears. In any cases, it may be useful to restore the convective thermal exchange inside fluids such as liquid oxygen. In this paper, the restoration of heat exchange by the way of creation of magnetic convection is numerically studied.

1. Introduction

An inhomogeneous magnetic field $\vec{B}$ distribution give volumic forces proportional to $\nabla(\vec{B}^2)$ and to the magnetic susceptibility χ:

$$\vec{f} = \frac{1}{2\mu_0}\chi\nabla(\vec{B}^2) \tag{1}$$

where μ_0 is the vacuum permeability.

The magnetic susceptibility is temperature dependent, in a way different for diamagnetic fluids (where χ varies as the mass per unit volume ρ), or for the paramagnetic fluids obeying to a Curie law (χ varies as $1/T$). This explains that a thermal convection may be induced by a convenient field distribution. This fact has been used for promoting [1] or controlling [2] classical gravity convection in fluids.

The $\nabla(\vec{B}^2)$ field obeys to general constraints because $\vec{B}$ obeys to Maxwell equations. As an example, $\nabla(\vec{B}^2)$ field cannot be a constant vector in any point of a 3D domain [3]. This is very important, thus for example a constant vector such as gravity cannot be perfectly compensated in any point of a 3D domain. That means also there exists a fundamental difference between classical terrestrial, gravity induced, convection, and the magnetic convection in zero-gravity space: the force distribution is constant (in little domains) for terrestrial gravity induced convection, and never constant in every point for magnetic forces.

Equation (1) shows that magnetically induced convection in zero-gravity space is directly depending on the magnetic field distribution. A first particular cylindrical distribution had been considered in a preceding paper [4], and applied to the modification of natural convection of water (diamagnetic fluid). This field distribution was characterized by a radial force field, with a constant value of force modulus inside the channel containing the fluid.

In the present paper, another field distribution is considered, leading to a greatly inhomogeneous force field in the channel. This force field is applied to thermal convection in a paramagnetic fluid such as liquid oxygen in space zero-gravity conditions.

2. Problem formulation. Governing equations

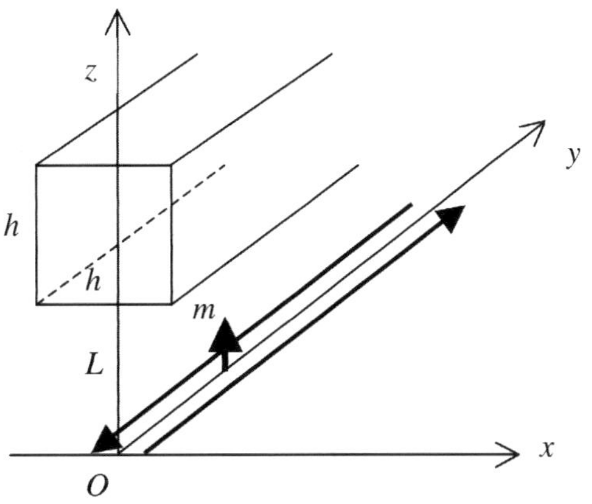

Figure 1. Problem formulation

Paramagnetic fills a long cylindrical channel of square cross-section ($h \times h$). Two opposite walls of the cylinder parallels to the x-axis are maintained at constant different temperatures. At the two remaining cylinder walls $x = \pm h/2$, the temperature varies according to the linear law. Magnetic field is produced by a line of permanent magnets the size of which is small in comparison with h, or by a bifilar line of conductors parallel to y-axis and fed by currents in opposite direction (fig.1). The magnets or the currents are located on the axis z at a distance L from the cylinder and their magnetic moment m per unit length is parallel to z-axis.

The field distribution is of the well-known cylindrical dipolar type, giving for the two components of $\nabla(\mathrm{B}^2)$ in the plane (x,z):

$$\frac{\partial(\mathrm{B}^2)}{\partial x} = -4\Phi_0^2 \frac{x}{r^6}, \quad \frac{\partial(\mathrm{B}^2)}{\partial z} = -4\Phi_0^2 \frac{z}{r^6}, \tag{2}$$

where $x^2 + z^2 = r^2$, and Φ_0 is a magnetic constant of the line (unit T.m^2) easily calculable, either for a line of permanent magnets or for a bifilar line of currents. It appears that the force field is central in the plane (Ox, Oz), and it varies rapidly (at the power -5) with the distance r from O. This quick variation explains that convection lines, when L is low, may be very different than observed in the classical gravitational convection.

The temperature dependence of the magnetic susceptibility of paramagnetic makes ponderomotive force to be non-gradient such that in the presence of both temperature gradient and non-uniform magnetic field the liquid cannot remain motionless. The governing equations for the flow are:

$$\begin{aligned} &\frac{\partial \mathbf{v}}{\partial t} + (\mathbf{v}\nabla)\mathbf{v} = -\frac{1}{\rho}\nabla p + \nu\Delta\mathbf{v} + \frac{1}{2\mu_0}\chi\nabla(\mathrm{B}^2), \\ &\frac{\partial T}{\partial t} + \mathbf{v}\nabla T = \kappa\Delta T, \quad \mathrm{div}\,\mathbf{v} = 0. \end{aligned} \tag{3}$$

We chose the following scales: for the length h, for the velocity ν / h, for time h^2/ν, for the temperature the difference in hot and cold wall temperatures ϑ, for the magnetic field induction $\mathrm{B}_0 = \mathrm{B}(z = L) = {}_0 / \Phi L^2$, for the magnetic susceptibility $\chi_0 = \chi(z = L)$; the temperature T_0 at lower wall is taken as the reference temperature. Then, we obtain for two-dimensional problem applying curl_y operation to the first equation of (3), introducing vorticity φ and streamfunction ψ by the relations $\varphi = \mathrm{curl}_y \vec{\mathrm{v}}$, $\mathrm{v}_x = -\partial\psi / \partial z$, $\mathrm{v}_z = \partial\psi / \partial x$:

$$\frac{\partial\varphi}{\partial t} + \frac{\partial\psi}{\partial x}\frac{\partial\varphi}{\partial z} - \frac{\partial\psi}{\partial z}\frac{\partial\varphi}{\partial x} = \Delta\varphi + \mathrm{Tm}\left(\frac{\partial\chi}{\partial x}\frac{\partial B^2}{\partial z} \quad \frac{\partial\chi}{\partial z}\frac{\partial B^2}{\partial x}\right),$$
$$\frac{\partial T}{\partial t} + \frac{\partial\psi}{\partial x}\frac{\partial T}{\partial z} - \frac{\partial\psi}{\partial z}\frac{\partial T}{\partial x} = \frac{1}{\mathrm{Pr}}\Delta T, \quad \varphi = -\Delta\psi \quad \mathrm{curl}_y \vec{\mathrm{v}}. \tag{4}$$

where $\mathrm{Tm} = \chi_0 B_0^2 h^2 / (2\mu_0 \quad ^2)$

Substituting the expressions for the derivative of $(\vec{\mathrm{B}}^2)$ from (2) and taking into account that the dimensionless magnetic susceptibility is $\chi = 1/(1+\alpha T)$ where $\alpha = \vartheta / T_0$ we have

$$\frac{\partial\varphi}{\partial t} + \frac{\partial\psi}{\partial x}\frac{\partial\varphi}{\partial z} - \frac{\partial\psi}{\partial z}\frac{\partial\varphi}{\partial x} = \Delta\varphi + 4l^4\mathrm{Tm}\frac{1}{r^6}\frac{\alpha}{(1+\alpha T)^2}\left(x\frac{\partial T}{\partial z} \quad z\frac{\partial T}{\partial x}\right),$$
$$\frac{\partial T}{\partial t} + \frac{\partial\psi}{\partial x}\frac{\partial T}{\partial z} - \frac{\partial\psi}{\partial z}\frac{\partial T}{\partial x} = \frac{1}{\mathrm{Pr}}\Delta T, \quad \varphi = -\Delta\psi \quad \mathrm{curl}_y \vec{\mathrm{v}}. \tag{5}$$

The boundary conditions at the rigid wall of container are no-slip conditions

$$\psi\big|_\Gamma = 0, \quad \left.\frac{\partial\psi}{\partial n}\right|_\Gamma = 0 \tag{6}$$

The thermal boundary conditions correspond to the perfectly conductive boundaries, the horizontal boundaries $z = l$, $z = l + 1$ are maintained at fixed different temperatures and at the vertical boundaries $x = \mathrm{m}l / 2$ the linear temperature distribution is fixed:

$$T(x,z) = A(z - l) \tag{7}$$

where $A = -1$ for the heating from below and $A = 1$ for the heating from above. Most of the calculations were performed for the fixed values of Prandtl number and parameter α: $\mathrm{Pr} = 1$, $\alpha = 0.03$. The parameters Tm and l were varied.

The problem was solved numerically by finite difference method. Uniform mesh with 41x41 nodes was used in most of the calculations.

3. Numerical results

The calculations show that at small values of the parameter Tm, independently of the direction of imposed vertical temperature gradient and on the distance l, there arises convective flow of two-vortex structure (see, fig.2). The boundary between the vortices coincides with z-axis. The flow direction is such that in the central part of the cylinder the fluid moves from the cold wall to the hot one.

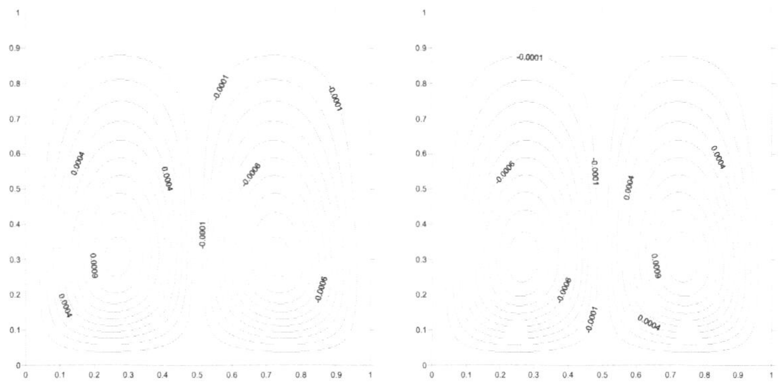

Figure 2. Streamlines of convective flow at $\mathrm{Tm}=10^3$, $l=1$ for the cases of heating from below (left) and above (right)

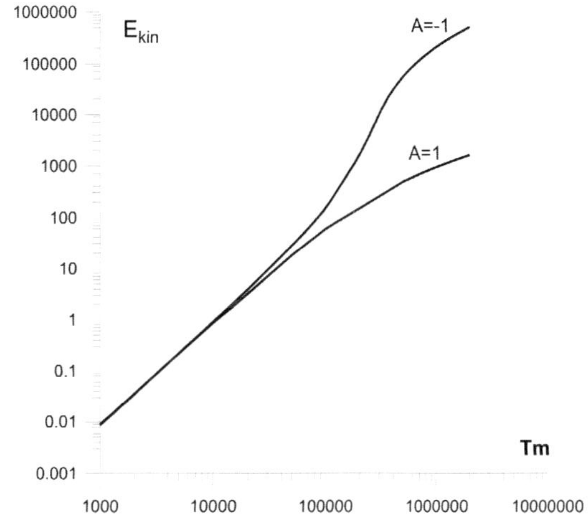

Figure 3. Kinetic energy of steady flow versus Tm for $l=1$ in the cases of heating from below and above

Fig.3 presents the dependences of the convective flow intensity on the parameter Tm at $l=1$ for the cases of the heating from below and from above. As one sees in both case the flow intensity grows monotonically with Tm . At small values of Tm the flow intensities are nearly the same, however with the growth of Tm , at some value of $\mathrm{Tm}=\mathrm{Tm}_*$ the flow intensity for heating from below demonstrates sharp increase and for heating from above the behaviour remains the same which results in the large difference in the flow intensities for $A=-1$ and $A=1$ at $\mathrm{Tm}>\mathrm{Tm}_*$. The flow structure for two cases at $\mathrm{Tm}>\mathrm{Tm}_*$ is presented in fig.4.

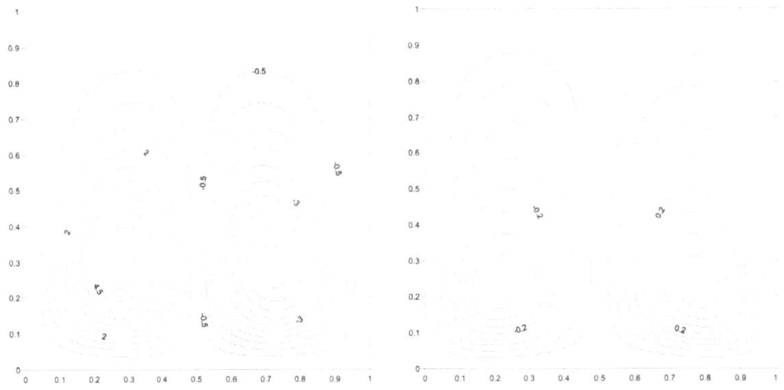

Figure 4. Streamlines of convective flow at $\mathrm{Tm} = 10^6$, $l = 1$ for the cases of heating from below (left) and above (right)

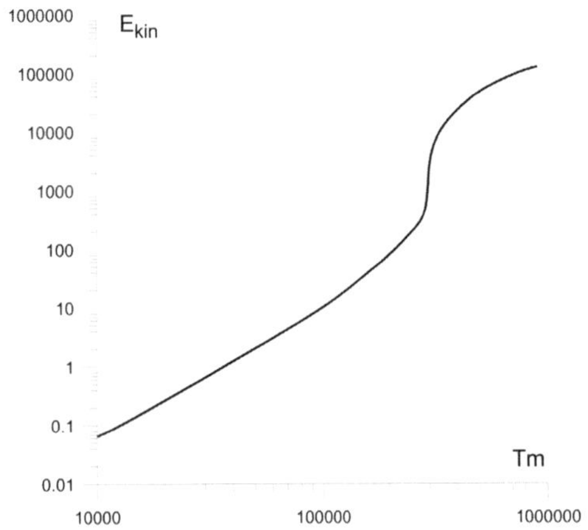

Figure 5. Kinetic energy of steady flow versus Tm for $l = 4$ in the case of heating from below

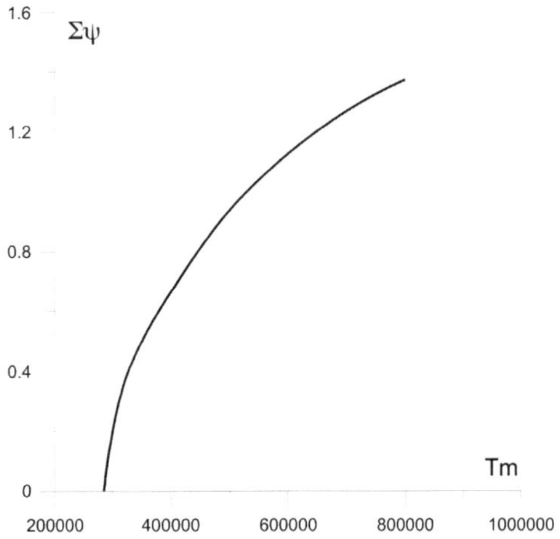

Figure 6. The sum of streamfunction over the fluid volume versus Tm for $l = 4$ in the case of heating from below

At larger distances between the magnets and the cylinder with the increase of Tm spontaneous breakdown of the symmetry occurs accompanied by the branch of two asymmetric solutions (see, figs.5 and 6).

4. Conclusions

We have theoretically and numerically studied the behavior of a paramagnetic fluid filling a square cylindrical channel, for zero-gravity space conditions. A particular magnetic field distribution is applied to the fluid, allowing the initiation of thermal convection when a temperature gradient is present. Magnetic dipolar sources are located "below" the channel, at a variable distance of it.

The analysis show that the streamlines of convective flow are related to a specific number Tm, the value of which depends on the magnetic field, the magnetic susceptibility and the viscosity of the fluid, and the size of the channel.

Numerical results show that for low value of Tm, and low distance between the channel and the magnetic line, in the central part of the channel the fluid moves from the cold wall to the hot one, with a two-vortex structure. Behavior is the same, either for heating from below (in the side of the magnetic sources) or for heating from above. Increasing Tm above a particular value leads to a net dissymmetry, with a sharp increase of the flow intensity for heating from below, instead the flow increase monotonically for heating from below.

For higher values of the distance between the channel and the magnetic line, leading to a better homogeneity of the magnetic force in the channel, asymmetric solutions for the flow appear, with another critical value for Tm number.

These results show that in zero gravity space conditions, the magnetic force distributions in the channel are highly influent on the shape of streamlines. Further work includes taking in account other realistic magnetic sources.

References

[1] K Nakatsuka, Y Hama, J Takahashi 1990 J. *of Magnetism and* Magnetic Materials **85**(2) 207.
[2] D Braithwaite, E Beaugnon, R Tournier 1991, Nature, 1991, **354**,
[3] L. Quettier, H. Felice, A. Mailfert, D. Chatain, D. Beysens 2005 *The European Physical Journal, Applied Physics*, **32**, 167.
[4] L Quettier, O Vincent-Viry, A Mailfert, F P Juster *The European Physical Journal, Applied Physics* **22**, 69-73.

Numerical modelling of admixture transport in a turbulent flow at river confluence

T. Lyubimova[1], A. Lepikhin[2], Ya. Parshakova[1], A. Tiunov[2], V. Konovalov[1], N. Shumilova[1]

[1] Institute of Continuum Media Mechanics, Ural Branch of RAS, Perm, Russia
[2] Mining Institute, Ural Branch of RAS, Perm, Russia

Correspondence to: T. Lyubimova (lubimova@psu.ru)

Abstract. The paper is concerned with the development of the hydrodynamic model of the Chusovskoy water intake located in the confluence zone of two rivers with essentially different hydrochemical regimes and in the backwater zone of the Kamskaya hydroelectric power station. The proposed model is used for numerical simulation in the framework of two-and three-dimensional approaches for the annual average, minimal and maximal values of the water flow rates in two rivers. The data for water mineralization in the water intake zone have been obtained. The recommendations for optimization of the water intake structure have been formulated.

1. Introduction

The objective of this paper is to develop a hydrodynamic model of the Chusovskoy water intake system, supplying most of Perm districts numbering about one million inhabitants with potable water. Numerous field observations have shown that in the region of the Chusovskoy water intake located in the confluence zone of the Chusovay and Sylva rivers with essentially different hydrochemical characteristics, water of the Sylva river is characterized by a higher degree of hardness and mineralization than water of the Chusovaya river. Moreover, the Chusovskoy intake system is located in the backwater zone of the dam of the Kamskay hydroelectric power station (HPS). Thus, the hydrodynamic regime of the water basin in the region of the water intake is defined by a rather complicated combination of the hydrological regimes of the Chusovaya and Sylva rivers, the water level of the Kamskiy water storage reservoir and the operating regime of the Kamskay HPS. It should also be noted that the rivers' regimes are characterized by high seasonal variability. Due to these factors and the limited possibilities of water monitoring, a numerical simulation of the above hydrological system is a matter of considerable importance for the environmental and engineering applications. Optimization of the water intake structure using the data of mathematical simulation may provide an alternative to the high-priced and time-consuming technological methods of lessening water hardness.

Simulation of the processes of the selective water intake from large water basins in the context of water-protective problems has been the subject of numerous studies by foreign and Russian researchers [1-7]. Simulation is generally done in the framework of a two-dimensional approach using the system of shallow water equations. A number of software products have been developed to realize numerically the two-dimensional Saint Venant equations.

A difference in the chemical composition of water is responsible for its density difference. Therefore, as the waters of two rivers are stirred, the more dense Sylva river water sinks to the bottom and the less dense Chusovaya river water rises to the surface. Since the non-uniform distribution of water density over the depth of the river is of fundamental importance for the examined problem, a numerical simulation in the framework of the two-dimensional approach including averaging over the depth is conceptually incorrect at least for part of the hydrological system, where the waters of two rivers are actively mixed.

On the other hand, in view of large dimensions and complicated geometry of the computational domain, a solution of this problem in the framework of a three-dimensional approach will require extensive computational resources.

To alleviate the problem, we used a combination of two- and three- dimensional computational schemes. The numerical simulations in the vicinity of the water intake head and the bottom barriers were made based on the three-dimensional model, whereas the hydrodynamics of the rest of the computational domain was computed based on the two-dimensional model.

To realize three- and two- dimensional simulations we used two different software products. The two-dimensional approach required application of the specialized hydrological software packet SMS v.10 produced by the American company AQUAVEO LLC (Donnell, 2009). For simulation of a three-dimensional problem we used the software packet for computational hydrodynamic Fluent v.6.3.26.

2. Two-dimensional model

Since the resource requirements for a two-dimensional simulation are relatively low we were able to examine the whole region of the Kamskiy water storage basin, which extends for about 100 km. The characteristic dimensions of the examined area are as follows: the length of the region from the Kamskaya HPS to the settlement of Palazna is ~ 40 km, from Kamskaya HPS to the place of the confluence of the Chusovaya and Sylva rivers ~23 km., from the confluence of the Chusovaya and Sylva rivers upstream the Chusovaya river ~ 22 km., from the confluence of the Chusovaya and Sylva rivers upstream the Sylva river ~ 17 km; the width of the water storage reservoir in the examined area ranges from 1.0 to 3.5 km and the average depth is about 15 m.

To set correctly the morphometry of the considered water basin (see fig.1) we used the technique of matching the data obtained from a detailed echo-sounding survey using GPS coupled sensors with satellite imagery. Raw data processing was accomplished using the Arc GIS software packet.

A two-dimensional (in a horizontal plane) model for the examined region of the water storage basin was constructed by making use of the software product SMS v.10 of the American company AQUAVEO LLC developed and manufactured by request and with participation of the USA Center of Hydraulic Investigations. The basis for this model is a set of modules (RMA2, RMA4, FESWMS, and others), which allow us to solve different problems. A brief outline of the RMA2 module which was used to analyze a two-dimensional flow [8] is given below.

RMA2 is a general –purpose module developed for a wide class of problems, in which the vertical acceleration is inessential and the velocity vectors have the same direction throughout the water basin depth at any time instant. It is used to calculate the eminences of water surface and horizontal velocity components in the laminar or turbulent flows with free surfaces in the two-dimensional domains.

The RMA2 module is designed to solve the Reynolds-type equations for turbulent flows derived from the Navier-Stokes equations by the finite element method. Friction is calculated by the Manning or Chezy formulas and the coefficients of the turbulent viscosity are used to determine the specific features of turbulence. It can be applied equally well to the stationary and non-stationary problems.

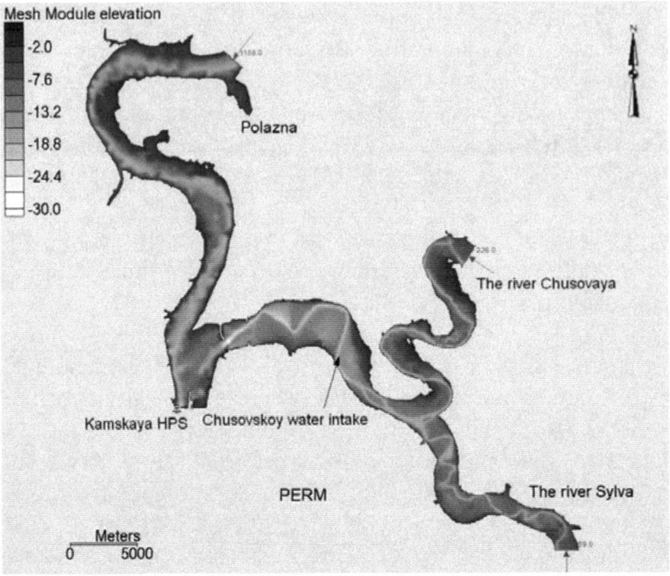

Figure 1. Bottom morphometry of the examined area of the Kamskiy water storage basin.

As the boundary conditions of the model we used the annual average flow rate of water in both rivers at the entrance to the water storage basin and the water level at the exit from the Kamskaya HPS, which is 108.5m. The data for the annual average flow rate and mineralization of water in the Chusovaya and Sylva rivers are given in Tables 1,2 . Figs. 2,3 show the calculated fields of the velocity vector and distribution of water mineralization in the confluence region of two rivers and the location of the Chusavskoy water intake system.

Water flow rate in Kamskiy water storage basin near the settlement of Polazna, Q, m^3 /s	Water level in the Kamskiy water storage basin near the Kamskaya HPS, H, m	Water flow rate in the Chusovaya river, Q, m^3 /s	Water flow rate in the Sylva river, Q, m^3 /s
1158.0	108.5	226.0	159.0

Table 1. Data for the flow rate and water level in the rivers of the Kamskiy water storage basin.

Mineralization of water in the Kamckiy water storage basin near the settlement of Polazna, C, mg / l	Mineralization of water in the Sylva river, C, mg / l	Mineralization of water in the Chusovaya river, C, mg / l
300.0	650.0	200.0

Table 2. Data for mineralization of water in the rivers of the Kamskiy water storage basin.

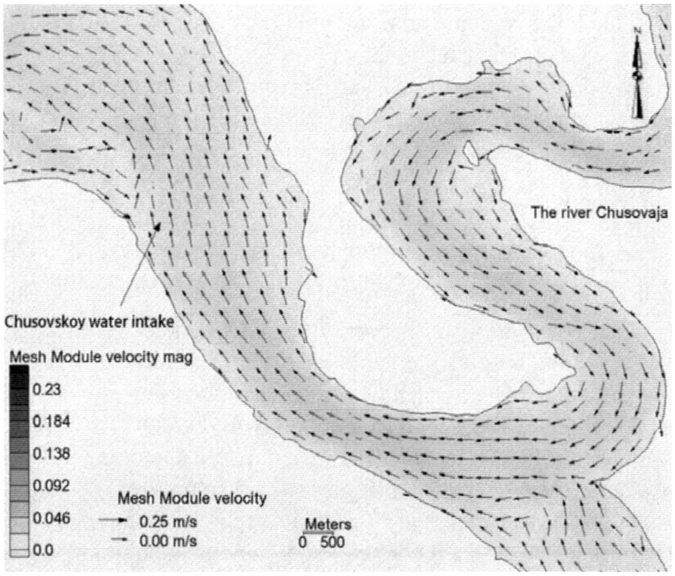

Figure 2. The vector field of velocity.

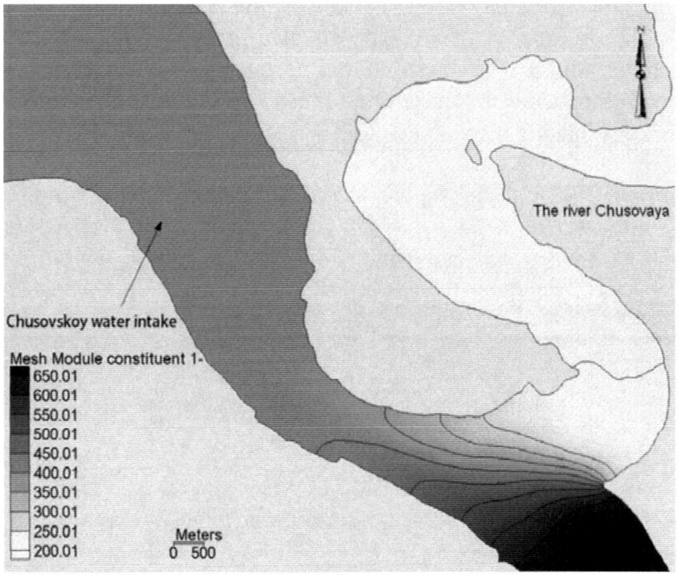

Figure 3. The field of distribution of water mineralization.

3. Three-dimensional model

As it has been mentioned above, our investigations would be far form complete, if only the results of numerical simulations made in the framework of two-dimensional model were not supplemented by the results of simulation of the three-dimensional flow region, which takes into account the vertical stratification of water mineralization and, accordingly, stirring of the Sylva and Chusovaya river waters due to the difference in water density.

The computations were made by applying the hydrodynamics software packet to one of the semi-empirical turbulence models. Since a tree-dimensional simulation requires extensive computational resources it was restricted to the region extending upstream for about 10 km from the confluence of the Sylva and Chusovaya rivers to the Chusovskaya water intake system.

The rivers' depth was taken to be constant over the whole length of the computational domain and was equal to 20 meters, which held 20 nodes of the mesh with adaptive refinement toward the bottom of the river. All in all, the mesh constructed with the aid of Gambit 2.3.16 packet consisted of 200000 nodes.

The flow was assumed to be isothermal and all changes in the water density were assumed to be caused by the difference in water mineralization found by the law $\rho = \rho_0 + 0.702\, kg/m^3 \times s$ where $\rho_0 = 1000.196\, kg/m^3$ and s is the mass concentration of the dissolved salts measured in promille.

At the inlet zone of the computational domain, (the place where the Chusovaya and Sylva rivers entered the domain) we used the boundary conditions of the VELOCITY-INLET type (the prescribed flow velocity was determined from the water flow rate) and for the outlet zone downstream from the Chusovskoy water intake we used the boundary conditions of the OUTFLOW type. The free surface of the river waters satisfied the boundary conditions of the WALL type with zero tangential stresses and other boundaries satisfied WALL-type conditions for a ''solid wall".

The choice of the computational model is of critical importance for anticipated results. Out of the turbulence models of the Fluent software packet we chose the Reynolds Stress model and used the Coupled scheme with discretization of the SECOND UPWIND type providing the second-order accuracy. The under relaxation factors used in our computations were standard.

We carried out preliminary simulations to debug the scheme for construction of an adequate numerical solution by applying the Fluent packet to the examined problem. First, we found a stationary solution to the problem neglecting the effect of gravity. The convergence of the obtained solution was achieved already after several hundred iterations. The solution showed that the vertical stratification of water mineralization was absent. Then, the stationary solution was refined in the framework of the non-stationary approach using the time step of 1 second and the second-order time approximation.

The non-stationary simulation lasted for a hundred thousand seconds and revealed a redistribution of salt concentration in the waters of two rivers. Since water of the Sylva river is heavier than that of the Chusovaya river it drops lower and while doing this is mixed with the lighter water of the Chusovaya river.

Flow rate type	Water flow rate in the Chusovskaya river, Q, m^3/s	Water flow rate in the Sylva river, Q, m^3/s
Annual average	226.0	159.0
Minimum	16.0	16.8
Maximum	5250.0	3750.0

Table 3. Data for the water flow rate in the Chusovaya and Sylva rivers.

The simulations were made for three values of the water flow rate in the Chusovaya and Sylva rivers: the minimal, maximal and annual average (see table 3). The mass concentration of salts dissolved in water was assumed to have the same value for three variants of the flow rate (table 2). Fig.4 shows the profiles of distribution of water mineralization through the depth of the river obtained in the vicinity of the water intake head.

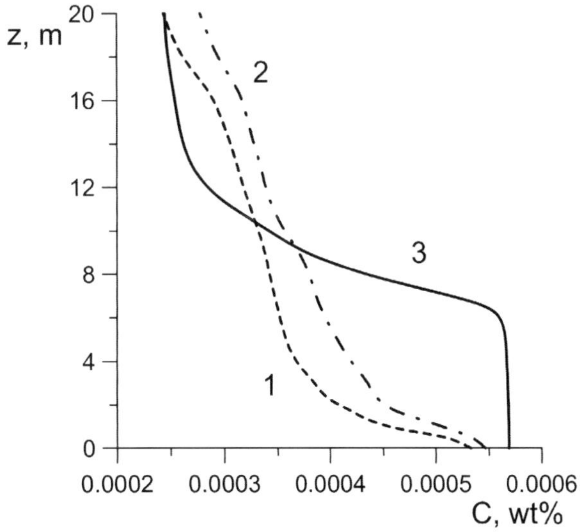

Figure 4. Profiles of distribution of water mineralization through the depth of the river. 1 – annual average flow rate; 2 – minimal flow rate; 3 – maximal flow rate.

4. Conclusions

The results of the two-dimensional simulations have shown that at typical values of the flow rates of the Sylva and Chusovaya rivers, their waters after confluence are intensively mixed in the horizontal direction. In the water intake zone, the admixture concentration in the horizontal direction is practically homogeneous. The three-dimensional simulations have supported this conclusion and revealed essential vertical inhomogeneity of admixture concentration through the water depth: water mineralization near the bottom a few times higher than water mineralization near the surface. Hence, the selective water intake from the near-surface layers can essentially reduce hardness of potable water consumed by the inhabitants of Perm.

References

[1] Lepikhin A.P., Nemkovskiy B.B., Onyanov V.A., Kapitanova E.N. The selective water intake (based on the example of the Chusovaya water intake Perm'). Water Supply and Sanitary Technology, 1988, N 3, pp.27-28.

[2] Vasiliev O.F., Kvon V.M., Lytkin Yu.M., Rozovskiy I.L. Stratified flows // Latest results in science and technology . Hydrodynamics, v. 8, Moscow: RSRI, 1975, p.74-131.

[3] Bocharov O.B., Vasilievв O.F. Semi-empirical model of the selective water intake from a stratified basin // Water resources, 1992, №3, p.16-21.

[4] Belolipetskiy V.M., Kostiuk V.Yu. numerical simulation of flow stratification in the Boussinesq approximation // Numerical methods of continuum mechanics. 1980, v. 17, №4, Novosibirsk, ITAM SB RAS, p.48-60.

[5] Belolipetskiy V.M., Kostiuk V.Yu. Shokin Yu.I. Mathematical modeling of the stratified fluid flows. Novosibirsk. Nauka. SB, 1991, p.176.

[6] Belolipetskiy V.M. Numerical simulation of hydrophysical processes in water basins// Computational technologies. 1998, v. 4, p.15-23.

[7] Nigber M.S., Mitra A.K. Computation of a downflow of the stratified liquid through a drain tube with the aim of determining the selective intake conditions //Transactions of the Americal Society of .. Theoretical foundations of enginering designs Тр. Амер. Об-ва имитацион. механ. Сер. Теоретические основы инженерных расчетов. 1988, №3, p.247-254.

[8] Donnell, Barbara P., Letter, Joseph V., McAnally, W. H., and others, "Users Guide for RMA2 Version 4.5," [09 Sept] 2009, [http://chl.wes.army.mil/software/tabs/docs.htp].

Journal of Physics: Conference Series **416** (2013) 012029 doi:10.1088/1742-6596/416/1/012029

Pressure-driven peristaltic flow

S V Mingalev, D V Lyubimov and T P Lyubimova

Physics Department, Perm State University, 614690 Perm, Bukireva street, 15, Russian Federation

E-mail: SMingaliov@gmail.com, lyubimov@psu.ru, lubimova@psu.ru

Abstract. The peristaltic motion of an incompressible fluid in two-dimensional channel is investigated. Instead of fixing the law of wall's coordinate variation, the law of pressure variation on the wall is fixed and the border's coordinate changes to provide the law of pressure variation on the wall. In case of small amplitude of pressure-variation on the wall A, expansion wave propagates along the length of channel and the wave results in the peristaltic transport of fluid. In the case of large A, the channel divides into two parts. The small pulsating part in the end of the tube creates the flow as a human heart, while the other big part loses this function. The solution of problem for the first peristaltic mode is stable, while the solution for the second "heart" mode is unstable and depends heavily on boundary conditions.

1. Introduction

When expansion wave propagates along the length of the tube, fluid contained in the tube is transported in the direction of the wave propagation. This kind of transport, called peristaltic, is used by many organs such as urethra, male reproductive system and gastro-intestinal tract.

Shapiro, Jaffrine and Weinberg [1] investigated the linearized problem of the peristaltic transport in two-dimensional channel and tube contained incompressible fluid. They considered also the wave length to be infinite. They found backward (reflux) time-mean flow near the tube periphery and the phenomenon of trapping: under certain conditions a bolus of fluid lying about the axis is transported with the wave speed. Jaffrine [2] continued this investigation and took into account effects of Reynolds number and wave number, when these are small but not negligible. The experiments of Weinberg, Eckstein and Shapiro [3] showed that the theory of Jaffrin [2] is valid up to Reynolds number of about 10. Takabatake and Ayukawa [4] investigated numerically the problem of peristaltic in case of two-dimensional channel. They found that validity of the perturbation solutions by Jaffrin is restricted within a narrower range than that which he had predicted. They also found the reflux near the central axis at large Reynolds number. Then Takabatake, Ayukawa and Mori [5] studied numerically the peristaltic transport and its efficiency in cylindrical tubes.

The interaction of peristalsis with elastic properties of the wall was investigated by Mittra and Prasad [6]. They used dynamic boundary conditions considering Newtonian fluid and solved this problem under the approximation of small amplitude ratio [7]. Muthu et al. [8] investigated this problem for micropolar fluids and found critical values of the parameters involving wall characteristics, which cause reversal mean flow. Muthu et al.[8], Mittra and Prasad [6] considered wall to have uniform cross section. Sankad and Radhakrishnamacharya [7] extended the analysis of Muthu et al. [8] to two-dimensional channel of non-uniform cross section. Carew and Pedley

[9] used viscoelastic active body equations with activation parameter which was specified in the form of a traveling wave, as a boundary condition [10]. Since the hallow organ's walls move under force produced by the muscle, the kinematic boundary conditions are changed to dynamic.

In the present study, the law of wall's coordinate variation isn't determined *a priori*. It is found from the initially definite law of pressure-variation on the wall. This way is based on the fact that some hollow organs change diameter under the signals of baroreceptors (sensors that detects the pressure). The effects of various parameters on flow rate and stability of solution have been studied.

2. The governing equations and method of solution

Consider a two-dimensional flow of incompressible Newtonian fluid with kinematic viscosity ν and density ρ in a channel of uniform thickness $2h$, with a sinusoidal wave of pressure traveling along the walls of the channel with speed c, amplitude a and wave length λ. We use the following non-dimensional variables, quantities and parameters

$$x' = \frac{x}{h}, y' = \frac{y}{\lambda}, t' = \frac{tc}{\lambda}, {\xi'}_1 = \frac{\xi_1}{h}, {\xi'}_2 = \frac{\xi_2}{h}, u = \frac{u'}{c}, v = \frac{v'}{c}, p = \frac{p'h^2}{\nu c \rho \lambda}, \tag{1}$$

$$\mathrm{Re} = \frac{hc}{\nu}, \delta = \frac{h}{\lambda}, A = \frac{a}{\rho c^2}, \tag{2}$$

where x and y are directed normal and parallel to the channel's wall respectively, $-1 - \xi_1$ is position of the first wall, $1 + \xi_2$ is position of the second wall, u and v are the velocity components in the y and x directions respectively, p is pressure, Re is the Reynolds number, δ is the wave number, A is the amplitude of pressure-variation on the wall, the non-dimensional equation are

$$\delta^3 \left(\frac{\partial v}{\partial t} + v\frac{\partial v}{\partial x} + u\frac{\partial v}{\partial y} \right) + \frac{1}{\mathrm{Re}}\frac{\partial p}{\partial x} - \frac{\delta^2}{\mathrm{Re}} \left(\frac{\partial^2 v}{\partial x^2} + \delta^2 \frac{\partial^2 v}{\partial y^2} \right) = 0, \tag{3}$$

$$\delta \left(\frac{\partial u}{\partial t} + v\frac{\partial u}{\partial x} + u\frac{\partial u}{\partial y} \right) + \frac{1}{\mathrm{Re}}\frac{\partial p}{\partial y} - \frac{1}{\mathrm{Re}} \left(\frac{\partial^2 u}{\partial x^2} + \delta^2 \frac{\partial^2 u}{\partial y^2} \right) = 0, \tag{4}$$

$$\frac{\partial v}{\partial x} + \frac{\partial u}{\partial y} = 0. \tag{5}$$

The non-dimensional boundary conditions are:

$$v = -\frac{\partial \xi_1}{\partial t}, u = 0, p = \delta A \mathrm{Re} \cos\left(t - y\right) \text{ for } x = -1 - \xi_1, \tag{6}$$

$$v = \frac{\partial \xi_2}{\partial t}, u = 0, p = \delta A \,\mathrm{Re} \cos\left(t - y\right) \text{ for } x = 1 + \xi_2. \tag{7}$$

When the law of wall's coordinate-variation is fixed, the governing equations aren't changed, but boundary conditions should be written as

$$v = -\Xi \sin(t - y), u = 0 \text{ for } x = -1 - \xi_1, \tag{8}$$

$$v = \Xi \sin(t - y), u = 0 \text{ for } x = 1 + \xi_2, \tag{9}$$

where Ξ is the amplitude of wall's coordinate-variation.

The perturbation solution of equation (3)-(7) in the term of wave number δ ($\delta << 1$) is:

$$u = \frac{A\mathrm{Re}\delta}{2} \sin(t-y)(x^2-1) + \frac{A\mathrm{Re}^2\delta^2}{24} \left[\cos(t-y)(x^2-1)(x^2-5) + 8A\sin^2(t-y)\right] + O(\delta^3), \tag{10}$$

$$v = \frac{A\mathrm{Re}\delta}{6}x(x^2-3)\cos(t-y) - \frac{A\mathrm{Re}^2\delta^2}{120}\left[(x^2-5)^2 - 80A\cos(t-y)\right]x\sin(t-y) + \mathrm{O}(\delta^3), \quad (11)$$

$$\xi_{L,R} = \xi = -\frac{A\mathrm{Re}\delta}{3}\sin(t-y) + \frac{A\mathrm{Re}^2\delta^2}{30}\left[5A - 2\cos(t-y)(5A\cos(t-y)-2)\right] + O(\delta^3), \quad (12)$$

$$p = A\mathrm{Re}\delta\cos(t-y) + \frac{A\mathrm{Re}\delta^3}{2}(x^2-1)\cos(t-y) + O(\delta^4). \quad (13)$$

These series are divergent not only for large value of wave number δ but also for large value of Re and A. For this reason the finite element method is used for the solution of equations (3)-(7). In the case of series the channel is infinite in the direction, but the finite element method doesn't allow simulation of infinite channel. In this case the channel's length is considered to be 8π and the solution should satisfy the additional soft boundary conditions at the ends of channel. These conditions are

$$\left.\frac{\partial p}{\partial x}\right|_{y=4\pi} = \left.\frac{\partial p}{\partial x}\right|_{y=-4\pi} = 0, \quad \left.\frac{\partial v}{\partial x}\right|_{y=4\pi} = \left.\frac{\partial v}{\partial x}\right|_{y=-4\pi} = 0, \quad \left.\frac{\partial u}{\partial x}\right|_{y=4\pi} = \left.\frac{\partial u}{\partial x}\right|_{y=-4\pi} = 0.$$

3. Flow rate

From the perturbation solution (10)-(13) flow rate is

$$Q = \frac{1}{2\pi}\int_0^{2\pi}\int_{-1-\xi}^{1+\xi} u\,dx\,dt = \frac{A^2\mathrm{Re}^2\delta^2}{3} + \frac{A^2\mathrm{Re}^2\delta^4}{9}\left[\left(\frac{65}{24}A^2 - \frac{31}{63}\right)\mathrm{Re}^2 - \frac{22}{5}\right] + O(\delta^6). \quad (14)$$

The series (12) for ξ allows to write the amplitude of wall's coordinate variation as

$$\Xi = \frac{A\,\mathrm{Re}\,\delta}{3} + \frac{A^2\mathrm{Re}^2\delta^2}{6} + O(\delta^3). \quad (15)$$

After finding A from (15):

$$A = \frac{3\Xi}{\delta\,\mathrm{Re}}\left[1 - \frac{3}{2}\Xi\right] + O(\delta^2),$$

and substituting it in (14), the flow rate can be rewritten as

$$Q = 3\Xi^2\left[1 - 3\Xi\right] + \mathrm{O}(\delta^4). \quad (16)$$

In the case when the law of wall's coordinate-variation is fixed (equations (3)-(5) and boundary conditions (8),(9)), from the perturbation solution we find flow rate

$$Q = 3\Xi^2 + \left[\frac{\Xi^2}{5\delta^2} + \frac{9\Xi^4}{4\delta^2} - \frac{\Xi^2\mathrm{Re}^2}{1575\delta^4}\right]\delta^4 + \mathrm{O}(\delta^5). \quad (17)$$

The comparison of (16) and (17) shows that both problems give the similar flow rate for small value of Re, δ and A. This conclusion is the result of the fact that one problem can be transformed into the other problem, when the solutions are stable. The law of wall's coordinate-variation can be found from the equations (3)-(5) with boundary conditions (6) and (7). Then this law can be fixed instead of cosinusoidal law in boundary conditions (8), (9) and as a consequence both problems give identical flows.

Finally, expansion wave propagates along the length of the channel and the peristaltic transport is observed for small value of Re, δ and A. Note that the expansion wave has the similar period 2π as the law of pressure-variation on the wall.

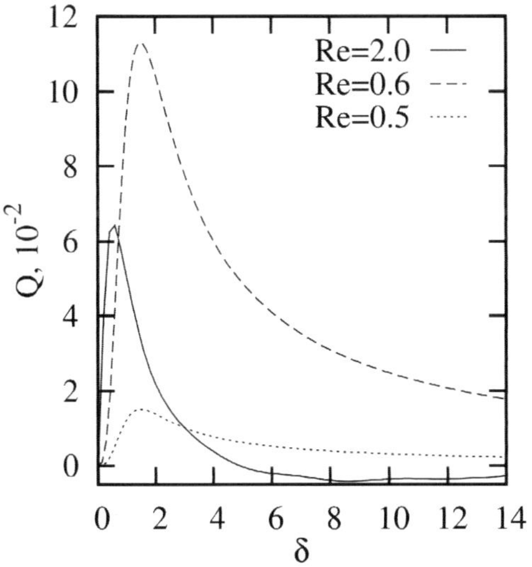

Figure 1. The dependence of flow rate on the wave number.

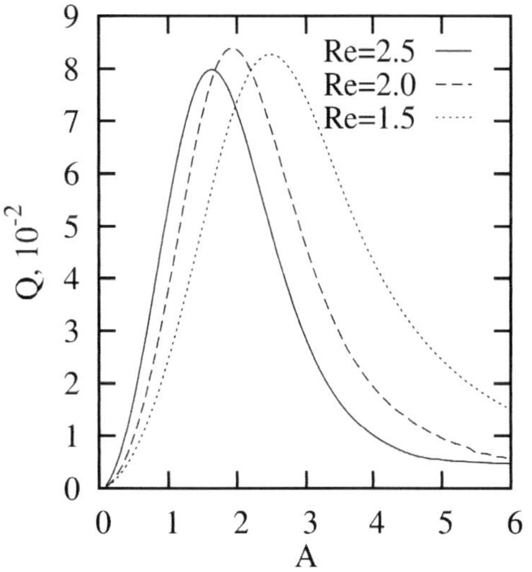

Figure 2. The dependence of flow rate on the amplitude of pressure-variation on the wall.

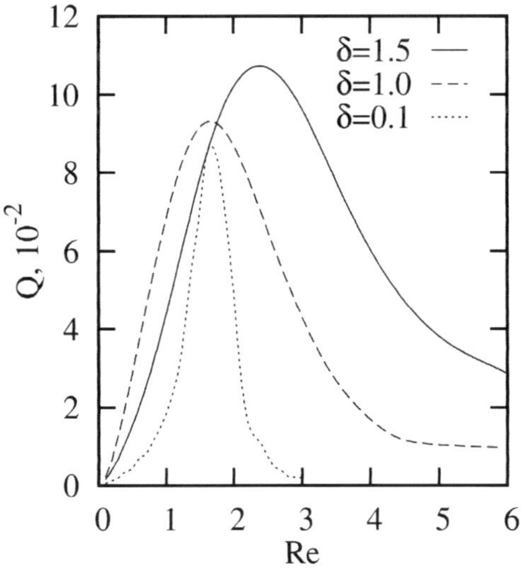

Figure 3. The dependence of flow rate on the Reynolds number.

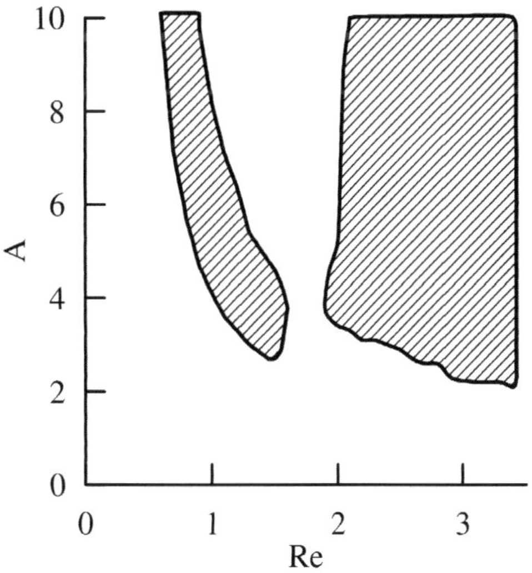

Figure 4. The neutral curve for $\delta = 1.5$.

The difference between the cases of different boundary conditions appears for large value of Re, δ or A. When the law of pressure-variation on the wall is fixed and the parameter is large, the width of the channel decreases (figure 5a, $H(A)$). The middle positions of walls after the relaxation time fall to an optimal value and stop to change (figure 5b). The optimal channel's middle width decreases with growth of the parameters. The figures 1, 2 and 3 show how flow rate depends on Re, δ and A. When the parameter is small, the flow rate increases with growth of the parameter. When the parameter is large, the flow rate and the amplitude of coordinate-variation (figure 5a, $\Xi(A)$) in the central part of walls decrease with growth of the parameter. The flow in this case is created by small pulsating part of walls at the channel's end. The

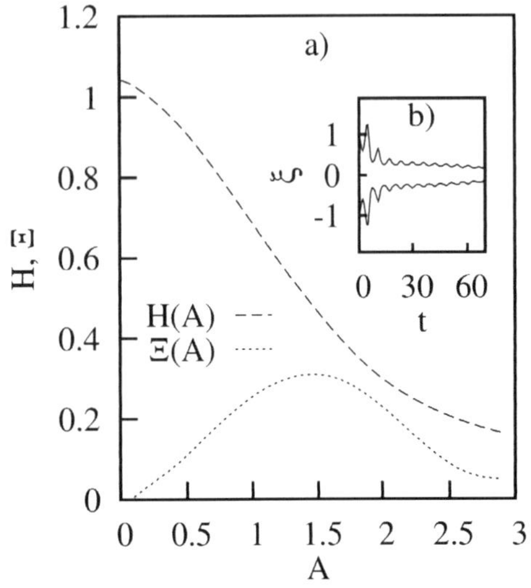

Figure 5. a) The dashed line shows the dependence of second wall's middle coordinate $H(A)$ on the amplitude of pressure-variation on the wall (Re = 3.6, $\delta = 1.5$). The dotted line shows the dependence of amplitude of wall's coordinate-variation $\Xi(A)$ on the amplitude of pressure-variation on the wall (Re = 3.6, $\delta = 1.5$).
b) The solid line shows the position of walls changes far from the end of the channel when the system is unstable ($A = 2.5$, Re = 2.5, $\delta = 1.5$).

amplitude of coordinate-variation grows in this part of the wall in opposite to its behavior in the central part of channel.

The pump's efficiency is determined by the power of walls which can be written as

$$P = -2\int p|_{x=1+\xi}\frac{\partial\xi}{\partial t}dy = -2A\,\mathrm{Re}\,\delta\int\frac{\partial\xi}{\partial t}\cos(t-y)dy. \tag{18}$$

When the parameter is small, the expression for power (18) in term of δ is

$$P = \frac{2\pi A^2\mathrm{Re}^2\delta^2}{3}\left[1-\delta^2\left(\frac{17}{105}\mathrm{Re}+\frac{4}{5}-\frac{7A^2\mathrm{Re}^2}{12}\right)\right]+O(\delta^4). \tag{19}$$

The energy lost to pump fluid in term of δ can be written as

$$\Lambda = \frac{P}{Q} = 2\pi - \frac{\delta^2\pi}{3}\left[\frac{23A^2\mathrm{Re}^2}{12}-\frac{4\mathrm{Re}^2}{315}-4\right]+\mathrm{O}(\delta^4). \tag{20}$$

The series (20) shows that Λ tends to 2π , when the parameters vanishes. Moreover, it shows that Λ decreases with growth of A. When the parameters are large, the numerical simulation shows that energy lost Λ increases with growth of A in opposite to its behavior for small parameters.

4. Solution stability

When the parameters are large, the chaotic behavior is observed in the system (figure 4). Note that the flow doesn't become turbulent and its structure doesn't significantly change but the behavior of walls changes (figure 7 and 6). The instability leads to the transition from one law of wall's coordinate-variation (figure 7) to the other (figure 6). Figure 6 shows, that channel divides into two parts. The small pulsating part at the end of the channel creates the flow as a human heart, while the other big part loses this function. When the solution becomes unstable, the peristaltic mechanism of fluid transport loses importance. The transition from stability to instability leads to decrease of the efficiency of fluid pumping and the energy lost becomes increasing with growth of A.

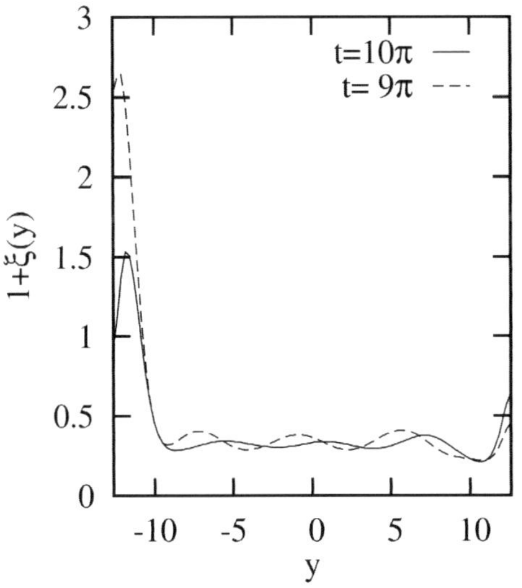

Figure 6. The position of wall in different time for $A = 2.5$, $\mathrm{Re} = 2.5$, $\delta = 1.5$, when the system is unstable.

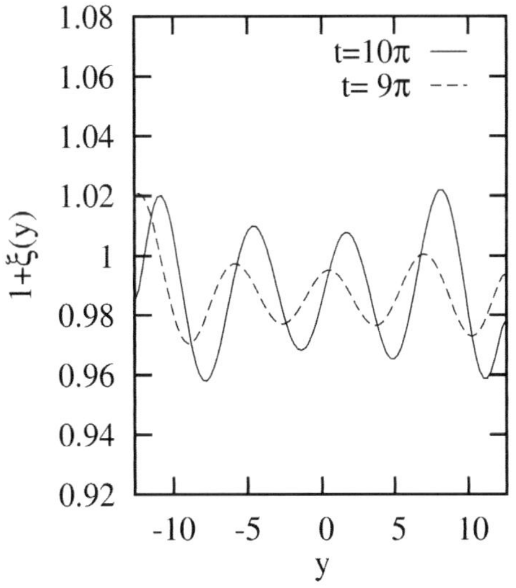

Figure 7. The position of wall in different time for $A = 0.1$, $\mathrm{Re} = 2.5$, $\delta = 1.5$, when the system is stable.

5. Conclusion

The transition from fluid's motion, which is created by peristaltic mechanism, to the motion, which is created by "heart", has been found. This transition is caused by the lost of solution's stability and the rise of dependence on soft boundary conditions. The channel with heart is a bad pump, because the transition leads to increase of energy lost and decrease of flow rate.

When the parameters Re, δ or A are small, the flow rate increases with growth of the parameters. If the parameters Re, δ or A are large, the flow rate decreases with growth of the parameters. The change in dependents is caused by the lost of solution's stability.

Acknowledgments

The research described in this publication was made possible in part by Award No. RUX0-009PE-06 of the U.S. CRDF and Award No. 10-01-96057 of Russian Foundation for Basic Research.

References

[1] Shapiro A H, Jaffrine M Y and Weinberg S L 1969 *J.Fluid Mech* **37** 799–825
[2] Jaffrine M Y 1971 *Int.J.Eng.Sci.* **11** 681–99
[3] Weinberg S L, Eckstein E C and Shapiro A H 1971 *J.Fluid Mech* **49** 461–79
[4] Takabatake S and Ayukawa K 1982 *J.Fluid Mech* **122** 439–65
[5] Takabatake S, Ayukawa K and Mori A 1988 *J.Fluid Mech* **193** 267–83
[6] Mittra T K and Prasad S N 1973 *J. Biomech.* **6** 681–93
[7] Sankad G C and Radhakrishnamacharya G 2010 *IJAMM* **6** 94 – 107
[8] Muthu P, Rathish K and Chandra P 2003 *ANZIAM J.* **45** 245 – 60
[9] Carew E and Pedley T 1997 *J.Biomech* **119** 66 – 76
[10] Bykova A A and Regirer S A 2005 *Fluid Dynamics* **40** 3 – 23

Journal of Physics: Conference Series **416** (2013) 012030 doi:10.1088/1742-6596/416/1/012030

Gaseous bubble dynamics in a pulsational viscous flow

T Lyubimova[1], A Cherepanova[1], L Filippov[2]

[1] Institute of Continuous Media Mechanics, Ural Branch of RAS, Perm, Russia
[2] Université de Lorraine, Rue du Doyen Marcel Roubault, F-54501 Vandoeuvre-lès-Nancy, France

Email: lubimova@psu.ru

Abstract. The paper is devoted to the investigation of gaseous bubble dynamics in a pulsational viscous flow. Behaviour of the bubble suspended in a viscous liquid of different density and located near the rigid wall of oscillating container is studied. The effect of the bubble and host liquid characteristics and vibration parameters on the interaction force is examined.

1. Introduction

If a solid particle is suspended in a liquid which fills a vibrating container then a non-zero average vibrational force acting on the particle from a pulsating liquid arises if densities of particle and host liquid are different. This force could lead to a paradoxical behavior of a solid particle in a pulsational flow under non-zero gravity: the particle more dense than the liquid could rise and less dense particle could sink [1-5]. For the translational vibrations, the average vibrational force acting on the particle from the pulsating fluid is subjected to the fast decrease with the growth of distance between the particle and wall [1-5]. In the case of deformable inclusion translational vibrations of container could also influence the average shape of inclusion, flattening it in the direction of vibrations [6]. The goal of the present paper is to study the effect of viscosity on the dynamics of deformable inclusion in a pulsating liquid.

2. Governing equations. Numerical method

Let us consider the dynamics of a deformable inclusion suspended in a viscous liquid of different density in a vibrating container. Investigation is made numerically by level-set method proposed in [7-9]. This method is based on the representation of a two-layer system as one medium with the parameters depending on a marker-function. In this case, the interface diffuses into transition layer with sharply changing parameters. The capillary force is approximated by volumetric force which is introduced in the following way:

$$\vec{F}_c = -\alpha K(\phi)\delta(\phi)\nabla\phi, \quad K(\phi) = \operatorname{div}\vec{n}, \quad \vec{n} = \nabla\phi/|\nabla\phi| \qquad (1)$$

where $\vec{n}$ is the unit vector normal to the interface, ϕ is the marker-function, $K(\phi)$ is the local curvature of the interface, $\delta(\phi)$ is the Dirac delta-function, α is the surface tension coefficient. Taking into account the force $\vec{F}_c$ we may perform numerical simulation of the behaviour of two-phase system in the framework of single-phase approach.

In the present paper the behaviour of a two-phase system consisting of gaseous bubble and host liquid in a closed cavity subjected to the translational vibrations with frequency ω and amplitude b is studied applying the approach described above. Both the gaseous and liquid phase are assumed to be incompressible. The governing equations describing the behaviour of a fluid with non-uniform density in the reference frame of oscillating container are written down in the dimensionless form as:

$$\frac{\partial(\rho\vec{v})}{\partial t}+(\vec{v}\nabla)\rho\vec{v}=-\nabla p+\frac{1}{\mathrm{Re}}\mathrm{Div}\,\bar{\bar{\sigma}}-a\rho\cos t\,\vec{j}-\frac{1}{\mathrm{We}}\vec{F}_c, \tag{2}$$

$$\frac{\partial\phi}{\partial t}+\vec{v}\nabla\phi=0,\quad \sigma_{ij}=\eta\left(\frac{\partial v_i}{\partial x_j}+\frac{\partial v_j}{\partial x_i}\right),\ \rho=\bar{\rho}+(1-\bar{\rho})H(\phi),\quad \eta=\bar{\eta}+(1-\bar{\eta})H(\phi),$$

where $\vec{j}$ is the unit vector of the vibration axis, $H(\phi)$ is the Heaviside function.

The equations contain the following dimensionless parameters: density and viscosity ratios $\bar{\rho}=\rho_g/\rho_l$, $\bar{\eta}=\eta_g/\eta_l$, $R=R_b/l$, Reynolds number $\mathrm{Re}=\omega l^2/\nu_l$, Weber number $\mathrm{We}=\rho_l\omega^2 l^3/\alpha$, dimensionless vibration amplitude $a=b/l$ (l is linear dimension of the cavity).

On the rigid walls of container the no-slip conditions are imposed.

We assume that in the absence of vibrations the inclusion has the shape of long circular cylinder, the container is a long cylinder of square cross-section and the axes of the container and inclusion are parallel to each other. This allows to consider the problem in two-dimensional formulation.

To solve momentum equation projection method allowing to exclude the pressure is used. When solving transport equation, the problem of numerical diffusion arises which does not allow to perform the calculations for long enough period of time. This difficulty is avoided applying reinitialization procedure. Most of the calculations was performed using the mesh 100x100, which was selected from the results of test calculations made with different number of nodes.

3. Numerical results

First series of calculations was made for the physical parameters which correspond to the two-phase system of liquid deuterium and its saturated vapour at temperature $T=T_c-500\,\mathrm{mK}$ close to the critical point $T_c=33\,\mathrm{K}$. The compressibility and phase change effects were neglected. In this case the density of liquid phase is $\rho_l=44.93\,\mathrm{kg/m^3}$, the density of gaseous phase is $\rho_g=18.46\,\mathrm{kg/m^3}$, the viscosities of both media are the same $\nu=1.07\times10^{-7}\,\mathrm{m^2/s}$, the surface tension coefficient is $\alpha=2.89\times10^{-5}\,\mathrm{N/m}$. The linear dimension of container is $l=1.5\,\mathrm{mm}$ and the bubble radius is $R=l/10$. In most of the calculations the frequency of vibrations was taken to be 38.5 Hz, and their amplitude 0.375 mm. The values of the dimensionless parameters which correspond to the parameter values indicated above are: $\bar{\rho}=\bar{\eta}=0.42, \mathrm{Re}=5080,\ We=303,\ a=0.25$. Initially the bubble is at rest and its mass center is located at the distance equal to two bubble radii from one of the horizontal walls of container (fig.1). The direction of vibrations is parallel to that wall and initial acceleration of the bubble is determined by the vibrational acceleration.

Numerical results for the parameters and initial conditions indicated above are presented in fig. 2. As one can see from this figure, under action of vibrations the bubble performs forced translational oscillations with the frequency equal to the vibration frequency. Additionally, accelerated average motion of the bubble to the nearest wall of container is observed. This motion is related to the appearance of average vibrational attraction force. The formula for average vibrational attraction force in the case of non-deformable inclusion was obtained analytically in [10-11] neglecting the viscosities of media. Appearance of the vibrational attraction of inclusion to the nearest wall is attributed to the Bernoulli effect. Since the parameter values used in the first series of our calculations well correspond to the validity range of the high frequency approach used in [10-11] it was expected to observe the behaviour similar to the one described in [10-11]. The calculations confirmed this hypothesis. The results of our calculations made at different vibration amplitudes and frequencies and the inclusion radii well correspond to [10-11].

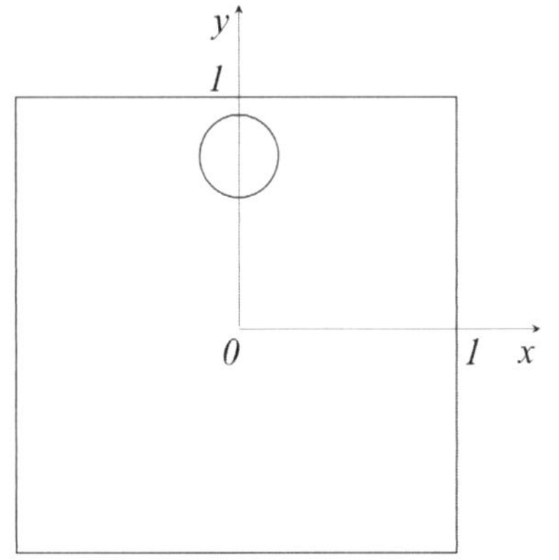

Figure 1. Problem configuration

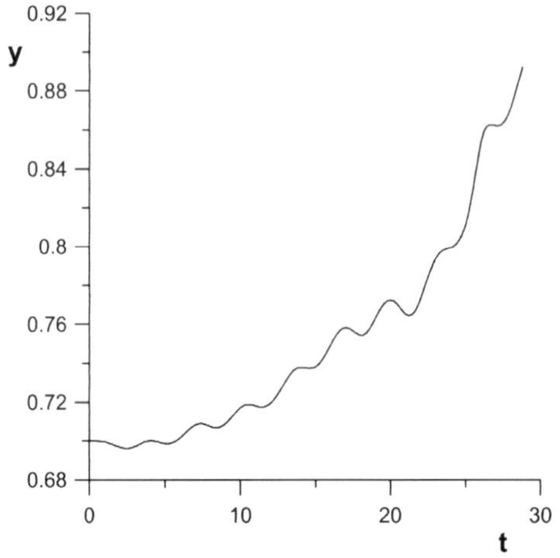

Figure 2. Temporal evolution of the dimensionless vertical coordinate of the bubble centroid for $\bar{\rho} = 0.42$, $\bar{\eta} = 0.42$, $We = 303$, $\mathrm{Re} = 5080$, $a = 0.25$ (vertical coordinate of wall equals to 1)

The goal of the second series of the calculations was to investigate the effect of viscosity on the interaction between the bubble and wall. These calculations were made for the same parameter values as in the first series except for the viscosity which is varied in the range $1.07 \times 10^{-7}\ \mathrm{m^2/s}$ - $1.07 \times 10^{-5}\ \mathrm{m^2/s}$. The results are presented in figure 3. As one can see, at low viscosities (curve 1 and curve 2) the bubble is subjected to the strong attraction to the nearest wall. For the viscosity ten times greater than the lowest value (curve 3), the attraction force is substantially weaker and only small displacement of the bubble to the rigid wall is observed. With further growth of the viscosity, attraction is replaced by the repulsion of the bubble from the wall (curves 4 and 5).

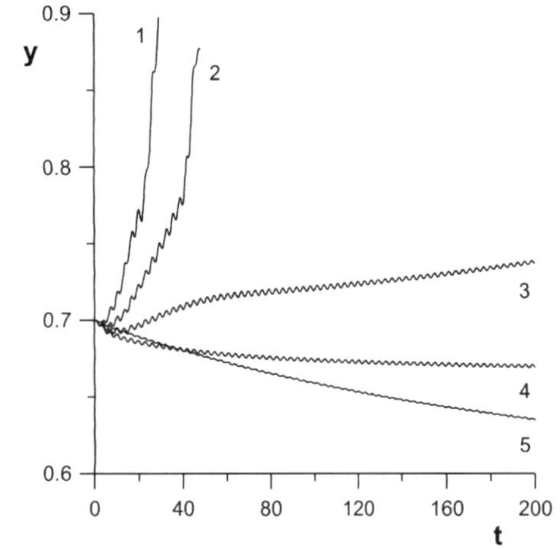

Figure 3. Temporal evolution of the dimensionless vertical coordinate of the bubble centroid for $\bar{\rho} = 0.42$, $\bar{\eta} = 0.42$, $We = 303$, $a = 0.25$: 1 - $\mathrm{Re} = 5080$, 2- $\mathrm{Re} = 1016$, 3 - $\mathrm{Re} = 508$, 4 - $\mathrm{Re} = 254$, 5 - $\mathrm{Re} = 50.8$

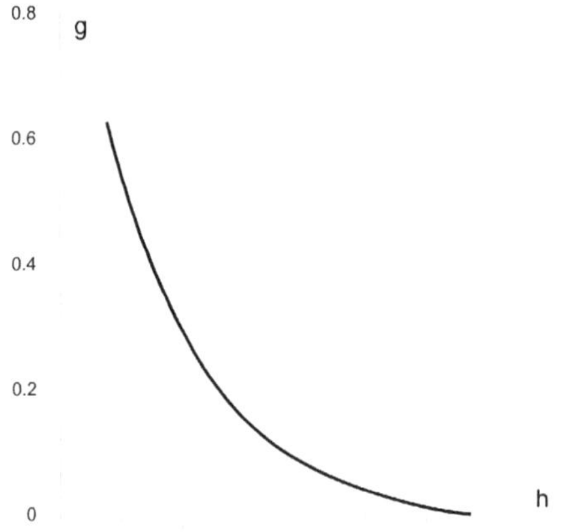

Figure 4. Dimensionless repulsion force versus the dimensionless distance from the wall for $\bar{\rho} = 0.42$, $\bar{\eta} = 0.42$, $We = 303$, $a = 0.25$, $\mathrm{Re} = 101.6$

Journal of Physics: Conference Series **416** (2013) 012030 doi:10.1088/1742-6596/416/1/012030

To define the dependence of the vibrational repulsion force on the distance from the rigid wall we used the balance method. For that we studied the dynamics of the bubble under the coupled action the vibrations and gravity. For our configuration the buoyancy force is directed to the upper wall. As a result, at any initial conditions a quasi-equilibrium state is attained at which the bubble performs small stationary oscillations around a position where the vibrational repulsion force is balanced by the buoyancy force. The calculations show that the repulsion force grows with the decrease of distance from the wall but remains finite even for the very small distances (fig.4). The magnitude of the repulsion force grows with the increase of the viscosity.

4. Conclusions

The paper focuses on the viscosity effect on the interaction of gaseous bubble suspended in liquid of different density with oscillating rigid wall. Numerical simulation shows that under vibrations the bubble could be either attracted to the rigid wall (at low viscosities) or repelled from the wall (at high viscosities). Appearance of the vibrational attraction of inclusion to the nearest wall is related to the Bernoulli effect: increase of the pulsational flow velocity between inclusion and the wall leads to the lowering of the pressure in this area which finally results in the attraction of inclusion to the wall. Repulsion of inclusion from the wall at high viscosities might be explained by the pulsational flow deceleration and consequently the growth of the pulsational pressure between inclusion and the wall.

References

[1] Chelomei V N 1983 *Dokl. AN SSSR* **270** 62

[2] Lyubimov D V, Lyubimova T P, Cherepanov A A 1987 On a motion of solid body in a vibrating fluid. *Convective Flows*, Perm, pp.61-70.

[3] Lyubimov D V., Cherepanov A A , Lyubimova T P 1992 The motion of solid body in a liquid under the influence of a vibration field. *Reviewed Proc. Of the First Int. Symp. On Hydromechanics and Heat/Mass Transfer in Microgravity, Gordon and Breach*, pp. 247-251.

[4] Lugovtsov B A, Sennitskii V L 1986 *Dokl. AN SSSR* **289** (2) 314.

[5] Lyubimov D V , Cherepanov A A, Lyubimova T P , Roux B 2001. *J. Physique IV* **11**, 83

[6] Lyubimov D V, Cherepanov A A , Lyubimova T P, Roux B 1996. *Microgravity Quarterly*, **6** 69

[7] D V Lyubimov, T P Lyubimova. 1990 Modelling in *Mechanics*, 4(21), **N 1**, pp.136-140.

[8] Brackbill J U, Kothe D B, Zemach C A 1992 *J. Comp. Phys.*,**100** 335.

[9] Sussman M, Smereka P, Osher S 1994 *J. Computational Physics* **114** 146.

[10] Lyubimov D V, Cherepanov A A, Lyubimova T P 1992 The motion of solid body in a liquid under the influence of a vibrational field. *Reviewed Proc. of the First Int. Symp. on Hydromechanics and Heat/Mass Transfer in Microgravity, Gordon and Breach*, pp.247-251.

Interest of new communicating material paradigm: An attempt in wood industry

J Jover[*], A Thomas[*], JM Leban[**], D Canet[***],
Research center for automatic control of Nancy, science faculty - BP 239
54506 Vandoeuvre les Nancy – France.
** Wood Quality department, national institute of agronomical research, Champenoux Center, Champenoux, France.
*** Departement of NMR metrology, Henry Poincare University, science faculty - BP 239, 54506 Vandoeuvre les Nancy – France.

E-mail: jeremy.jover@univ-lorraine.fr

Abstract. This paper present a new paradigm in which the wood material could become communicating. We decide to use Nuclear Quadrupole Resonance to mass marking the wood. We imagine a new method to create identification codes. At first, we examine the feasibility of this mass marking method by impregnating wood to obtain a specific marking signal. In parallel, we study the interest to abide information provided by this marker to control the supply chain. We model the supply chain (e.g. the information/decisional flow) to understand which information is important and how to use it.

1. Introduction

More and more, with a view to increase productivity or traceability, supply chain manager mark piece. The auto-ID technology, and in particular RFId, are more and more extended in the way to realize this product marking. Nowadays, advances in reading of these marks make possible to read many tag's names in the same times at long distance. In another way the storage capacity increases: on the one hand RFId chips have more memory, and in the other hand we know how associate chip names to a data base where they are reference. Some research projects showed the relevance of the communicating product concept and search for spread out in industry. So it's possible to increase the productivity of the workshop by using communicating products. For example tags are put on the kanban in order to become it active [1]: kanban label can "discuss" with the other labels to be better ordain.

But these products marking are discrete. That is to say one product is marked with only one tag. This is really practical for finished products but that means products could not be modified. May be we could imagine a new paradigm wherein it is the material that becomes communicating instead of the product. In this case we could cut the material and conserve the information associate with each element. Thus the material has its personal information like physical characteristic and origins with it. Under this new paradigm, our research group has studied the pertinence and the feasibility of this vision in wood industry and textile. This paper deals only on the wood mass marking. We look for how mass marking the wood and we project us in the exploitation of this marking for piloting the wood supply chain. To understand the effect of the information loss on the supply chain, we modelize all the wood supply chain and the information exchange. This paper will follow the present plan: first we will focus on the tractability techniques in wood industry. Secondly we will study the marking we want to experiment, using nuclear quadrupole resonance (NQR). Then we will modelize the information/decisional flow in order to study the interest of this information for piloting the production. Finally we will describe a first demonstrator which will confirm our concept.

Published under licence by IOP Publishing Ltd

2. The traceability in wood indutry

2.1. The importance of the tractability in the wood industry

Even though the wood was the oldest material used in the history, it was every time considered by people like an easy material in which the economic issue is weak. But today the mind change: wood industry has to be improved. Wood industry tries more and more to setup systems of material traceability. Many reasons are put forward: the increase of the outlaw wood exploitation [2], the need to improve the wood utilization [3] [4] and the important information loss all along the wood life cycle [4]. Some actions were realized by the manufacturers to increase wood traceability. For example:

- Certifications of the wood origins like PEFC (Pan European Forest Certification) or FSC (Forest Stewardship Council). These certifications give the proof that the woods come from forests which are durably managed and the supply chain respects specific rules of management.
- The European project Indisputable key [5] which search improvement ways for the wood traceability and methods for the information conservation width.

2.2. The information conservation in wood industry

The problems which are mentioned before have one common point: the wood industry can't conserve the information of the tree all along the supply chain. Finally the wood supply chain is sequence of information loss. A European project [5] shows that the leakage is between every important steps 'figure 1'. The solutions proposed by this project seek to solve the information loss between forest and log selection. Other papers like [3] look for solution between log selection and log sawing. Few publications deal with the information loss after sawing because the impact is less important, the information can be finding easily by measurement on the wood (humidity, class of quality, etc.).

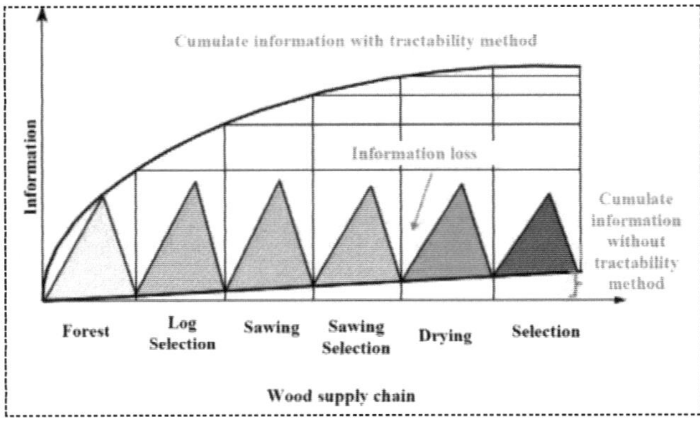

Figure 1. Schematization of information loss at each step during wood supply chain (source: the Indisputable Key).

The researchers try to put in place new identification and traceability methods for wood. The techniques used until today for marking logs are describe in publication [2]. The first markings put in place were paint mark or hammering, used by the log owners to recognize their own. While they are cheap these marks have a real big problem: you can reproduce it easily. The logs owner think to a new method that could be harder to reproduce: the bar code print on a plastics stamp which is hammering on the log. These stamps are unique and their reproductions are harder than the paint marking. But this marking can be easily removed from the log. To improve the security and reading possibility, the idea was to put in the log an RFId or magnetic tag which can be read in a dirty, dusty or damp environment. The project [5] submits an idea of RFId tag: this tag was cast in a resin wedge. Then the wedge is hammering at the end of the log. Another solution was to print a bar code directly at the end of the log.

But all these markings are discrete (i.e. one mark for one object). Furthermore they mark wood between forests and sawmill. The mark can be easily removed from the log by sawing the end of the log because it is not attach to the totality of the log. That's why other technics have been found. The use of the micro wave [6] allowed having a biometric code of the wood piece. An emission of micro wave goes through the piece of wood and gives a biometric mark of the wood. But this biometric mark changes in the time because wood could be manufactured or changed by the time. Another way to have a mark of the wood is the use of the X-ray: you obtain a picture of all the piece of wood. By the similitude you can recognize the piece. With this method the traceability can be possible by reconstruction of the log, and also the tree. But this technic repulse sawyer because the quantity of pictures to save is huge.

In the tree there is a unique code that could be used to identify it: the DNA [7]. This is the same technique than for human. It is used in the cooperage industry to identify the origins of the tree. But this method of identification is so complicated to integrate in an industrial process [2].

2.3. Issue

We understand that a mass marking is useful. However how marking the wood in the mass? If the code is unique for one tree, how to code the pieces cut in the tree? How to use this information to pilot the supply chain?

3. Utilization of the nuclear quadrupole resonance like a tag.

As one of the imperatives of this present project is the mass marking and as it will be probably (other while necessarily) a molecular marking, the non-destructive and non-invasive identification technique will use the radio frequency (not absorbed by the material). In addition this technique has to use a spectroscopies method which gives an ambiguous identification. We are obliged to use the nuclear magnetic resonance (NMR), the electron paramagnetic resonance (EPR) or the nuclear quadrupole resonance (NQR). NMR and EPR need a static magnetic field which complicates the equipment without given a specific determination. So we have to use the NQR which use a null field and the frequency response is characteristic to the molecule integrated to the material.

3.1. Principe of the NQR

The NQR result from nucleus quantum properties with a spin upper than ½. The nucleus is made up of protons and neutrons. The distribution of them in the core creates a distribution of the charge that could be spherical or with an axial symmetry 'figure2'. The spin is associated to the distribution: if the spin is ½ the distribution is spherical, if the spin is up than ½ the distribution has an axial symmetry. This symmetry allowed the apparition of a nuclear quadrupole moment. This moment can interact with the electric field gradient present in the nucleus environment and provoke a gap of the energy level. Between this energy level, transition could be induce by radio frequency as in our case.

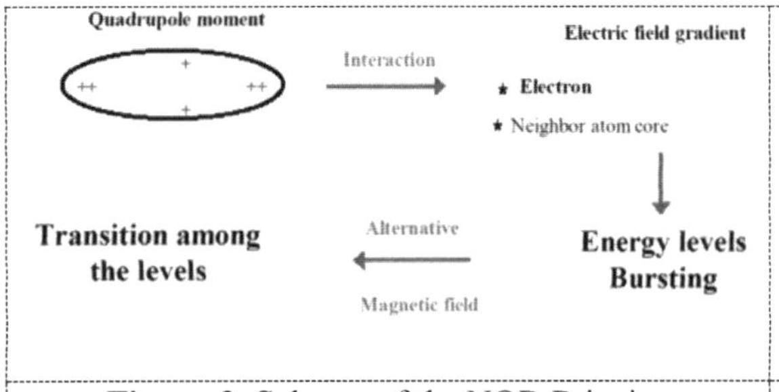

Figure 2. Scheme of the NQR Principe

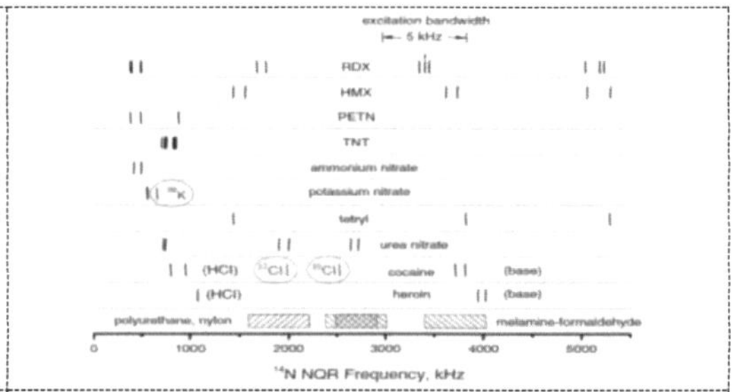

Figure 3. Representative RQN frequency for some explosives et narcotics. [8]

The frequency of this transition is a direct measure of the quadrupolaire interaction, and depends of the electronic distribution around the core. So it's possible to characterize the molecule. The land mine could be detected by this way [8]. A spectrum is established for explosive 'figure 3'. This is like the molecule's fingerprint.

3.2. Principe of the NQR

The objective is to create an identifying code that could be integrated in trees. So we imagine mixed some molecules you know the frequency responses. By using a transmitter/receiver and when the product is integrated in the tree, you have a code that is the identifying of the tree.
But NQR has a mediocre sensibility and temperature has a big influence on the frequency response [9]. As the NMR metrology laboratory of the henry Poincare University master the measurement of the sodium nitrite (NaNO2) we use it to see the effect of the wood on the NQR signal. We realize two experiences:

- NaNO2 and sawdust in order to be sure that wood doesn't influence NQR signal
- A piece of wood impregnate with a solution of sodium nitrite so as to know if the nitrite signal is present after drying

3.3. Results and conclusion

We test two different ratios during the first test: firstly, we put fifty percent of wood and NaNO2, secondly we reduce the quantity of NaNO2 to twenty five percent. We note that the sodium nitrite's signal is present in the both cases so wood doesn't affect NQR signal 'figure4'. Some parasitic signals appear come from the preparation on the nitrite but disrepair little by little (piezoelectric effects).
The second experience shows us that one part of the nitrite has been correctly crystallized because we see the sodium nitrite ray. But an abnormal ray appears in the two spectrums 'figure5'. If we made the parallel with the explosive detection [10] which show that the ray width is more important than the crystal lattice have defects, this ray is a part of the sodium nitrite which is more amorphous or interacts with the wood.

We can conclude that it will be possible to mark wood using NQR response like code but nowadays the technique isn't efficient.

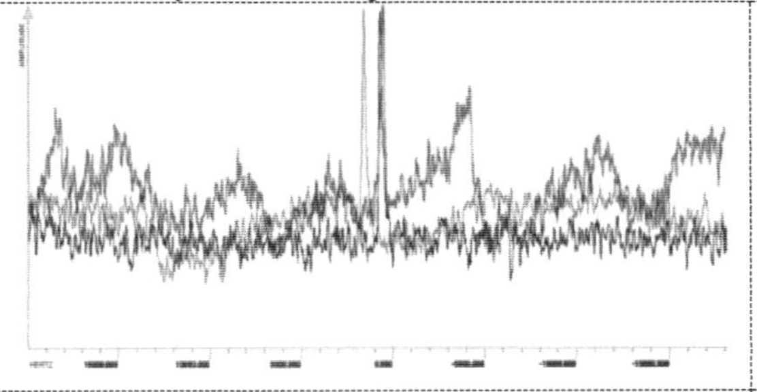

Figure 4. Sodium Nitrite response Spectrum at 4,64MHz. Blue Spectrum: 100% de $NaNO_2$ (powder), reference (8 scans). Red spectrum: 50% $NaNO_2$, 50% sawdust (32 scans). Green spectrum: 25% $NaNO_2$, 75% sawdust (128 scans).

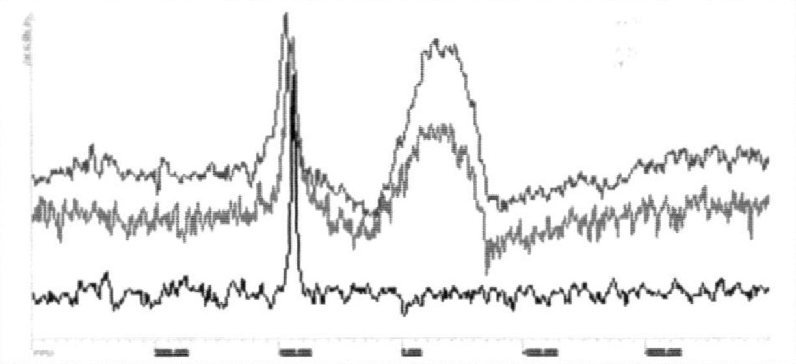

Figure 5. Sodium Nitrite response Spectrum at 4,64MHz. Black spectrums: 100% de NaNO2, reference (8 scans). Red and blue spectrum: impregnate wood with two different NaNO2.(40 000 scans)

4. Modeling the flow of information

In order to see the interest of the information in the wood supply chain, we model them. Like this we can understand the utility of the information at each activity.

4.1. Information flow Model in the wood supply chain.

Few models describing all the wood supply chain exist; often it is part of this supply chain. Some projects show the information's loss [5] or the exchange between the activity sectors [11]. These

Journal of Physics: Conference Series **416** (2013) 012031 doi:10.1088/1742-6596/416/1/012031

projects show the most important information is at the beginning of the supply chain (from forestry to end of sawing). So we decided to break up this part of the supply chain.
The forest inventories give us all the information relative to the forestry or the logging industry [12] and some paper explain the utility of this information [13]. We have used the books of FCBA [14] to find the information used to pilot sawmills. When the data was gathering and the supply chain modeled, we have studied all activity to identify the information at the input and the output (A1).

4.2. Study of the model: the output information of the forestry and logging industry.
All the data use in the forestry and the logging industry are data which refer to the growth of the tree. There are two industrial utilities for that data: first, with that data you can know the tree's origin. Secondly, they could be used by the sawyer in order to determine the potential of a plot. Rules exist between the tree's growth and the quality of the tree. So sawyer can use an electronic data exchange, like a database, to know the plots and have ratios. With them he can determine if the plots will be good for him or not. (I.e. if the sawyer cut wood for pallet, he has to choose plot with a good ratio in pour quality wood).

4.3. Study of the model: the output information of sawmill.
In sawmill, the information has another utility. This information enables the sawyer to tune his materiel. When you have a hardwood you know the stress on the blade is bigger so you tune your saw cut at lower speed. So if you could find the information relative to the tree you can tune the machine in real time. But there is a big problem with the unique code: how to identify the pieces cut in the log if they have all the same code? This is our future research: we would like to study which way is the best to have a code for the "children" (pieces of wood cut in the log)? If it was possible to code the "children", we can imagine that the sawyer have a data base where he can look all his storage. When a customer wants a product, he describes the destination of the product and the sawyer can deliver the product which is the closest from his need.

4.4. Future researches
As we said just before, our future researches are to find the best way to code the "children". It looks easy to create a code, but what is the best one to include in an industrial process?

5. The demonstrator
While the NQR marking is not ready, we would like to continue the research on the generation of the code for the "children" pieces. So we decide to simulate the mass marking material. So we decide to build a demonstrator which simulates a log mass marking. Another utility of this demonstrator is to show the benefit of marking the tree in the mass. You can simulate a log's storage and when you scan the storage, you can find which log is the most appropriate to cut function of the orders: you know the information relative to the logs and the orders and the computer find which log is the best one for the situation. So we made a requirement for our demonstrator.

As we want to simulate a log which must be cut and mass mark, the demonstrator have to be easy to disassemble and have all the parts marked by RFId. We have decided to use piece of wood with little magnets on it to make the assembly easiest. On each piece of wood we have to put an RFId tag 'figure6'. If all the RFId tags have the same code, we can simulate mass marking wood. This wood is not marked itself but we don't want to cut piece of wood, just disassemble our tree. For the second utility we have just to change the RFId tags with other ones that have many different codes.

At the same times, we made some model of an informatics program and we asked to an informatics engineer to program the software 'figure7'.

Figure 6. Piece of wood composing our tree without the RFId tags

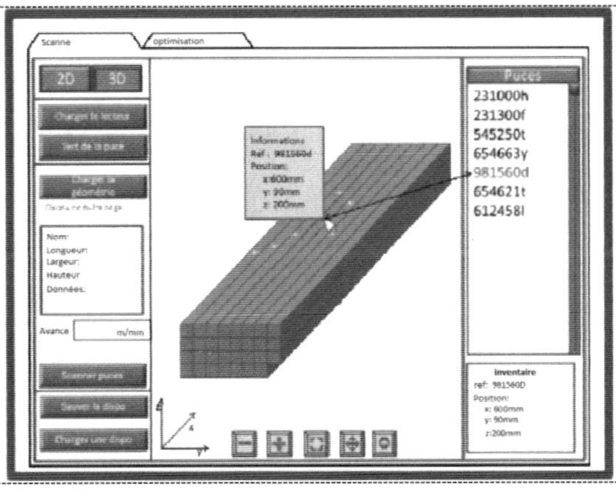

Figure 7. Model of the software

6. Conclusion

Finally the wood mass marking stays a futurist project. But with this project we show that it's not impossible to mark the wood. NQR marking could offer possibilities: with other products, best research on the NQR stability... so researches must be done on this subject.

Concerning the model, it enables us to understand the utility of each data present in the wood supply chain. We understand the benefit from this marking. The next research that our research group has to do is the generation of the "children" piece code with a reflection on the best way to obtain it.

For this we have to finish our demonstrator. Perhaps, the mass marking will be efficient in wood. For the moment the textile mass marking is more concrete than the wood mass marking and researches are bring to light better results.

7. Thanks

The authors gratefully acknowledge the financial support of the CPER 2007-2013 "Structuration du Pôle de Compétitivité Fibres Grand'Est" (Competitiveness Fibre Cluster), through local (Conseil Général des Vosges), regional (Région Lorraine), national (DRRT and FNADT) and European (FEDER) funds. The authors want to thank the INRA (National Institute in Agronomical Research) and, particularly, Jean-Michel LEBAN and Charline FREYBURGER for the use of their X rays scanner.

8. References

[1] Klein, T. (2008). Le kanban actif pour assurer l'interopérabilité décisionnelle centralisé/distribué. Application à un industriel de l'ameublement. Thesis of the university Henry Poincare.

[2] Dykstra, D.P., Kuru, G., Taylor, R., Nussbaum, R., Magrath, W., and Story, J. (2002). Technologies for wood tracking: Verifying and Monitoring the Chain of Custody and Legal Compliance in the Timber Industry, World Bank / WWF Alliance for Forest Conservation and Sustainable Use.

[3] Chiorescu, S., and Grönlund, A. (2004). The Fingerprint Method: Using Over-bark and Underbark Log Measurement Data Generated by Three-dimensional Log Scanners in Combination with Radiofrequency Identification Tags to Achieve Traceability in the Log Yard at the Sawmill, Scandinavian Journal of Forest Research, 19: 4, 374 — 383

[4] Ginet, C., and Golja, R. (2007). Traçabilité et échanges électroniques pour la filière forêt bois papier. Fiche information–forêt n°750 d'Afocel

[5] The Indisputable Key. (2009). www.indisputablekey.com, European project.

[6] Charpentier, P. and Choffel, D. (2003). The feasibility of intrinsic signature identification for the traceability of pieces of wood, Forest Products Journal, 53(9) : 40-46

[7] Deguilloux, M.F., Pemonge, M.H., and Petit, R. J. (2003) DNA-based control of oak wood geographic origin in the context of the cooperage industry, Annals of Forest Science 61 (2004) 97–104
[8] Garroway, A.N. (1994). Explosives detection by Nuclear Quadrupole Resonance, Cargo Inspection Technologies, San Diego.
[9] Ferrari, M. (2006). Temperature effects on 14N Nuclear Quadrupole Resonance (NQR) parameters, Poster.
[10] Buess, M.L. and Caulder, S.M. (2004) Appl. Magn. Reson. 25, 383.
[11] Jover, J. ; (2010). Le bois communicant sur pied, rapport de projet de fin d'étude pour le diplôme d'ingénieur ENSTIB.
[12] Inventaire Forestier National. (2010). Documentation relative aux données brutes « arbre », Campagne 2008.
[13] Pignard, G. (1993). La base de données dendrométriques de l'inventaire forestier national. Revue forestière française, XLV, n° spécial.
[14] FCBA (Institut technologique Forêt, Cellulose, Bois, Ameublement) (2001). Manuel de scierie: techniques et matériels. (Paris: FCBA).

9. Apendice A

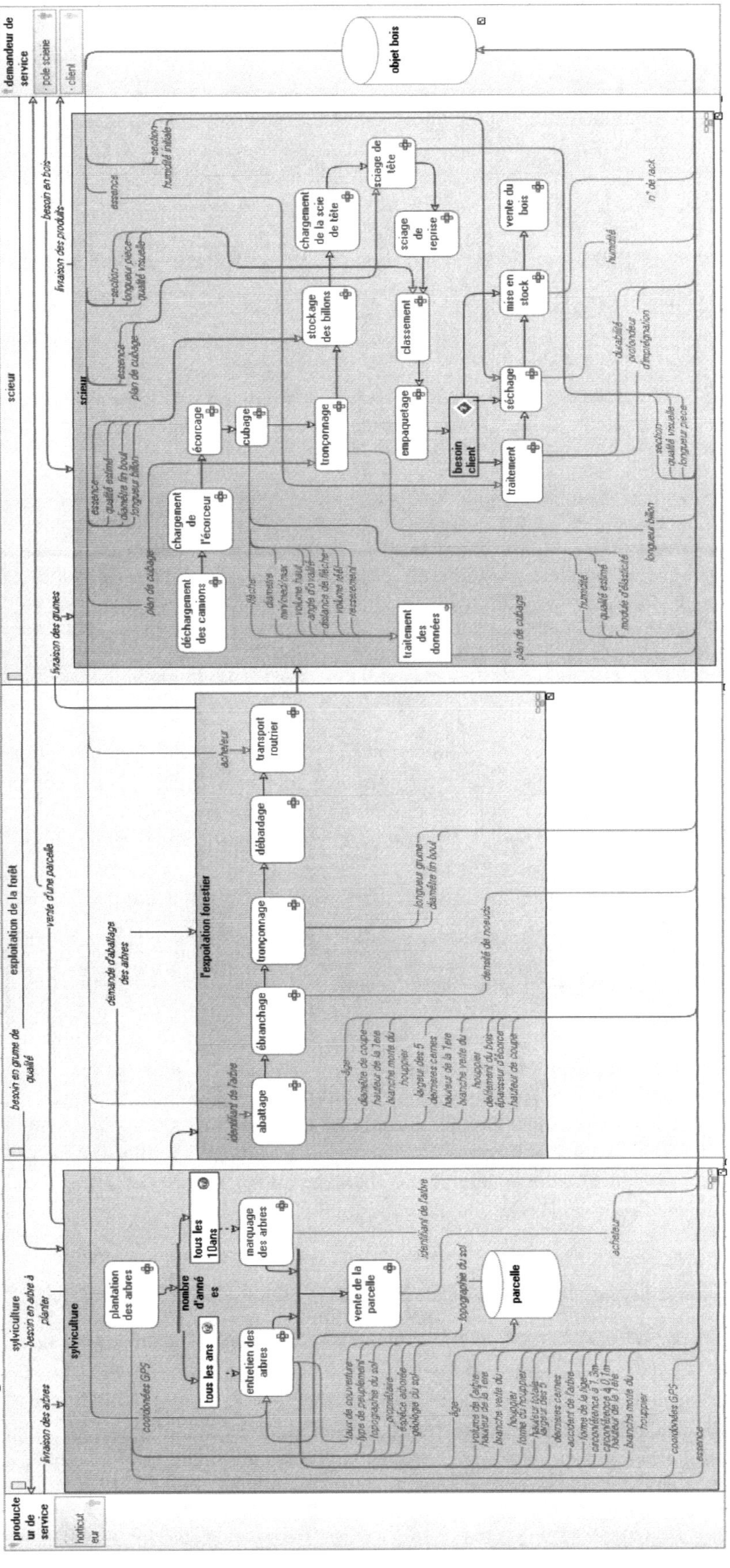

A1. Model of the information exchange in the wood supply chain using the wood object data base

Journal of Physics: Conference Series **416** (2013) 012032 doi:10.1088/1742-6596/416/1/012032

Fibre-optic current sensor exploiting Sagnac interferometer with Lyot-type fibre depolarizers

S M Kostritskii, A A Dikevich, Yu N Korkishko, V A Fedorov

RPC Optolink, Proton, Proezd 4806 bd.5, Zelenograd, 124498, Moscow, Russia

E-mail: skostritskii@optolink.ru

Abstract. A Sagnac interferometer-type fibre-optic current sensor using single-mode fibre was proposed. To prevent non-reciprocal phase error, depolarizers are used in the Sagnac coil. This configuration makes the current sensor less costly and rather accurate. We fabricated the current sensor for an electric power substation by using the proposed technique and confirmed its characteristics. The dynamic range of 50 dB and ratio error within ±0.2% were satisfied. These results have demonstrated that the Sagnac interferometer-type current sensor using a single-mode fibre is suitable as an instrument for electric power plants.

1. Introduction

Until now, measuring current in electrical equipment was performed using a current sensor with an iron core and copper wire windings. By introducing high-performance protection systems and high-voltage power supplies, attention has been paid to current measurement using fibre-optic current sensors. This makes the size smaller because the sensor fibre is easily isolated. Moreover the environmental influence is minimized because insulating oil or SF_6 gas is not needed.Recently various fibre-optic current sensors have been proposed. In these sensors we think the Sagnac interferometer type fibre-optic current sensor [1,2] is the most promising because the technology used for fibre-optic gyroscopes can be applied without any change. It is highly reliable and applicable for electrical equipment. Nevertheless, the Sagnac loop is configured in the Sagnac interferometer-type fibre-optic current sensor using polarization-maintaining (PM) fibres. Using PM fibres require highly accurate axial alignment during fusion splicing, and installing the fibre-optic current sensor in a substation is complicated. Further, the PM fibres themselves are costly. Here, we propose a Sagnac interferometer-type fibre-optic current sensor, which enables fibre-optic transmission using a single-mode fibre. The basic concept is the same as that of a gyroscope using a depolarizer [3]. First the basic configuration and principles are shown. Next, the results of the characteristics tests are described. It was confirmed that the proposed current sensor has the accuracy required for application to electrical equipment.

2. System configuration

Using the scheme with counter-propagation of light, there is an opportunity to distinguish the birefringence-caused effects (for they are reciprocal) from the Faraday rotation-caused (this one being non-reciprocal). Although this scheme allows in principle to decrease the influence of the fiber birefringence, all the approaches, which are available now, failed to find an appropriate optical scheme to avoid strong reciprocal-effects-based errors. This problem can be solved by compensation of polarization noise with aid of spectral averaging of state of polarization (SOP).

The basic configuration of our fibre-optic current sensor is a Sagnac interferometer using the SM isotropic fiber-loop and depolarizers. A 1.55 μm super-luminescent diode (SLD) is used as the light source. Light polarization, which is transmitted through the Sagnac loop, is a well-known source of noise. To suppress this noise, a highly accurate gyroscope is configured with a Sagnac loop using PM fibres. Meanwhile, a depolarizer is installed as a part of the Sagnac loop in the depolarized-type gyroscope, too. Light travelling through the Sagnac loop clockwise or anti-clockwise is depolarized to prevent noise due to polarization. This method is also applicable to fibre-optic current sensors. The first depolarizer is placed between the SM coupler and the multifunction integrated-optical chip (MIOC) to stabilize the amount of light passing through the MIOC. The depolarizer is of Lyot type. It is configured by fusion splicing two PM fibres with a 1:2 length ratio at 45° to the optical axis. To detect a phase difference due to the Faraday effect, light passing through the sensing fibre coil must not have polarization oscillating due to birefringence. Two fibre depolarizers are placed at both ends of the sensor fibre to obtain a stable SOP in average over SLD spectrum with a low degree of polarization (DOP).

The SLD used is λ = 1.55 μm, and $L_c \leq 40$ μm, and if the PM fibre beat length is 3 mm, $\Delta\beta = 2000$ rad m^{-1}, and so it is desirable PM fibre length for a shorter arm of depolarizer L to be 15 cm or longer. In this experiment, L was ≥ 50 cm. The sensing fibre birefringence reduces fibre-optic current sensor sensitivity. According to the design, the maximum measured current is 100 kArms. Sensitivity can be improved by increasing the number of sensing coil fibre turns, but because of the constraint of the maximum measured current, a ten-turns coil is used. The fibre coil diameter is 200 mm. Even if the lengths of depolarizers and connecting fiber cable are added to this, the total length is less than 100 m. Then, as the time difference between the clockwise and anti-clockwise light passed through the MIOC is insufficient, sufficient modulation cannot be performed. To solve this problem, a compensating extra fibre coil, 400 m in length, was added. The extra coils are configured using single-mode fibres for 1.5 μm band.

The performance (accuracy specification) was tested according to IEC-60044-8 with a view to commercial production. The dynamic ranges of existing wound current sensors are narrow, and measurement has traditionally been performed using two-core current sensors, one for the control of small current, and one for the protection of large current.The accuracy classes most frequently used are 0.5 for control, and 5P for protection. Use of fibre-optic current sensors makes it possible to widen the dynamic range. With the proposed fibre-optic current sensor, the aim is to simplify the configuration to one sensor only from the existing two sensors by simultaneously satisfying the 0.5 class for control and the 5P class for protection using one sensor. The current sensors most frequently used in Russia have a higher error tolerance than the 0.5 and 5P classes, which are specified by IEC-60044- 8, so that the proposed current sensor would be suitable for use in Russia, too. The rated frequency is thought to be 50 Hz.

3. Experimental results and discussion

The accuracy of the fibre-optic current sensor was measured in various environments to check the stability. The accuracy of the current and voltage metre used for electrical equipment is generally shown by ratio error, and can be defined using the following formula [4]:

$$\varepsilon(\%) = [\,(K_r\, U_s - I_p) / I_p] \times 100$$

where K_r is the rated transformation ratio, I_P is the RMS value of the actual primary current and U_S is the RMS value of the optical current sensor output. Attention must be paid to the fact that I_P and U_S are the values of the actual current and not of the rated current. The ratio error is measured by an automatic instrument transformer test set by using a balanced bridge.

The measurement results are all within the ±0.2% range of ratio error, indicating that the fibre-optic current sensor has excellent linearity within ±0.2%, Fig. 1. The tolerable error of the 0.5 class sensor, which is defined by IEC 60044-8, is shown in the diagram using solid lines. The current sensors most commonly used for protection are those of this class in the world. The accuracy of the proposed fibre-optic current sensor satisfies the standards with sufficient margin.

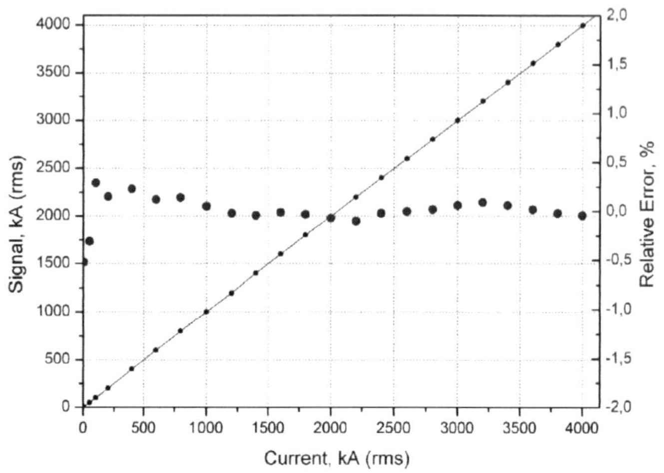

Figure 1. Output signal of fibre-optic current sensor versus primary current.

The sensitivity of the fibre-optic current sensor is temperature dependent. The main reason for the sensitivity change is the temperature dependence of the SLD wavelength. As the temperature dependence of the SLD is about 0.2 nm×K^{-1}, the SLD wavelength is shifted about 20 nm by the 100 °C temperature change. The sensing fibre's Verdet constant is proportional to $1/\lambda^2$ (where λ is the SLD wavelength), and so 20 nm wavelength shift brings about 5% sensitivity change of the current sensor. To compensate for the temperature dependence of the current sensor, we install a semiconductor sensor to measure the SLD operating temperature. The signal processing unit has a ROM lookup table that memorizes data of the SLD operating temperature and the current sensor sensitivity. The signal processing unit calculates temperature compensated current data from GYRO circuit output and semiconductor sensor output. The temperature dependence of the Verdet constant of the sensing fibre is 0.69×10^{-4} K^{-1} [5]. It is also compensated by using a semiconductor sensor. The temperature difference of the sensor fibre and SLD operation temperature is designed to be less than 30 K. The sensitivity change of the fibre-optic current sensor because of the temperature dependence of the Verdet constant is less than 0.2%. After temperature compensation, the actual measured values are all within the ratio error range of ±0.2%, indicating excellent temperature stability.

Because fibre-optic current sensors are not subject to saturation such as iron core magnetic saturation, excellent linearity can be maintained even at large current. When the power systems are fully operational, they operate at the rated current for each system, for example, 4000 A, but if an accident such as grounding or a short circuit occurs, then the current is overloaded. As it is necessary to detect such current accidents, the fibre-optic current sensor must be able to measure the area up to 20 times of the rated value [6]. In the case of protection systems using a conventional wound current sensor, iron core saturation needs to be taken into consideration, and therefore, errors of 5% are tolerated for class 5P defined by IEC 60044-8.

The large-current experiment was performed also. The sensing coil is housed at the top of the magnetic polarizer. The electronic circuit is housed in a water-proof case at the ground potential. The sensing coil and electronic circuit are connected using a single-mode fibre. A maximum current of 100 kArms, which passes through the centre of the sensing coil using a shunt generator, is supplied to the conductor. The input current simulates the waveform in the event of a transmission line accident, and the waveforms in an exponential factor with a time constant of 0.03 s at an ac signal of 50 Hz are superposed. The input current was measured using a shunt resistor with excellent linearity even at large current. The top diagram shows the input current waveform, and the bottom diagram shows the fibre-optic current sensor output waveform. There is a good agreement between the two waveforms.

The ratio error between the waveforms is 1% or less, which is less than the 1% accuracy of the shunt resistor used for measurement. Linearity of the input–output characteristics of the current sensor up to 100 kA was examined by this measurement, but nonlinearity of the fibre-optic current sensor was not measured.

The dynamic range of the fibre-optic current sensor was 50 dB in the range from the noise level at 0 Arms of 10 Arms to 100 kArms. Therefore conventional two-core current sensors, which are used for control and protection by changing the sensitivity, can be replaced by only one fibre-optic current sensor. If the optical current sensor is used for ac measurement, then dc drift is not taken into account. But to measure the effects of non-reciprocal phase shift, the DC drift of the fibre-optic current sensor was measured. If a non-reciprocal phase shift exists, the dc component will be the output to the fibre-optic current sensor even if the primary current is zero, and this dc component will change over time. Such DC component changes were 10 Arms or less when measurement was performed for 1 h. This is 0.05 mrad when converted to phase difference. If non-polarization is insufficient, the polarized light will fluctuate due to birefringence changes, causing non-reciprocal phase shifts. These results indicate that the depolarizers are working effectively. Consequently, it was confirmed that the performance of the fibre-optic current sensor using single-mode fibre is equal to, or superior to, that of the previously reported fibre-optic current sensor using PM fibre.

4. Summary

We proposed a Sagnac fibre-optic current sensor using single-mode fibre, and checked its performance. It was demonstrated that error due to polarization fluctuations can be completely eliminated by using depolarizers. It is thought that the realization of single-mode fibre transmissions makes installation of fibre-optic current sensors easier and less costly. The input–output and temperature characteristics satisfied the ratio error within 0.2%. This performance is equal to, or superior to, that of the previously reported Sagnac interferometer-type fibre-optic current sensor using PM fibre. Considerable flexibility exists for tailoring the number of fiber turns on the sensing head to center the current sensing range around the customer's desires. Fibre-optical current sensor design has evolved from the first prototypes made by RPC Optolink in 2007 [6], which in turn were a short design step away from the fiber optic gyroscope in production at RPC Optolink for commercial applications since 2002.

5. References

[1] Frosio G and Dandliker R 1994 *Appl. Opt.* **33** 6111
[2] Blake J 1996 *Proc. SPIE Fiber-optic and laser sensors* **XIV** 166
[3] RPC Optolink web site: www.optolink.ru
[4] Takahashi M 2004 *Meas. Sci. Technol.* **15** 1637
[5] Williams P A 1991 *Appl. Opt.* **30** 1176
[6] Kostritskii S M, Dikevich A A, Korkishko Yu N, Fedorov V A, Prilutskii V E, Ponomarev V G and Morev I V 2007 *Photon Express* **6(62)** 177

Journal of Physics: Conference Series **416** (2013) 012033 doi:10.1088/1742-6596/416/1/012033

Gated luminescence in as-grown and reduced undoped $LiNbO_3$ crystals

S M Kostritskii[1], M Aillerie[2], S Margueron[2] and P Bourson[2]

[1]RPC Optolink, Proton, Proezd 4806 bd.5, Zelenograd, 124498, Moscow, Russia
[2]LMOPS, University Paul Verlaine of Metz and Supelec, 2 rue E. Belin, 57070 Metz, France

E-mail: skostritskii@optolink.ru

Abstract. The results of comparative study of the gated two-color luminescence and intrinsic optical absorption in as-grown and chemically reduced nominally pure $LiNbO_3$ crystals are reported. A powerful laser beam at 514.5 nm was used as gating radiation, and a laser weak beam at 632.8 nm was used to probe the luminescence. The dependences of luminescence spectra on the power of gating radiation and crystal composition have been investigated. The luminescence intensity correlates with the bipolaron absorption band in the chemically reduced crystal. We conclude that this gated luminescence at 800-950 nm is caused by the radiative transitions involving polaron states and it may be directly correlated with the two-color photorefractive sensitivity used for gated holographic recording.

1. Introduction

Lithium niobate ($LiNbO_3$) has been extensively studied for holographic data storage media since the discovery of photorefractivity in this material [1]. One of the critical issues in the implementation of holographic data storage is to fix the stored information. Due to the dynamic nature of photorefractive effect, the stored information is gradually erased during readout. Several techniques have been developed to overcome this volatile readout, including thermal fixing and electrical fixing [2]. Thermal processes or applications of high external electric fields required for these methods might be impractical. One promising method is to use two-color recording, in which a gating beam (usually of higher photon energy than the recording beams) is present during recording for sensitizing the material. This gating light just aids the recording process but is not at the recording wavelength, thereby making the readout at a longer wavelength nonvolatile. In addition, the stored information can be erased with the gating beam (or gating beam plus one of the recording beams). This approach is attractive in that it is an all-optical process.

The holographic storage system that uses the two-photon gated recording method has been recently reported: High diffraction efficiency is achieved when a cw near-IR laser (800 nm) is used for writing and a green/blue incoherent light for gating [3,4]. There are many hints that short-living (order of ms at room temperature) small (single) polarons strongly influence photorefractive sensitivity. At room temperature almost all of these polarons are paired forming the bipolarons in equilibrium. Bipolarons can be broken into single polarons again by illumination within visible range. The dissociation of bipolarons changes the optical absorption spectrum of the crystals. The broad absorption band near 500 nm has been attributed to the presence of the bipolarons and a narrower band at 760 nm has been

[1] To whom any correspondence should be addressed.

assigned to the formation of single polarons under illumination with proper wavelength [4,5]. The second important source of small polarons is two-photon band-to-band excitation of an electron, forming a pair of the electron and hole polarons within bandgap [6]. To optimize the crystal properties for two-color holography, it is necessary to study the efficiency of small polarons excitation with recording red/near-IR light in different crystals. The usual approach consists of direct measurement of photovoltaic current density at two-color excitation [3,7]. However, this approach is rather sophisticated technically. As alternative approach, we propose to use the measurements of two-photon luminescence, as it has been established [5], that two-photon photorefractive sensitivity is correlated with this photoluminescence (PL) intensity, as both effects have the same microscopic origin: the photoexcitation of small polarons.

2. Experimental techniques

Nominally pure $LiNbO_3$ crystals with different composition were used in this study. Some samples of the congruent crystal were annealed in pure hydrogen atmosphere at 500 °C or in vacuum 10^{-3} Torr at 500÷700 °C for various times in range from 0.5 to 11 hours. As can be seen from the corresponding absorption spectra, the strength of the bipolaron absorption band depends on the reduction temperature in a strongly nonlinear way. This fact makes it difficult to control the reduction process and may lead to arbitrarily different results for slightly altered experimental conditions. Since the dependence of bipolaron concentration on the reduction conditions in nominally pure $LiNbO_3$ is not known in general, the most convenient characterization is given by the properties of the bipolaron band itself [2]. We will therefore refer to the steady-state absorption at λ =500 nm, i.e., the maximum of the bipolaron absorption band, as the degree of reduction. The two near-stoichiometric samples have slightly different composition, as it were fabricated by various techniques: double cruicible CZ (DCCZ) growth and vapor-transport-equilibrium (VTE) treatment. According to our previous study, the deviation (Δx_c = 50 - x_c) from true stoichiometric composition is larger in the DCCZ sample than in the VTE one by factor of 4.1, while composition parameter $x_c = \{[Li_2O]/([Li_2O] + [Nb_2O_5])\}\times 100\%$ has been found to have almost same value of 49.96±0.05 % in the both crystal, according to the Raman data [5].

To investigate the contribution of small polarons to PL, the samples was irradiated by a focused beam of 514.5 nm from an argon-ion laser for pumping (gating), or/and by another focused laser beam of 632.8 nm from a helium-neon laser for probing. The on/off statuses of these two beams were controlled by two shutters. To focus both pump and probe beams, a high-NA microscope objective (×50) was used. This objective is part of a confocal microscope within a "LabRam" spectrometer. Therefore, a miscroscope confocal hole allows extract luminescence and Raman signal from an area, where overlapping between pump and probe beams is largest. To calibrate a spectral dependence of sensitivity of the LabRAM Jobin-Yvon spectrometer in red and near-IR ranges (640-1000 nm), a NIST calibrated halogen lamp were used as the standard quasi-white light source. This calibration of the both photodetector and monochromator responses has provided possibility to use the LabRam in obtaining of appropriate information about the PL bands in the near-IR range.

3. Experimental results and discussion

The PL spectra excited by probe beam of a He-Ne laser in as-grown nominally pure $LiNbO_3$ crystals contain a weak band, which has intensity depending strongly on the crystal composition and degree of residual chemical reduction. This broad luminescence band is observed to be centred around 800-820 nm in the raw spectra measured by the LabRAM spectrometer [5]. Calibrating the spectrometer's spectral sensitivity, we conclude that these PL spectra consist of the two broad bands centred at 835 and 950 nm, Fig. 1. Note that this result is in satisfactory agreement with the previously reported data on position of the PL bands (840 and 953 nm) [8].

Integral PL intensity, I_{PL}, as well as ratio between intensities of two PL bands depends drastically on composition of as-grown crystals: at decrease of x_c., i.e. increase of intrinsic defect concentration, the sharp growth of absolute intensity of 835-nm PL band is observed at slight increase of 950-nm band intensity, e.g. ratio I_{PL} (835nm)/I_{PL}(950nm) ≈ 1.6 in as grown congruent crystals.

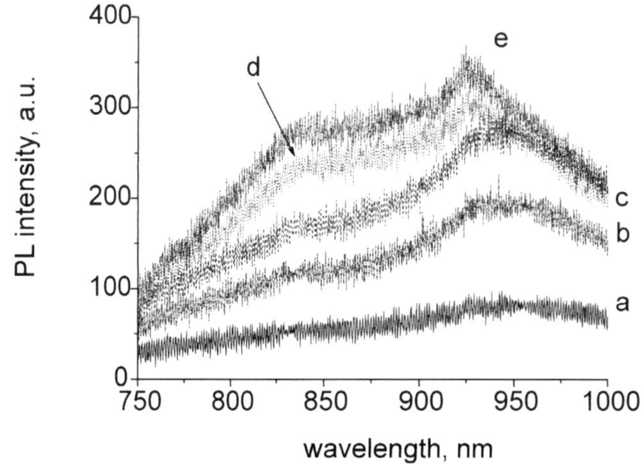

Figure 1. PL spectra of a slightly reduced DCCZ crystal excited by: single 10-mW-514.5-nm beam (a), single 10-mW-632.8-nm beam (b), both beams (c-e) at constant power (10 mW) of 632.8-nm beam and different powers of pump 514-5-nm beam: 20 mW (c), 40 mW (d) and 50 mW (e).

These bands were observed also in PL spectra measured at two-color excitation, when second green pump beam of Ar^+-laser is used to increase PL excitation efficiency. Green and red beams were aligned to overlap inside the crystal so that the change of PL in the presence of green light could be monitored. As seen in figure 1, as-grown crystals show significant increase of PL intensity in the presence of green light. The green-induced increment of PL intensity disappears immediately, when the green pump beam was turned off. Note, that one-color PL excited by single green pump beam has much smaller intensity in comparison with one-color PL excited by red probe beam, Fig. 1.

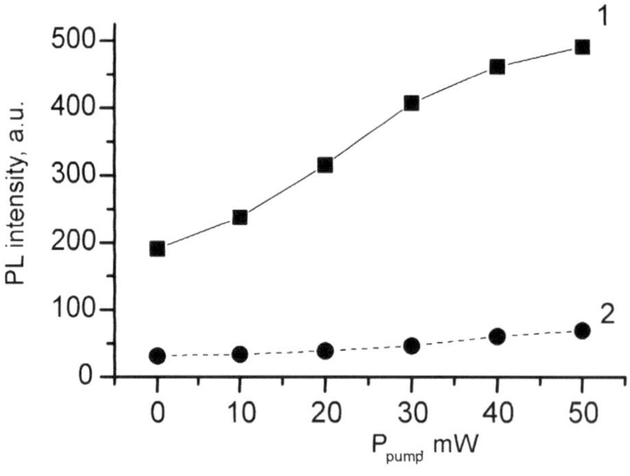

Figure 2. PL intensity versus pump power P_{pump} in the as-grown congruent (1) and near-stoichiometric VTE crystals (2).

Figure 2 shows the PL intensity variation with the pump power at a constant probe 632.8-nm-beam power (10 mW) in the as-grown congruent and near stoichiometric VTE crystals, i.e. in samples with different residual lithium deficiency. I_{PL} demonstrates a marked growth with power P_{pump} of the pump 514.5-nm-beam in the both crystals. It may be regarded as a clear evidence of the two-color (gated) excitation of this luminescence. The slope of power dependence of PL intensity, $\partial I_{PL}/\partial P_{pump}$, as well as the maximum value of I_{PL} increase with Δx_c, i.e. there is clear correlation between $\partial I_{PL}/\partial P_{pump}$ and intrinsic defects concentration. The dependence of I_{PL} on P_{pump} tends to saturation at higher pump powers in congruent crystal, Fig.2(1), while no the tendency to saturation is observed in VTE crystal, Fig.2(2). The spectral shape and absolute intensity of PL differ slightly with the specimens of as-grown congruent $LiNbO_3$ crystal.

Additional chemical reduction via annealing in vacuum or H_2 atmosphere has been established to introduce a large amount of extra luminescence at 835 nm, presumably corresponding to the large bipolarons concentration in these crystals. This growth of PL intensity correlates closely with a marked increase of the bipolaron absorption band [1,5,9], having its center at ~ 500 nm. Both I_{PL} and ΔI_{PL} (measured at 835 nm) increase monotonically with the increase of the optical absorption coefficient α at 500 nm within α range from 0.2 to 3 cm^{-1}. Thus, the maximum value of normalized increment of PL intensity (ΔI_{PL} / I_{PL}) is evaluated to be on the order of 0.1 in the slightly and moderately reduced congruent crystals at P_{pump} = 50 mW, Fig. 2. The dependence of I_{PL} on α shows an evident saturation in the more deeply chemically reduced crystals with larger α , while ΔI_{PL} sharply decreases with the further increase of α and $\Delta I_{PL} = 0$ in the reduced crystal with the largest value of α (8.1 cm^{-1}) among studied samples. Besides, the PL spectra of deeply reduced crystals have pronounced difference from the spectra of all other crystals, as the 950-nm band is not observed, Fig. 3.

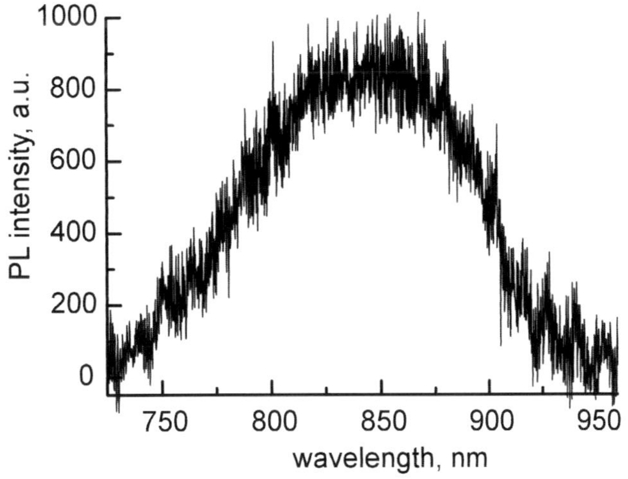

Figure 3. PL spectrum obtained by two-color excitation at P_{pump}= 20 mW in the strongly reduced congruent crystal with $\alpha = 8.1$ cm^{-1}.

It has been found that the some important features of the PL spectra of the reduced nominally pure $LiNbO_3$ can be consistently explained by photoexcitation of the ($Nb_{Li}^{3\bullet}$-$Nb_{Nb}^{'}$) bipolarons [5,8]. Bipolarons may be considered as pairs of small polarons stabilized by a joint lattice distortion and covalence. They can be broken into single (small) polarons by illumination with a proper wavelength via a photo-dissociation process. At room temperature single polarons are short-living (≤ ms) because of their small activation energy (0.3 eV) for annihilation. The kinetics of the dissociation-annihilation process may be described by a second order reaction [9]: B ⇔ P + P. The existence of these single polarons in continuous wave (cw) illuminated crystals manifests itself in the luminescence spectra, as the polarons can be further excited by absorption of a photon with energy corresponding to the single polaron absorption band centred at 780 nm. Electrons elaborated in this way relax to the bottom of the conduction band, captured at a locally distorted site and after relaxes either vibrationally or radiatively to a level of a localized polaron defect below the band edge, i.e. within the band gap. Along the latter non-phonon relaxation process a photon with smaller energy is emitted, giving the luminescence signal at $\lambda > 800$ nm. Such a radiative (non-phonon) transition from a locally distorted site to the ground polaron state is possible if the configuration coordinate of the excited quasi-localized state in the conduction band would be the same as that of the ground state of a polaron [10]. Of course, the probability of this process is small and, indeed, the polaron luminescence is relatively weak and can be observed only at relatively high excitation intensities specific for strongly focused laser beams obtained by the high-magnification objectives used in the confocal microscope of a LabRam spectrometer [5,8].

With densities N_B of bipolarons and N_p of polarons the kinetics can be described by the following rate equation taking into account generation and recombination of polarons [11]:

$$dN_p/dt = g'N_B J_{pump} - rN_P^2 \tag{1}$$

If the concentration of bipolarons changes only slightly during illumination, $\Delta N_B << N_B$, the term $g'N_B$ in Eq. (1) can be replaced by a constant g. At cw laser intensities typical for our luminescence experiments this is a good approximation and the term gJ_{pump} can be used instead of $g'N_B J_{pump}$ in Eq. (1) as generation rate [11]. The concentration N_P can be evaluated from data on light-induced changes of optical absorption α_{li}, as $\alpha_{li} = sN_p$ with $s = s_P - 0.5s_B$, where s_P denotes the absorption cross section of polarons and s_B that of bipolarons. Solving the rate equation (1), one obtains for steady-state absorption changes during illumination:

$$\alpha_{li}^{s} = s(gJ_{pump}/r)^{1/2} \tag{2}$$

where $1/r \sim \tau_P$, and τ_P is the polaron lifetime limited by fast thermal relaxation to bipolaron states or other deep traps. Combined solution of equations, describing the dissociation-annihilation process and the luminescence gives the following results for the dependence of polaron luminescence intensity I_{lum} on probe laser intensity J_{prob}:

$$I_{lum} = A\gamma\{sN_B\tau_P(1 - \exp(-J_{pump}/J_s)\}^{1/2} J_{prob} \tag{3}$$

where J_s is specific saturation parameter: In the case $J_{pump} > J_s$, N_P has the same order of magnitude as N_B, and a pump intensity law of I_{lum} is caused by a significant depletion of bipolaron states, relating to a decrease of instant bipolaron concentration N_B in the crystals illuminated by a high-intensity pump beam.

At the same time, such experimental conditions (light wavelength and intensity) allow for appearance of marked generation of electron free polaron - hole polaron pairs due to the two-photon interband absorption [6,12]. According to the data on PL of polarons in $SrTiO_3$ [6], the sequent recombination of these polarons should be sufficient to give rise of luminescence in $LiNbO_3$, too.

The recent data [12] on competition of the two light-induced processes (photodissociation of bipolarons, and two-photon interband absorption) in small polarons generation has been used as a key argument for assignment of 835-nm and 950-nm bands to the different radiative transitions between polaron states. For example, significant increase of 835-nm band accompanied by evident suppression of 950-nm band with growth of α in deeply reduced crystals gives evidence of dominating role of the bound Nb_{Li}^{4+} polarons, appearing due to the bipolarons photodissociation, in PL at 835 nm. The PL at 950 nm is, hence, assigned to the recombination of free electron and O^- hole polarons. Note that existence of 835-nm band in the PL spectra of as-grown samples is strongly related to x_c, i.e. to intrinsic antisite defects Nb_{Li}^{5+} concentration, and, therefore, may be assigned either to a residual bipolarons content in as-grown crystals, since these crystals grown in standard conditions may be regarded as slightly reduced $LiNbO_3$ [13], or to formation of the bound Nb_{Li}^{4+} polarons due to electrons trapping at unoccupied Nb_{Li}^{5+} ions during hopping transport of electrons, occurs via free Nb_{Nb}^{4+} polarons [14]. Eventually, an involvement of impurity $Fe^{2+/3+}$ ions in the recombination and excitation process in nominally pure $LiNbO_3$ would yield much weaker effects than the polaronic one [12], in contrast to the situation reported for PL in the reduced iron-doped $LiNbO_3$ crystals [15,16].

Similar behaviour with respect to chemical reduction degree and crystal composition was observed for the two-photon gated holographic recording via bound polarons/bipolarons in the reduced nominally pure $LiNbO_3$ [3-4]. Thus, the data on PL intensity may be used for rough estimation of efficiency of the gated holographic sensitivity of nominally pure crystals with given composition and chemical reduction degree. Besides, PL spectroscopy gives valuable insight on charge transport governed by free polarons that must have some practical importance, as such a transport is very perspective to use for dynamic holographic recording with light in the infrared spectral range [14]. Two-color gated luminescence from the pre-exposed states has not been completely analyzed yet. The dependence of bleaching rate on the gating wavelength should also be taken into account. Optimization of the recording conditions might enable further improvements to be achieved.

4. Summary

We have reported the data on measurement of the gated two-photon luminescence upon background shorter-wavelength illumination in as-grown and reduced nominally pure $LiNbO_3$ crystals. The gating-light power, composition and bipolaron concentration dependences of this weak photoluminescence in the near IR-range allow for conclusion, that the light-induced generation of free and bound small Nb_{Nb}^{4+} and Nb_{Li}^{4+} electron polarons and bound small O^- hole polarons is responsible for this gated luminescence in the nominally pure $LiNbO_3$ crystals. It has been shown, that the data on the luminescence intensity may be used for rough estimation of the gated holographic sensitivity of the reduced nominally pure $LiNbO_3$ crystals.

5. References

[1] Von der Linde D, Glass A M and Rodgers K F 1974 *Appl. Phys. Lett.* **25** 155
[2] Amodei J J and Staebler D L 1971 *Appl. Phys. Lett.* **18** 540
[3] Hesselink L, Orlov S, Liu A, Akella A, Lande D and Neurgaonkar R 1998 *Science* **282** 1089
[4] Guenther H, Macfarlane R, Furukawa Y, Kitamura K and Neurgaonkar R 1998 *Appl.Opt.* **37** 7611
[5] Kostritskii S M, Sevostyanov O G, Bourson P, Aillerie M, Fontana M D and Kip D 2007 *Ferroelectrics* **352** 61
[6] Hasegawa T and Tanaka K 2001 *J Luminescence* **94-95** 15
[7] Liu Y, Kitamura K, Takekawa S, Nakamura M, Hatano H and Furukawa Y 2003 *OSA Trends in Optics and Photonics* **99** 55-60
[8] Harhira A, Guilbert L, Bourson P and Rinnert H 2007 *phys. stat. sol (c)* **4** 926
[9] Schirmer O F, Imlau M, Merschjann C and Schoke B 2009 *J. Phys.: Condens. Matter* **21** 123201
[10] Yakovlev V Y, Kabanova E V, Veber T and Paufler P 2001 *Phys. Sol. State* **43** 1580
[11] Jermann F, Simon M, Bower R, Krätzig E, Schirmer O.F 1995 *Ferroelectrics* **165** 319
[12] Merschjann C, Schoke B and Imlau M 2007 *Phys. Rev. B* **76** 085114
[13] Kostritskii S M and Sevostyanov O G 1997 *Appl. Phys. B* **65** 527
[14] Merschjann C, Berben D, Imlau M and Wohlecke M 2006 *Phys. Rev. Lett.* **96** 186404
[15] Parsons R, Cornish W D and Young L 1975 *Appl.Phys.Lett.* **27** 654
[16] Harhira A, Guilbert L, Bourson P and Rinnert H 2008 *Appl. Phys. B* **92** 555

AUTHOR INDEX

AUTHOR INDEX

Institute of Physics
Dirac House, Temple Back
Bristol BS1 6BE UK

proceedings
.com